CHEMISTRY OF
METALLOPROTEINS

CHEMISTRY OF METALLOPROTEINS

Problems and Solutions in Bioinorganic Chemistry

JOSEPH J. STEPHANOS
ANTHONY W. ADDISON

Published by John Wiley & Sons, Inc., Hoboken, New Jersey
Published simultaneously in Canada

For general information on our other products and services or for technical support, please contact our
Customer Care Department within the United States at (800) 762-2974, outside the United States
at (317) 572-3993 or fax (317) 572-4002.

Wiley also publishes its books in a variety of electronic formats. Some content that appears in print may
not be available in electronic formats. For more information about Wiley products, visit our web site
at www.wiley.com.

Library of Congress Cataloging-in-Publication Data:

Stephanos, Joseph J., author.
  Chemistry of metalloproteins: problems and solutions in bioinorganic chemistry/by Joseph J.
Stephanos, Anthony W. Addison.
     p.; cm. – (Wiley series in protein and peptide science)
  Includes bibliographical references and index.
   ISBN 978-1-118-47044-2 (paperback)
  I. Addison, A. W., author. II.   Title. III.   Series: Wiley series in protein and peptide science.
  [DNLM: 1. Metalloproteins–chemistry–Examination Questions. QU 18.2]
  QP551
  572′.6076–dc23
                                                                                    2013041995

Printed in the United States of America

10 9 8 7 6 5 4 3 2 1

CONTENTS

PREFACE

This book is an attempt to reveal the chemical concepts that rule the biological action of metalloproteins. The emphasis is on building up an understanding of basic ideas and familiarization with basic techniques. Enough background information is provided to introduce the field from both chemical and biological areas. It is hoped that the book may be of interest to workers in biological sciences, and so, primarily for this purpose, a brief survey of relevant properties of transition metals is presented.

The book is intended for undergraduates and postgraduates taking courses in coordination chemistry and students in biology and medicine. It should also be a value to research workers who would like an introduction to this area of inorganic chemistry. It is very suitable for self-study; the range covered is so extensive that the book can serve as a student's companion throughout his or her university career. At the same time, teachers can turn to it for ideas and inspirations.

The book is divided into seven chapters and covers a full range of topics in bioinorganic chemistry. It is well-illustrated and each chapter contains suggestions for further reading, providing access to important review articles and papers of relevance. A reference list is also included, so that the interested reader can readily consult the literature cited in the text.

It is hoped that the present book will provide the basis for a more advanced study in this field.

JOSEPH J. STEPHANOS
ANTHONY W. ADDISON

1

INTRODUCTION

The discipline of *bioinorganic chemistry* is concerned with the function of metallic and most of nonmetallic elements in biological processes. Also, it is the study of the chemistry, structure, and reactions of the metalloprotein molecules belonging to the living cell.

The precise concentrations of different ions, for instance, in blood plasma indicate the importance of these ions for biological processes, (Table 1-1).

Such elements fall into four broad classifications: the polluting, contaminating, beneficial, and essential elements.

- Polluting elements: Pb, Hg, and Cd
- Contaminating elements: vary from person to person
- Beneficial elements: Si, V, Cr, Se, Br, Sn, F, and Ni
- Essential elements: H, C, N, O, Na, Mg, K, Ca, P, S, Cl, Mo, Mn, Fe, Co, Cu, Zn, and I (Fig. 1-1).

Twenty-five elements are currently thought to be essential to warm-blooded animals (Table 1-2).

Essentiality has been defined according to certain criteria:

- A physiological deficiency appears when the element is removed from the diet.
- The deficiency is relieved by the addition of that element to the diet.
- A specific biological function is associated with the element.

Chemistry of Metalloproteins: Problems and Solutions in Bioinorganic Chemistry, First Edition.
Joseph J. Stephanos and Anthony W. Addison.
© 2014 John Wiley & Sons, Inc. Published 2014 by John Wiley & Sons, Inc.

TABLE 1-1   Ion Concentration in
Extracellular Blood Plasma

Ion	mM	Ion	mM
Na^+	138	SO_4^{2-}	1
Cl^-	100	Fe	0.02
K^+	4	Zn^{2+}	0.02
Ca^{2+}	3	Cu^{2+}	0.015
Mg^{2+}	1	Co^{2+}	0.002
HPO_4^{2-}	1	Ni^{2+}	0

Every essential element follows a dose–response curve, shown in Fig. 1-2. At lowest dosages the organism does not survive, whereas in deficiency regions the organism exists with less than optimal function.

The ten ions classified as trace metal are Fe, Cu, Mn, Zn, Co, Mo, Cr, Sn, V, and Ni, and the four classified as bulk metals are Na, K, Mg, and Ca. The nonmetallic elements are H, B, C, N, O, F, Si, P, S, Cl, Se, and I.

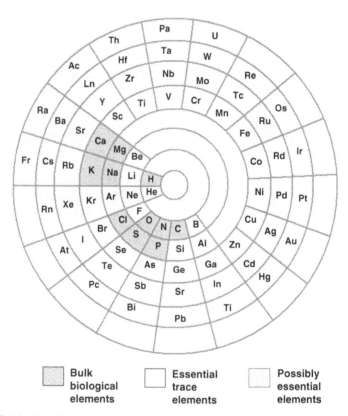

FIGURE 1-1   Distribution of elements essential for life (Cotton and Wilkinson, 1980). (See the color version of this figure in Color Plates section.)

TABLE 1-2   Percentage Composition of Essential Elements in Human Body

Element	Percentage (by Weight)		Element	Percentage (by Weight)
Oxygen	53.6		Silicon	0.04
Carbon	16.0		Iron, fluorine	0.005
Hydrogen	13.4		Zinc	0.003
Nitrogen	2.4		Copper, bromine	2×10^{-4}
Sodium, potassium, sulfur	0.10		Selenium, manganese, arsenic, nickel	2×10^{-5}
Chlorine	0.09		Lead, cobalt	9×10^{-6}

Biological Roles of Metal Ions

What are the general roles of metal ions in biological systems?

The general roles of metal ions in biological systems are summarized in Table 1-3. Metals in biological systems function in a number of different ways:

- Groups 1 and 2 metals operate as *structural* elements or in the *maintenance of charge, osmotic* balance, or *nerve* impulses.
- Transition metal ions that exist in single oxidation states, such as zinc (II), function as structural elements in superoxide dismutase and zinc fingers or as *triggers* for protein activity, e.g., calcium ions in calmodulin or troponin C.

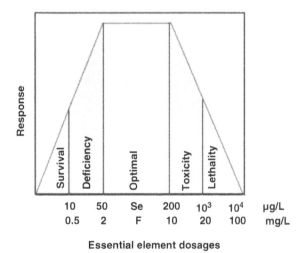

FIGURE 1-2   The dose-response curves of selenium and fluoride.

TABLE 1-3   Role of Metal Ions and Examples

Metal	Functions and Examples
Na^+, K^+	Charge transfer, osmotic balance, nerve impulses
Mg^{2+}	Structure in hydrolases, isomerases, phosphate transfer, and trigger reactions
Ca^{2+}	Structure, charge carrier, phosphate transfer, trigger reactions
Zn^{2+} (tetrahedral)	Structure in zinc finger, gene regulation, anhydnase, dehydrogenase
Zn^{2+} (square pyramidal)	Structure in hydrolases, peptidases
Mn^{2+} (octahedral)	Structure in oxidases, photosynthesis
Mn^{3+} (tetragonal)	Structure in oxidase, photosynthesis
Fe^{2+}	Electron transfer, nitrogen fixation in nitrogenase, dioxygen transport in hemoglobin and myoglobin
Fe^{3+}	Electron transfer in oxidases
Cu^+, Cu^{2+}	Electron transfer in type I blue copper proteins, oxidases and hydroxylases in type II blue copper proteins, hydroxylases in type III blue copper proteins, dioxygen transport in hemocyanin
Co^{2+} (tetrahedral)	Alkyl group transfer, oxidases
Co^+, Co^{2+}, Co^{3+} (octahedral)	Alkyl group transfer in B_{12}
Ni^{2+} (square planar)	Hydrogenase, hydrolases
Mo^{4+}, Mo^{5+}, Mo^{6+}	Nitrogen fixation in nitrogenose, oxo transfer in oxidases

- Transition metals that exist in multiple oxidation states serve as *electron carriers*, e.g., iron ions in cytochromes or in iron–sulfur clusters of the enzyme nitrogenase or copper ions in azurin and plastocyanin.
- As facilitators of *oxygen transport*, e.g., iron ions in hemoglobin or copper ions in hemocyanin.
- As sites at which enzyme *catalysis* occurs, e.g., copper ions in superoxide dismutase or iron and molybdenum ions in nitrogenase.
- Metal ions may serve multiple functions, depending on their location within the biological system, so that the classifications in Table 1-3 are somewhat arbitrary and/or overlapping.

PROTEINS: FORMATION, STRUCTURES, AND METALLOPROTEINS

This section is designed to introduce the chemistry of proteins. The text broadly includes where and how the proteins are formed, along with the structure and formation of metalloproteins.

Following the introduction of organelles and their functions within the cell, the discussion will be concerned with the general structure of deoxyribonucleic acid (DNA) and how the nucleus maintains its control of cell growth, division, and formation of [messenger, transfer, and ribosomal ribonucleic acid (mRNA, tRNA, rRNA)]. This is followed by how mRNA and tRNA master the formation of proteins within a cell. Then, primary, secondary, tertiary, and quaternary structures of the formed proteins and the factors that control each of these structures are discussed.

Specific points about the ligation of various metal ions to different amino acids within the proteins are made, and the binding stabilities of various metal ions toward different amino acids are arranged.

The general formulas, side chains, and corresponding names of the common natural α-amino acids, the formation of the peptide chain from the amino acids, and the physiological roles of proteins are described.

The chemistry of the prosthetic and cofactors is explored. Enough basic biochemistry is presented to enable the student to understand the discussions that follow.

Organelles and Their Functions

Identify the organelles and their functions within the cell.

- Cells are the building blocks of all living things.
- There are similarities in the appearance, chemical constituents, and activities of all cells (Fig. 1-3).
- Different structures within the cell are called *organelles*.
- Each organelle has an important, specific function in the cell.
- The *mitochondria* are responsible for conversion of food into usable energy (metabolism):
  - They contain enzymes for cell metabolism.
  - More than 50% of the energy produced by mitochondrial oxidation of carbohydrates is recaptured as adenosine diphosphate (ADP) and converted into adenosine triphosphate (ATP).

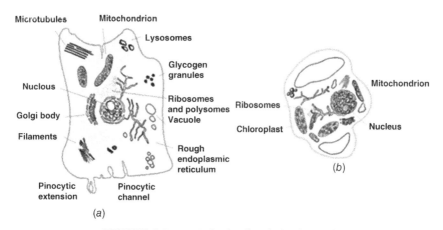

FIGURE 1-3 (*a*) Animal cell and (*b*) plant cell.

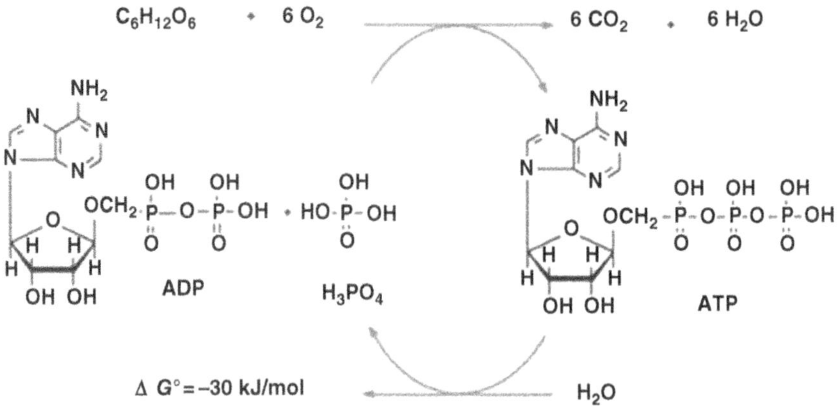

$$C_6H_{12}O_6 \quad + \quad 6\,O_2 \quad \longrightarrow \quad 6\,CO_2 \quad + \quad 6\,H_2O$$

$$\Delta\,G° = -30\ kJ/mol$$

SCHEME 1-1 Derived energy is trapped in adenosine triphosphate molecules (ATP).

- o The derived energy is trapped in ATP molecules (Scheme 1-1).
- o ATP can diffuse rapidly throughout the cell, delivering energy to sites where it is required for cellular processes.
- In green plants, *chloroplasts* contain chlorophyll molecules and other pigments.
  - o Chlorophyll and other pigments in chloroplasts absorb light energy from the sun and use it to produce ATP, glucose, and oxygen.
- *Ribosomes* are round particles (mRNA) that are sent by the nucleus to activate protein synthesis.
  - o The mRNA causes a specific protein molecule to be synthesized from the pool of amino acids present in the cell cytoplasm.
- The *nucleus*, or command station, contains information for the development and operation of the cell.
  - o This information is stored chemically in long molecular strands called DNA. A combination of DNA and protein forms fine strands of chromatin. When a cell is about to divide, the chromatin strands coil up and become densely packed, forming chromosomes.
  - o The number of chromosomes varies with the species: Humans have 23, the fruit fly has 4, corn has 10, and the mosquito has 3.

Structure of DNA

What is the general structure of deoxyribonucleic acids, DNA?

- Polymerization of nucleoside phosphates produces the nucleic acids, DNA and RNA.

Bases, phosphate, and deoxyribose → nucleic acid
Nucleoside
Nucleoside monophosphate
This H in DNA is OH in RNA
Anionic
B Base
Deoxyribose
Nucleic acid

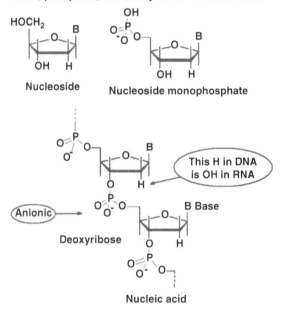

- DNA is a giant molecule with molecular weight of order 1 billion or more.
- The information is chemically stored by nitrogen-base molecules that are bonded to the sugar residues of the sugar–phosphate chain.
- There are four nitrogen bases:
  (a) Two purines, which are bicyclic molecules:

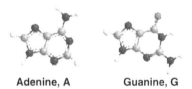

Adenine, A Guanine, G

  (b) Two pyrimidines, which are monocyclic:

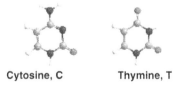

Cytosine, C Thymine, T

- The order in which they appear on the chain makes up the molecular message (Fig. 1-4).
- The DNA molecule is also capable of duplicating itself and dividing.

Phosphate   Deoxyribose   Nitrogen bases

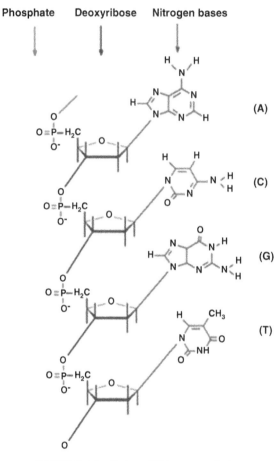

FIGURE 1-4 Order of N bases on chain.

- Under a microscope we can see the duplicated chromosomes divide equally as the cell divides.
- The DNA double strand forms when the bases on the two adjacent single strands form hydrogen bonds:

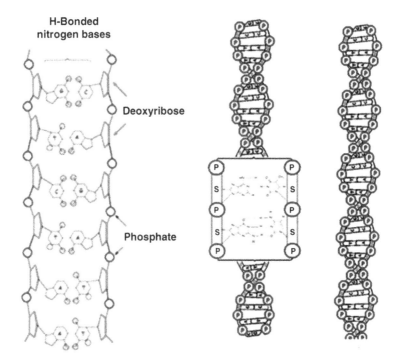

FIGURE 1-5 DNA double strand.

- Adenine and thymine form a hydrogen bonded pair, or *complementary base pair*.
- Cytosine and guanine also form a complementary base pair (Fig. 1-5).
- These complementary base pairs are conformed by the base ratios: G/C = 1 and A/T = 1 (Table 1-4).

TABLE 1-4 Nitrogen-Base Content of DNA from Different Organisms

Species Tissue Source	Calf Thymus	Crab All tissue	Algea (*Euglcna*) Chloroplast	Virus (Coliphaga ×174) Replicative Form
A	29.0	47.3	38.2	26.3
T	28.5	47.3	38.1	26.4
A/T	1.01	1.00	1.00	1.00
G	21.2	2.7	12.3	22.3
C	21.2	2.7	11.3	22.3
G/C	1.00	1.00	1.09	1.00

Note: Data in mole percent.

Cell Growth and Division

How does the nucleus maintain its control of cell growth and division?

- During ordinary cell division called mitosis, two new cells result from a single parent.
- Each daughter has the same number of chromosomes as the parent.
- If DNA is the molecular stuff of the chromosome, it must be able to reproduce itself.
- The DNA double helix rewinds and separates into two single strands (Fig. 1-6).
- As the unwinding occurs, the single strands act as templates for synthesis of new complementary strands.
- When the parent DNA double helix has completed its unwinding, two new DNA double-stranded molecules are formed.
- The process by which new DNA is formed is called *replication*.

Protein Synthesis

How can proteins be synthesized in cells?

- The order of the N bases on the DNA molecule determines the order of amino acids in the protein molecule.
- While DNA is in the nucleus, the proteins are synthesized on ribosomes outside the nucleus as follows:

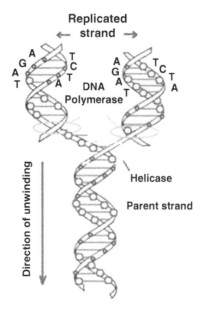

FIGURE 1-6   DNA double helix rewinds and separates into two single strands.

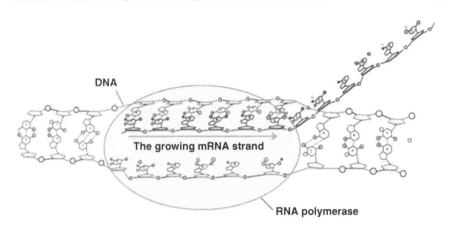

FIGURE 1-7　Synthesis of mRNA.

o As the DNA double helix unwinds, the N base segment becomes exposed.

o The DNA molecule serves as template for the synthesis of *mRNA* molecule.

o The synthesis of mRNA is analogous to the replication synthesis of DNA (Fig. 1-7).

o mRNA has structure similar to DNA but contains:

   • Ribose instead of deoxyribose

   • N-base uracil instead of thymine:

• After mRNA is synthesized, it is transported out of the nucleus and becomes attached to the ribosomes, where the protein syntheses begin (Fig. 1-8).

• At the ribosomes, the order of the bases on the mRNA determines the amino acid sequence in the protein molecule.

• The amino acid sequence is determined by a triplet code on the mRNA molecule.

• A group of three N bases represents a code for signifying a single amino acid (Scheme 1-2).

• The amino acids are brought to the mRNA at the ribosomes by much smaller RNA molecules called *tRNA*.

• Each tRNA has a triplet of bases, which is complementary to an amino acid code on mRNA.

• The tRNA molecules bring the amino acids to the ribosomes as they move along the mRNA strand, and the amino acids are knit into the growing protein chain.

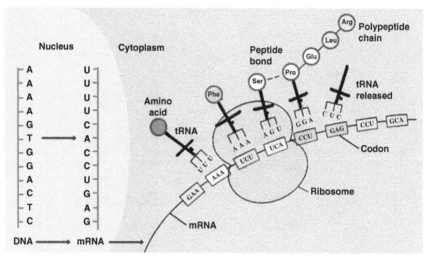

FIGURE 1-8   Protein synthesis.

- After the tRNA has discharged its amino acid passenger, it moves out into the cytoplasm, finds another amino acid, and returns to the ribosome surface.

Common Natural α-Amino Acids

Give the general formula, side chain, and corresponding name of the common natural α -amino acids.

- There are 20 common natural amino acids.

Gly :	GGU, GGC, GGA, GGG	Met :	AUA, AUG
Ala :	GCU, GCC, GCA, GCG	Tyr :	UAU, UAC
Val :	GUU, GUC, GCA, GUG	Asp :	GAU, GAC
Arg :	CGU, CGC, CGA, CGG	Asn :	AAU, AAC
Pro :	CCU, CCC, CCA, CCG	Glu :	GAA, GAU
Ser :	UCU, UCC, UCA, UCG	Gln :	CAA, CAG
Thr :	ACU, ACC, ACA, ACG	His :	CAU, CAC
Leu :	CUU, CUC, CUA, CUG	Lys :	AAA, AAG
Leu :	UUA, UUG	Phe :	UUU, UUC
Ile :	AUU, AUC	Arg :	AGA, AGG
Ser :	AGU, AGC	Chain termination :	UAA, UAG
Cys :	UGU, UGC		

SCHEME 1-2   Genetic codes.

- The general formula for an α-amino acids is

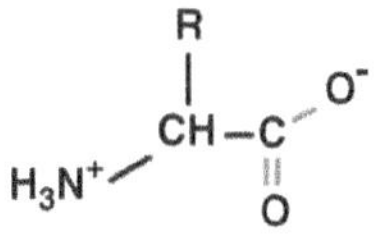

- They are summarized in Table 1-5.

TABLE 1-5   L-α-amino Acids

Name	Side Chain (R at pH = 7)	pK_a	Nature
Glycine, Gly, G	—H	2.35, 9.78	Structural, spacer
Alanine, Ala, A	—CH₃	2.35, 9.87	Hydrophobic
Valine, Val, V	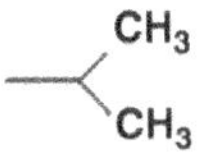	2.29, 9.74	Hydrophobic
Leucine, Leu, L	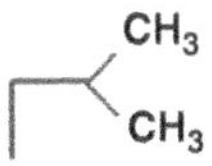	2.33, 9.74	Hydrophobic
Isoluecine, Ile, I	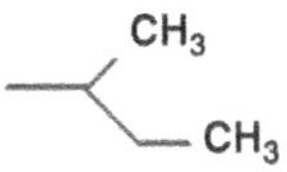	2.32, 9.76	Hydrophobic
Phenylalanine, Phe, F	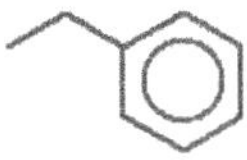	2.16, 9.18	Hydrophobic
Proline, Pro, P	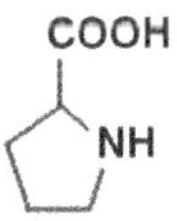	1.95, 10.64	Hydrophobic, structural
Tryptophan, Trp, W	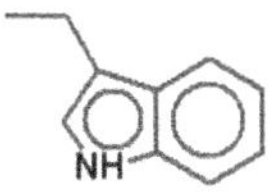	2.43, 9.44	Hydrophobic
Serine, Ser, S	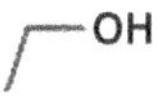	2.19, 9.21	Ambivalent
Threonine, Thr, T	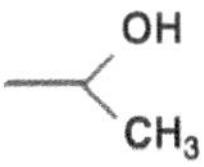	2.09, 9.11	Hydrophobic

(*continued*)

Table 1-5 (*Continued*)

Name	Side Chain (R at pH = 7)	pK_a	Nature
Methionine, Met, M	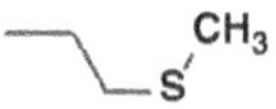	2.13, 9.28	Hydrophobic but weak Lewis base, soft
Tyrosine, Tyr, T	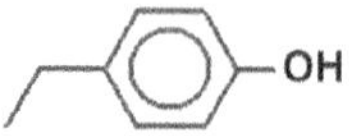	~10	Hydrophobic, but strong Lewis base, only when deprotonated
Aspartic, Asp, D	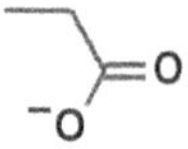	~5	Hydrophobic, Lewis base, anion
Asparagine, Asn, N	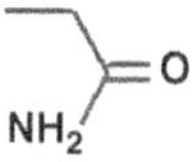	2.1, 8.84	Lewis base, anion
Glutamine, Gln, Q	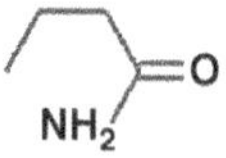	1.99, 3.90, 9.90	Polar, neutral
Glutamic, Glu, E	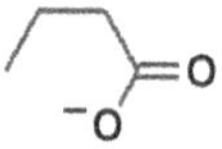	2.16, 4.27, 9.36	Lewis base, anion
Histidine, His, H	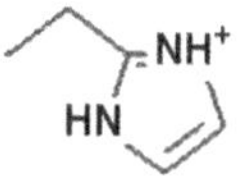	1.80, 6.04, 9.33	Hydrophobic, Lewis base
Cysteine, Cys, C		1.92, 8.35, 10.46	Lewis base, anion, soft
Lysine, Lys, K		2.16, 9.18, 10.79	Polar, cationic, protonated
Arginine, Arg, R	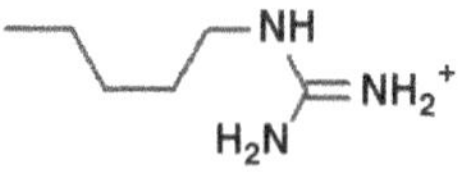	1.82, 8.99, 12.48	Polar, cationic, protonated

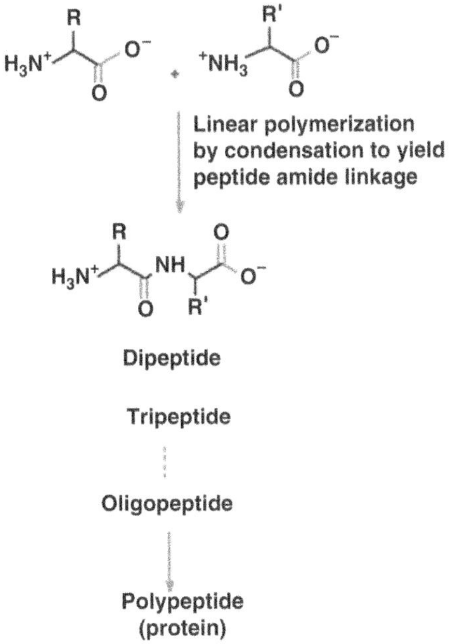

SCHEME 1-3 Polypeptide formation.

Peptide Chain Formation

How can the peptide chain be formed from the amino acids?

- Linear polymerization by condensation to yield amide peptide linkage (Scheme 1-3).
- All proteins are polypeptides.

Protein physiological functions

What are the physiological roles of proteins?

- The physiological roles of proteins are:
  - Structural: finger nails, hair, and skin
  - Transport: oxygen, electrons, and iron
  - Catalysis: enzymes responsible for all synthesis of proteins, DNA, and organics

Structural Features of Proteins

Define: primary, secondary, tertiary, and quaternary structures. And what are the factors that control each of these structures?

- The properties and functions of a particular protein depend on the sequence of the amino acids in the protein, or the *primary structure.*
  - The primary structure determines higher levels of structures.
  - These structural details are crucial to the biological role of a protein.
- The *secondary structure* arises from the relative disposition of atoms in the polypeptide "backbone":

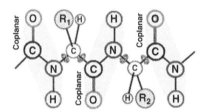

    The groups of four gray-shaded atoms are coplanar. Free rotation occurs about the bond connecting the carbon with the carbonyl and the nitrogen. Therefore, the extended polypeptide chain is a semirigid structure with two-thirds of the atoms of the backbone held in a fixed plane.

  - Examples of secondary structures:
    - (a) random coil
    - (b) α–helix (Fig. 1-9)
    - (c) β–pleated (Fig. 1-10), associated as (i) parallel and (ii) antiparallel
    - (d) reverse turns (Fig. 1-11)
    - (e) omega loops (Fig. 1-12)

    Both reverse turns and omega loops appear at the outer surface of the molecules.

- A *Tertiary structure* refers to the folding of the already secondary structured amino acids to form a three-dimensional (3D) structure. The overall 3D architecture of the polypeptide backbone:
  - Fibrous proteins: coils (Fig. 1-13).
  - Globular proteins: compact, ellipsoidal, spherical, until denatured. The folded tertiary, globular, structure of myoglobins is imposed over the helical secondary structure. Structures from X-ray diffraction are shown in Fig. 1-14.
  - Synthetic polypeptides have random or simply repetitive structures.

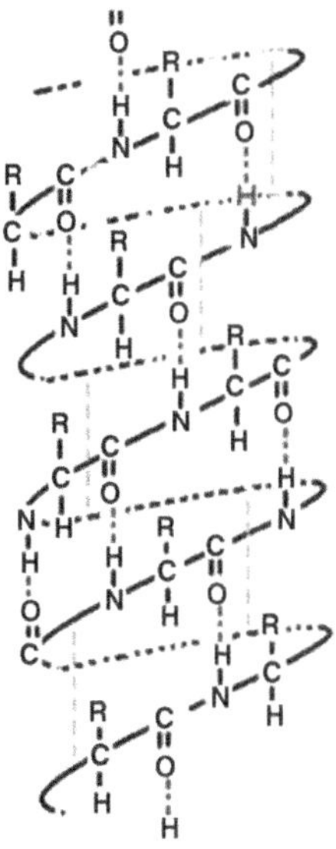

FIGURE 1-9 α-Helix structure of protein.

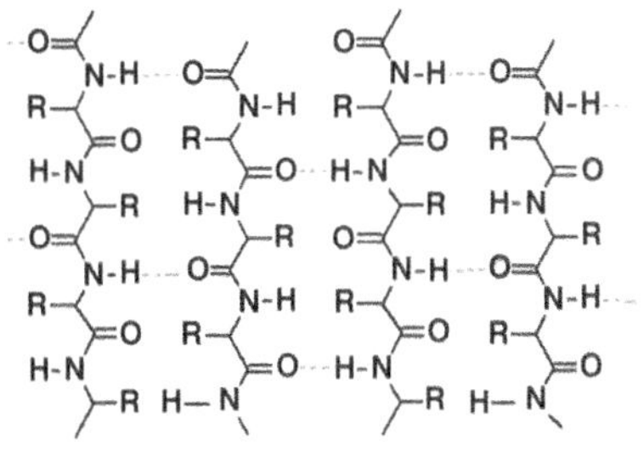

FIGURE 1-10 β-Pleated structure.

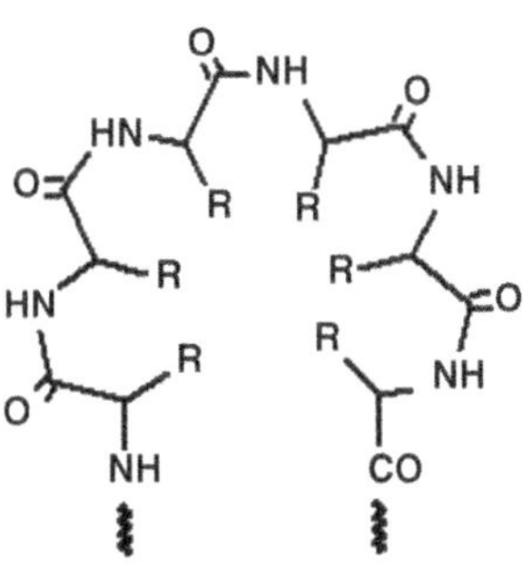

FIGURE 1-11 Reverse turn.

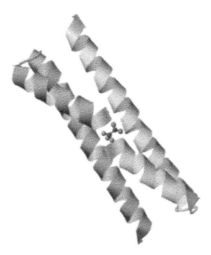

FIGURE 1-12   Omega loop.

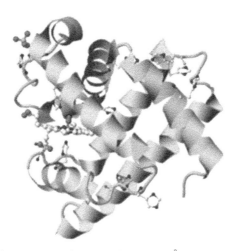

FIGURE 1-13   Fibrous oligomers, PDB 1G6U (Ogihara et al., 2001).

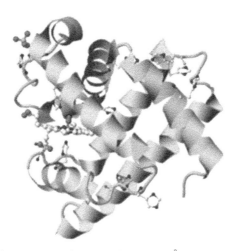

FIGURE 1-14   Tertiary structure of oxymyoglobin at 1.6 Å resolution, PDB 1MBO (Phillips, 1980).

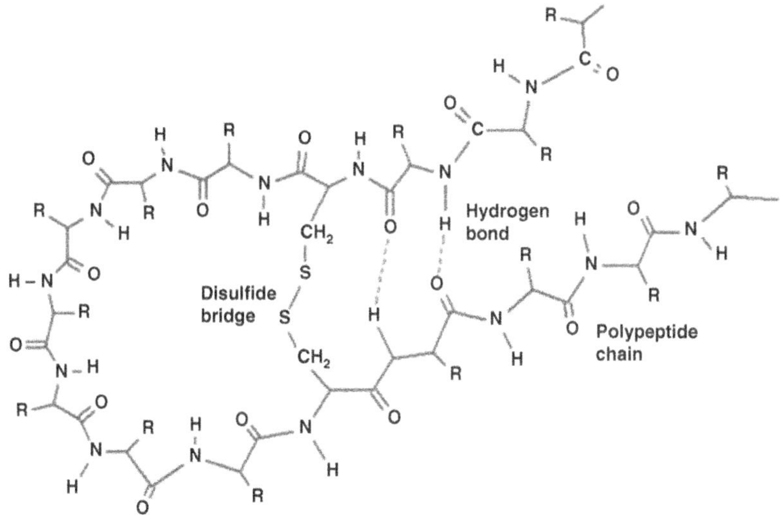

FIGURE 1-15 Disulfide linkages and H bonding.

Causes of Polypeptide Chain Folding

- Disulfide linkages (cysteine) (Fig. 1-15)
  - Disulfide linkage is a redox reaction:

$$R-S-S-R + 2H^+ + 2e^- \rightleftharpoons 2RSH$$

the standard redox potential,
$E^0 = -0.40$ V
for Cystine/Cysteine

  - To a large extent, various folding factors rely on "correct" positioning to relevant residues by other contributing folding factors, i.e., there is cooperativity, which causes an entropy gain.
- H-bonding
  - Hydrogen bonds within proteins are constant in time, whereas in fluid media, they are constantly breaking and re-forming.
  - Again, groups are in correct proximity (Fig. 1-15).
- A *hydrophobic interaction* arises from the unfavorable nature of interactions between water and nonpolar solutes (Fig. 1-16).
  - There is a common tertiary structure among proteins from diverse species. Similar folding conformations of distantly related cytochromes (Figs. 1-17 to 1-19) are noticed.
  - The interior packing is composed of generally dense, van Der Waals interactions, although voids are found (Scheme 1-4).

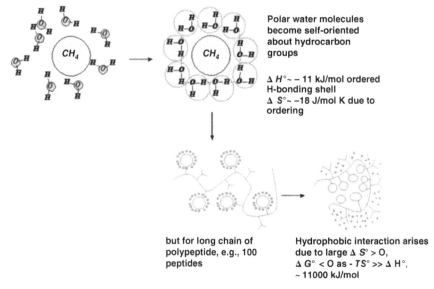

Polar water molecules
become self-oriented
about hydrocarbon
groups

$\Delta H° \sim -11$ kJ/mol ordered
H-bonding shell
$\Delta S° \sim -18$ J/mol K due to
ordering

but for long chain of
polypeptide, e.g., 100
peptides

Hydrophobic interaction arises
due to large $\Delta S° > O$,
$\Delta G° < O$ as $-TS° \gg \Delta H°$,
~ 11000 kJ/mol

FIGURE 1-16   Hydrophobic interaction and solvent entropy.

FIGURE 1-17   Cytochrome c, tuna, PDB 3CYT (Takano and Dickerson, 1980).

FIGURE 1-18   Cytochrome c_{553}, *Bacillus pasteurii*, PDB 1C75 (Benini et al., 2000).

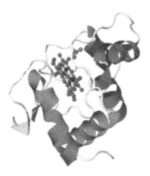

FIGURE 1-19 Cytochrome c_2, *Rhodobacter sphaeroides*, PDB 1CXC (Axelrod et al., 1994).

- ○ Ionized groups occur:
  - (i) On the outer surface (majority, 100%)
  - (ii) In clefts or inner sites, where they have a particular/special role in the protein's function
- *Unfolding* is caused by:
  - (i) Conformational entropy—more orientations accessible
  - (ii) Strain in folded state
- *Quaternary structure* refers to the aggregation of polypeptide chains into larger assemblies such as in hemoglobin (Fig. 1-20), and hemerythrin. In hemerythrin

$$8\text{He} \quad \rightleftarrows \quad \text{He}_8$$

Subunits (13,500)   Octamer (108,000)

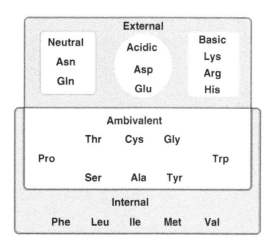

SCHEME 1-4 External, ambivalent, and internal amino acids.

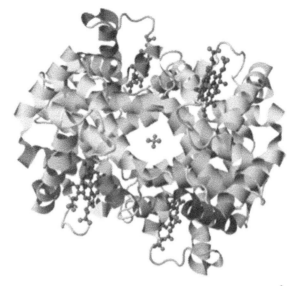

FIGURE 1-20 Quaternary structure of human oxyhemoglobin at 2.1 Å resolution, PDB 1HHO (Shaanan, 1983).

There are two types:

(i) Isologous: ⇆

(ii) ii. Heterologous: ⇉

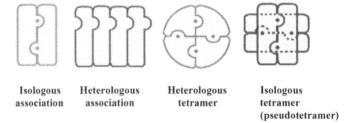

| Isologous association | Heterologous association | Heterologous tetramer | Isologous tetramer (pseudotetramer) |

- *Aggregation* is driven by:
  (i) Hydrogen bonding
  (ii) Hydrophobic interaction
  (iii) Salt bridges: Lys^+, Arg^+ vs. Glu^-, Asp^-

- Ionic moieties brought together, Coulombic attraction, so $\Delta H < 0$, solvent H_2O released, so $\Delta S > 0$. For example, the association of the four subunits of Hb_4 has Standard Gibbs free energy, $\Delta G^{\circ} = -60\,kJ, -15\,kJ/subunit$.

Metal Amino Acid Complexes

Arrange the binding stabilities of various metal ions toward different amino acids.

- M^{n+} binding group are:

$$\text{Less stable} \xrightarrow{\text{Met}^-\ \text{Tyr}^-\ \text{Glu}^-\ \text{Asp}^-\text{His}} \text{More stable}$$

Define:
(a) **Enzymes**
(b) **Metalloenzymes**
(c) **Coenzymes**
(d) **Cofactors**

- Enzymes:
  - Catalyze biological processes
  - Control rates of reactions
  - Promote certain geometries in the transition state, which lowers the activation energy for the formation of one product rather than the other
- *Matalloenzymes* are composed of:
  - A protein structure (called apoprotein/apoenzyme)
  - Small prosthetic group
  - Prosthetic groups that are a simple metal ion, a complex metal ion, or an organic compound
- *Coenzyme* reversibly combines with the enzyme for a particular reaction and then is released to combine with another.
- *Cofactors* are the prosthetic groups and the coenzymes.
  - Provide ability to *transfer* molecular groups or radicals that polypeptides cannot (e^-, phosphate, alkyl group, etc.), in enzyme-catalyzed reactions
  - Provide ability to *bind and transfer* molecules that polypeptide cannot (e^-, O_2)
  - Simple cofactors: Mg^{2+}, Fe^{2+}, Zn^{2+}, etc.
  - Several are nucleotide derivatives of ATP
    - (i) Nicotinamide adenine dinucleotide (NADH) (Scheme 1-5): a mild source of H^- as NADH. Note pervasive presence of phosphate ester links.
    - (ii) Ubiquinone (coenzyme-Q), CoQ_6, CoQ_{10} (Scheme 1-6).

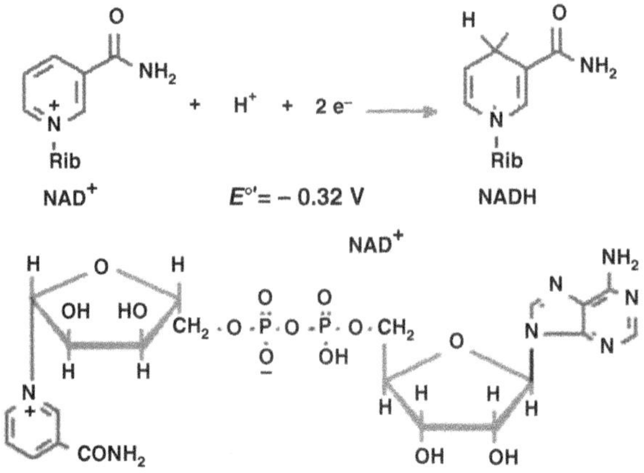

SCHEME 1-5   Nicotinamide adenine dinucleotide.

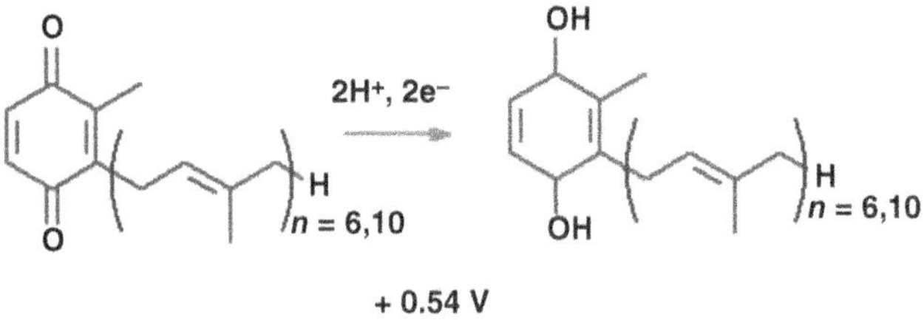

$E^{\circ\prime} = -0.32$ V

$+ 0.54$ V

SCHEME 1-6   CoQ$_6$ and CoQ$_{10}$.

(iii) Flavins, i.e., flavin mononucleotide (FMN) (riboflavin phosphate) (Scheme 1-7):

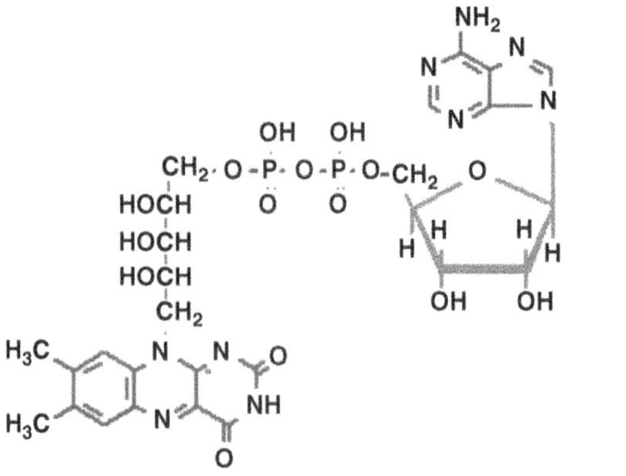

Riboflavin phosphate

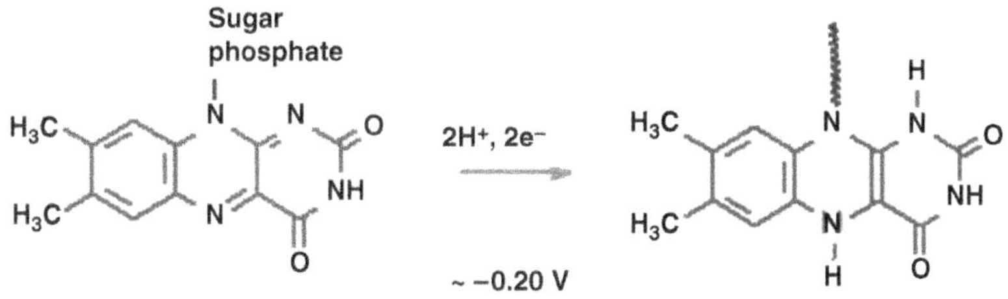

SCHEME 1-7 Flavin mononucleotide.

See also flavin adenine dinucleotide (FAD).

(iv) Tetrapyrrolic cofactors:

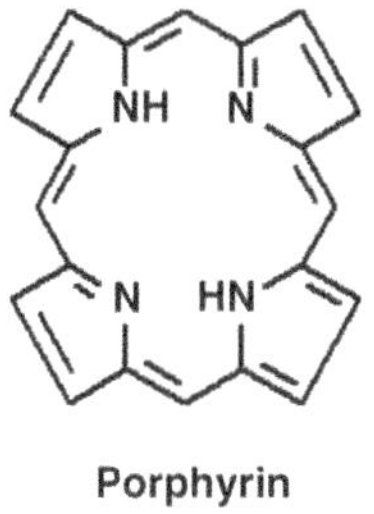

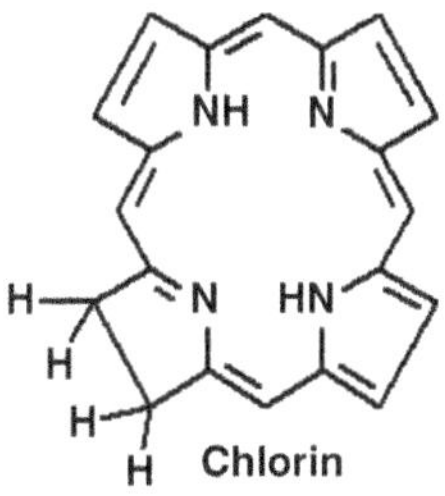

Porphyrin   Chlorin

(v) Phosphates

- Suitable as noncarbon "universal" component
- Carbon-based esters subjected to hydrolysis by digestive enzymes
- Must be readily available in environment, so $G°$ is not wasted in hunting and concentrating
- Used in presence of Cys–SH, so cannot be oxidized or reduced
- Used for persistent structures, so should be inert (slow reacting)
- Kinetic consideration:

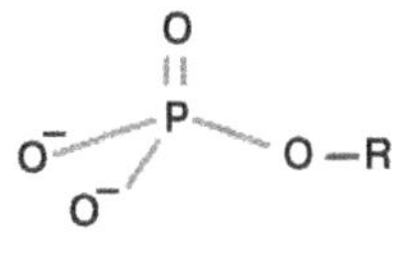

P^{5+} used as ester

Rate of hydrolysis depends on rate of P–O bond scissions

Comparison of P–O versus M–O Bond Scission Rates for Row 3 Elements

Compare M–O bond scission rates for row-3 elements in group-oxidation states in H_2O, and show the advantages of P–O.

- Use O^{17}or O^{18} isotopes as tags [(e.g. O^{17}, nuclear magnetic resonance (NMR)]:

- $t_{1/2}$ for O exchange at room temperature in seconds:

Na^+	Mg^{2+}	Al^{3+}	Si^{4+}	P^{5+}
10^{-9}	10^{-5}	10^0	10^3	10^{10}

Therefore, phosphates are better than silicate.

- Relative exchange rates: $Na^+ > Mg^{2+} > Al^{3+} > Si^{4+} > P^{5+}$

    → Slower
    → Size shrinks (r decreases)
    → Charge increases (q increases)

- Exchange rates (↓) decrease as q/r (↑) increase as well as the ionic potential gets higher.
- High ionic potential polarizes ligand, e.g., H_2O, and introduces covalency.

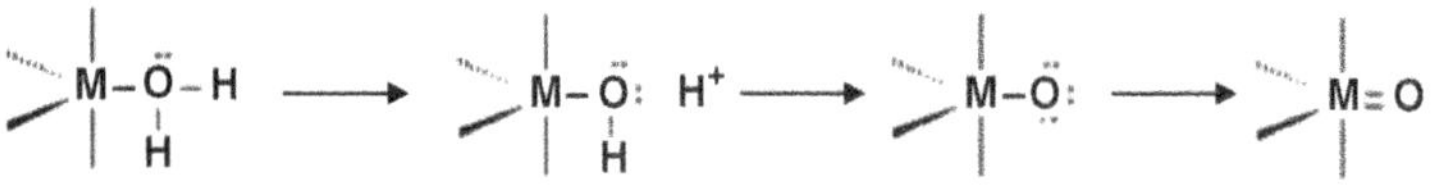

Consequently, H^+ is released, as increases of q/r lead to decreasing pK_a:

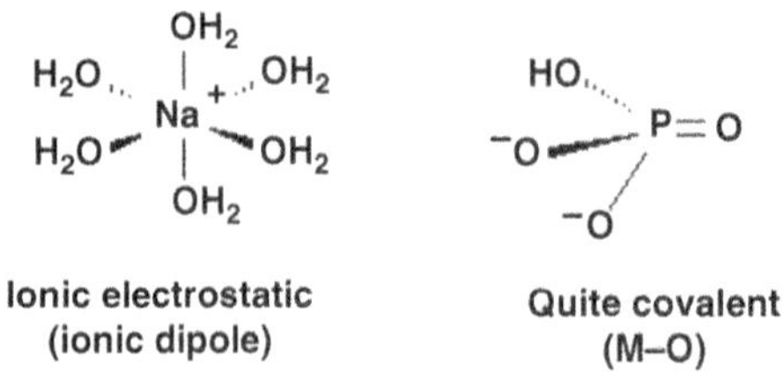

Ionic electrostatic   Quite covalent
(ionic dipole)   (M–O)

Redox Advantages: Sulfur versus Phosphorus

Consider all S and P possible anions, which will be the strongest oxidizing agent? And which will be the strongest reducing agent?

- Comparison of P and S:

$$2\ SO_4^{2-} + 4H^+ + e^- \longrightarrow S_2O_6^{2-} + 2\ H_2O \qquad E^\circ\ (mV) = -220$$
$$S^{6+} + e^- \longrightarrow S^{5+}$$
$$2\ SO_3^{2-} + 4H^+ + 2e^- \longrightarrow S_2O^{2-} + 2\ H_2O \qquad E^\circ\ (mV) = -86$$
$$S^{4+} + 2\ e^- \longrightarrow S^{2+}$$
$$S + 2H^+ + 2e^- \longrightarrow H_2S \qquad E^\circ\ (mV) = +142$$
$$S^0 + 2\ e^- \longrightarrow S^{2-}$$
$$SO_4^{2-} + 2H^+ + 2e^- \longrightarrow SO_3^{2-} + H_2O \qquad E^\circ\ (mV) = +172$$
$$cS^{6+} + 2e^- \longrightarrow S^{4+}$$
$$SO_4^{2-} + 8H^+ + 6e^- \longrightarrow S + 4H_2O \qquad E^\circ\ (mV) = +35$$
$$S^{6+} + 6e^- \longrightarrow S^0$$
$$S_2O_6^{2-} + 4H^+ + 2e^- \longrightarrow 2\ SO_3^{-2} + 4H_2O \qquad E^\circ\ (mV) = +564$$
$$S^{5+} + 2e^- \longrightarrow S^{4+}$$
$$PO_3^{3-} + 2H^+ + 2e^- \longrightarrow PO_2^{3-} + H_2O \qquad E^\circ\ (mV) = -499$$
$$P^{3+} + 2e^- \longrightarrow P^+$$
$$PO_2^{3-} + 4H^+ + e^- \longrightarrow P + 2H_2O \qquad E^\circ\ (mV) = -508$$
$$P^+ + e^- \longrightarrow P^0$$
$$PO_4^{3-} + 2H^+ + 2e^- \longrightarrow PO_3^{3-} + H_2O \qquad E^\circ\ (mV) = -276$$
$$P^{+5} + 2e^- \longrightarrow P^{3+}$$
$$P + 3H^+ + 3e^- \longrightarrow PH_3 \qquad E^\circ\ (mV) = -63$$
$$P^0 + 3e^- \longrightarrow P^{3-}$$

- Hot, concentrated H_2SO_4 is a strong oxidizing agent; dilute H_2SO_4 is not an oxidizing acid.

    SO_3^{2-}, sulfite ion is mild reducing agent

    HSO_3^-, hydrogen sulfite ion is mild reducing agent

    SO_4^{2-}, sulfate ion is oxidizing agent only in concentrated acid
- H_3PO_4, is not oxidizing agent

    $H_2PO_3^-$, dihydrogen phosphite ion is reducing agent in H^+ or OH^-

    HPO_3^{2-}, hydrogen phosphite ion is reducing agent in H^+ or OH^-

Bioenergetic Phosphate Derivatives

Give examples of the phosphate adducts that are bioenergetically important.

- Other PO_3^{3-} adducts are bioenergetically important:

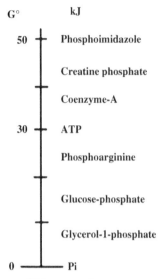

- ATP acts as G – currency of bioenergetics:

$$ATP + H_2O + (Mg^{2+}) \rightarrow ADP + Pi \qquad \Delta G^\circ = 30 \, kJ/mol$$

Corresponds to free energy available from transferring PO_3^- unit to H_2O.

REFERENCES

H. L. Axelrod, G. Feher, J. P. Allen, A. J. Chirino, M. W. Day, B. T. Hsu, and D. C. Rees, *Acta Crystallogr., Sect. D*, 50, 596–602 (1994).

S. Benini, A. Gonzalez, W. R. Rypniewski, K. S. Wilson, J. J. Van Beeumen, and S. Ciurli, *Biochemistry*, 39, 13115–13126 (2000).

F. A. Cotton and G. Wilkinson, in *Advanced Inorganic Chemistry*, 4th ed., Wiley, p. 1311, Hoboken, NJ (1980).

N. L. Ogihara, G. Ghirlanda, J. W. Bryson, M. Gingery, W. F. DeGrado, and D. Eisenberg, *Proc. Natl. Acad. Sci. USA*, 98, 1404–1409 (2001).

S. E. Phillips, *J. Mol. Biol.*, 142, 531–554 (1980).

B. Shaanan, *J. Mol. Biol.*, 171, 31–59 (1983).

T. Takano and R. Dickerson, *Proc. Natl. Acad. Sci. USA*, 77, 6371–6375 (1980).

SUGGESTIONS FOR FURTHER READING

Texts on Biochemistry

1. J. J. R. F. da Silva and R. J. P. Williams, *Biological Chemistry of Elements*, Oxford University Press, New York (1991).

2. L. Stryer, *Biochemistry*, Freeman, New York (1995).

3. D. Voet and J. G. Voet, *Biochemistry*, Wiley, New York (1995).

4. *Advances in Protein Chemistry*, Academic Press, New York.

5. P. D. Boyer, Ed., *The Enzymes*, 3rd ed., Academic, New York (1970–1986).

6. H. Neurath, Ed., *The Proteins*, 2nd ed., Academic, New York, vol. I to vol. V.

Texts on Bioinorganic Chemistry

7. I. Bertini, H. B. Gary, E. I. Stiefel, and J. S. Valentine, in *Biological Inorganic Chemistry*, University Science Books Sausalito, CA, (2007).

8. H-B. Kraatz and N. Metzler-Nolte, in *Concepts and Models in Bioinorganic Chemistry*, Wiley, Hoboken, NJ (2006).

9. W. Kaim and B. Schwederski, in *Bioinorganic Chemistry: Inorganic Elements in the Chemistry of Life*, Wiley, New York (1994).

10. E-I. Ochaia, in *Bioinorganic Chemistry: A Survey*, Elsevier, San Diego, CA (2008, 2011).

11. R. M. Roat-Malone, in *Bioinorganic Chemistry: A Short Course*, Wiley-Interscience, Hoboken, NJ (2002).

12. S. J. Lippard and J. M. Berg, in *Principles of Bioinorganic Chemistry*, University Science Books, Mill Valley, CA (1994).

13. L. Que, Jr., Ed., *Physical Methods in Bioinorganic Chemistry*, University Science Books, Mill Valley, CA (2001).

14. R. B. King, Ed., *Encyclopedia of Inorganic Chemistry*, Wiley, New York, (1994).

15. J. E. Macintyre, A. Exec. Ed., in *Dictionary of Inorganic Compounds*, Chapman and Hall, London (1992).

16. A. S. Brill, in *Transition Metals in Biochemistry in Molecular Biology, Biochemistry, and Biophysics*, A. Kleinzeller, G. F. Springer, and H. E. Wittman, Eds., Vol. 26, Springer-Verlag, Berlin (1977).

17. D. R. Williams, Ed., *An Introduction to Bio-Inorganic Chemistry*, C. C. Thomas, Springfield, IL (1976).

18. R. F. Gould, Ed., *Bioinorganic Chemistry*, ACS Advances in Chemistry Series No. 100; K. N. Raymond, Ed., *Bioinorganic Chemistry II*, ACS Advances in Chemistry Series No. 162, American Chemical Society, Washington, DC (1977).

19. A. W. Addison, W. R. Cullen, D. Dolphin, and B. R. James, Eds., *Biological Aspect of Inorganic Chemistry*, Wiley, New York (1977).

20. E. I. Ochiai, *Bioinorganic Chemistry: An Introduction*, Allyn and Bacon, Wiley, Rockleigh, Boston (1977).

21. H. Sigel, in *Metal Ion in Biological Systems*, Vols. 1–20, Marcel Dekker, Basel.

22. G. L. Eichhorn, Ed., *Inorganic Biochemistry*, Elsevier, Amsterdam (1973).

23. M. N. Hughes, in *The Inorganic Chemistry of Biological Process*, 2nd ed., Wiley, London (1981).

24. P. M. Harrison and R. J. Hoare, in *Metals in Biochemistry*, Chapman and Hall, London and New York (1980).

25. R. J. P. Williams and J. R. R. F. da Silva, in *New Trend in Bio-Inorganic Chemistry*, Academic, London (1978).

26. C. A. Mc Auliffe, Ed., *Techniques and Topic in Bioinorganic Chemistry*, Macmillan, London (1975).

2

ALKALI AND ALKALINE EARTH CATIONS

Alkali metals and alkaline earth cations constitute about 1% of human body weight, whereas the first transition series represents less than 0.01%.

The alkali metals have a single s electron in the outer shell with low ionization enthalpies. The chemistry of these metals is of their M^+ ions, which have a sphere-shaped and low polarizability. Alkali metal ions are weakly complexed by simple anions or monodentate ligands. Chelation is a necessary condition for significant complexation.

Sodium is the major component of the cations of extracellular fluid. It is largely associated with chloride and bicarbonate to regulate the acid–base equilibrium. The other important function of sodium is maintenance of the osmotic pressure of body fluid and consequently protection of the body against excessive fluid loss. It also functions in the preservation of normal irritability of muscle and permeability of the cell. Deficiency of sodium causes nausea, diarrhea, muscular cramps, dehydration, and Addison's disease.

Potassium is the principal cation of intracellular fluid. Within the cell, it functions like that of sodium. High intracellular potassium concentrations are necessary for some metabolic functions, including protein biosynthesis. A number of enzymes, such as pyruvate kinase, diol dehydratase, and yeast aldehyde dehydrogenase require K^+ for maximal activity. A prolonged deficiency of potassium may yield serious harm to the kidney. The symptoms of low serum potassium concentrations (hypokalemia) involve muscle weakness, irritability, and paralysis. High serum potassium (hyperkalemia) occurs in adrenal insufficiency (Addison's disease).

Chemistry of Metalloproteins: Problems and Solutions in Bioinorganic Chemistry, First Edition.
Joseph J. Stephanos and Anthony W. Addison.
© 2014 John Wiley & Sons, Inc. Published 2014 by John Wiley & Sons, Inc.

Alkaline earth cations are highly electropositive and have standard electrode potentials similar to those of alkali ions. All alkaline earth cations, M^{2+}, are smaller and less polarizable than the isoelectronic alkali ions, M^{1+}. The cations Mg^{2+} and Ca^{2+} show a tendency to form complexes with oxygen ligands; however, Mg^{2+} tends more than Ca^{2+} to be bound by nitrogen donors.

Magnesium is the fourth most abundant cation in the body. The body contains about 21 g of magnesium. Seventy percent is combined with calcium and phosphorus in the complex salts of bone; the remainder is in the soft tissues and body fluids. Magnesium affects muscle and nerve impulse and acts on some enzymes, particularly those of glycolysis. Convulsions and anesthesia are associated with deficiency and excess of magnesium, respectively.

Calcium is present in the body in larger amounts than any other mineral element. About 99% of calcium in the body is in the skeleton. Calcium is essential to muscle contraction, normal heart rhythm, nerve impulse, and activation of some enzymes. Diseases arising from deficiency of calcium are bone deformities and tetany, while diseases associated with an excess of calcium are cataracts, gallstones, and atherosclerosis.

The discussion will be concerned with the roles of these ions in the biological systems and the different mechanisms of their movement through the membrane's cell. The mechanism of the Na^+/K^+ ion pump ATPase and the roles of the metals' cations are explored.

Specific points about the types, examples, and selectivity of various ionophores to metal ions and the adapted stereochemistry are made.

This is followed by a discussion of how metal ions master information transmission through the nerve, DNA or RNA polymerase, Embden–Meyerhof pathway, pyruvate kinase, and phosphoglucomutase and the interconversion of ADP and ATP.

How calcium ions control muscle contractions is discussed and examples of calcium proteins are provided. Finally, we give a comparison between the biological roles of calcium and those of magnesium.

Na^+ and K^+ in Biological Systems

What are the roles of Na^+ and K^+ in biological systems?

- Sodium is the principal *extracellular* cation (Table 2-1).
- K^+ is the principal *intracellular* cation.

TABLE 2-1   Concentration of Na^+, K^+, and Cl^- in Seawater, Red Cell, and, Blood Plasma

	In Seawater (mM)	In Red Blood Cell (mM)	In Blood Plasma (mM)
Na^+	460	11	160
K^+	10	92	10
Cl^-	550	50	100
μ	650	85	160

- The precise concentrations of Na^+ and K^+ in and out of the cell indicate not only the biological role but a discriminatory mechanism, which controls the selective uptake of K^+ into the cell from plasma.
- Biological roles of Na^+ and K^+:
  1. Act as trigger and control mechanisms. The selective distribution of cations inside and outside the cell and breakdown of this selectivity can act as a trigger for some biological events.
  2. As principal biological cations with Cl^-, provide the ionic strength ($\mu = \Sigma\, m_i z_i^2$, where m_i is the molality (mol/kg) and z is the charge of positive or negative ions), which defines:
     ○ The operating environment for biological macromolecules (proteins, polysaccharides, and nucleic acids)
     ○ The conformations of these macromolecules
  3. Responsible for concentration differences (gradients) across biological membranes (e.g., inside vs. outside of cell; Table 2-1):
     ○ Store energy
     ○ Transmit information

Solvent and Solute Movement through Cell Membrane

Why are there different mechanisms for solvent and solute movement through the membrane's cell, and what are these mechanisms?

- Membranes differ in their composition, structure, and permeability and respond to different solvents.
- Some permit only solvent to pass through or the solvent may actually dissolve in the membrane as it passes through it.
- Some permit the passage of ions and small molecules.
- Some require a carrier molecule to encapsulate and transport the charged species.
- The mechanisms are:
  ○ Osmosis
    (a) Essential feature: Vapor pressures of the solvent on the two sides of a membrane are different
    (b) Principal moving components: small ions and water
  ○ Donan effect (passive transport)
    (a) Essential feature: Bulky charged species are confined to one side of a membrane and induce unequal distribution of small ions at equilibrium.
    (b) Principal moving components: small ions
  ○ Imbibition
    (a) Essential feature: Water is tightly bound by macromolecules.
    (b) Principal moving component: water

o Active transport

    (a) Essential feature: Solutes are transferred across a membrane into a region of higher free energy as a consequence of exergonic process occurring within the membrane.

    (b) Principal moving components: solutes.

o Other important small ions: Ca^{2+}, Mg^{2+}, HPO_4^{2-}, HCO_3^-

Osmotic Pressure

What is the relationship between osmotic pressure and molecular weight of the components that move through the membrane's cell?

- Solvent passes through a membrane from a dilute solution into a more concentrated one.

- If pressure is applied to the more concentrated solution, the flow of solvent can be slowed, halted, or reversed depending on the amount of pressure applied.

- The osmotic pressure Π of the solution is the minimum pressure that must be applied to the solution to prevent the flow of the solvent from pure solvent into the solution:

$$\frac{\Pi}{cRT} = A_1 + A_2c + A_3c^2 + \cdots$$

where c is the concentration, $A_1 = 1/M_n$, and M_n is the average molecular weight, defined as

$$M_n = \frac{\sum_i n_i M_i}{\sum_i n_i}$$

- For many dilute solutions it is sufficient to ignore the terms with $A_3, A_4, \cdots$, yielding

$$\frac{\Pi}{cRT} = \frac{1}{M_n} + A_2c$$

- A straight line will be obtained when Π/cRT is plotted against c; the intercept is the reciprocal of M_n.

- The slope is A_2: The larger is A_2, the more nonideal are the solutions examined.

Passive Transport (Donnan Effect) Concentration Gradients

In passive transport, assume the initial concentrations are as follows:

Protein Side of Membrane	Other Side of Membrane
$[\text{Protein}^{-z}] = y_0 = 50\,\text{mM}$	$[\text{Protein}^{-z}] = 0$
$[K^+] = x_0 = 100\,\text{mM}$	$[K^+] = x_0 = 100\,\text{mM}$
$[Cl^-] = ? = 50\,\text{mM}$	$[Cl^-] = x_0 = 100\,\text{mM}$

TABLE 2-2   Initial Ion Concentrations before Equilibrium

Protein Side of Membrane	Other Side of Membrane
$[Protein^{-z}] = y_0 = 50$ mM	$[Protein^{-z}] = 0$
$[K^+] = x_0 = 100$ mM	$[K^+] = x_0 = 100$ mM
$[Cl^-] = x_0 - zy_0 = 50$ mM	$[Cl^-] = x_0 = 100$ mM

Note: Membrane is permeable to H_2O, K^+, Cl^- but not to proteins.

Calculate:

(a) The equilibrium concentrations of $[Protein^{-z}]$, $[K^+]$, and $[Cl^-]$

(b) The potential that is generated by the concentration gradient

How can the cell dissipate the generated E_K^+?

- Large-molecule charged species are confined to one side of a membrane and cause unequal distribution of small ions at equilibrium.
- Principal moving components are small ions such as K^+ and Na^+, but protein molecules cannot move through the membrane.
- To illustrate the origin of the Donnan effect, consider the initial concentrations given in Table 2-2.
- Assume $x_0 = 100$ mM, $y_0 = 50$ mM, and $z = 1$. Then

$$\{G_{Cl} = G^{\circ}_{Cl} + RT \ln[50]\}_{\text{protein side}}$$

$$\{G_{Cl} = G^{\circ}_{Cl} + RT \ln[100]\}_{\text{other side}}$$

$$\Delta G_{Cl} = \{G^{\circ}_{Cl,RHS} - G^{\circ}_{Cl,LHS}\} + RT \ln[100/50]$$

where G is the Gibbs free energy and Cl^- has more free energy on the right-hand side (RHS), so move to reach equilibrium, to make $\Delta G_{Cl} = 0$.

- There will be a transfer of $[Cl^-]$ from the other membrane side to the protein side.
- Protein Pr^- cannot diffuse from the left-hand-side (LHS) to the RHS, so to keep electrical neutrality; K^+ diffuses with Cl^- to the LHS.
- Chemical potential must be equal on both sides of the membrane at equilibrium. This requires that the product $[K^+][Cl^-]$ must have equal values on both sides of the membrane. Therefore, the $[K^+]$ gradient developed.
- $[K^+]$ is not equal on both sides of the membrane, $[K^+]_{LHS} > [K^+]_{RHS}$, and opposes the $[Cl^-]$ gradient.
- The solutions on each side of the membrane would no longer be electrically neutral.

TABLE 2-3   Ion Concentrations at Equilibrium

Protein Side of Membrane	Other Side of Membrane
$[\text{Protein}^{-z}] = y_0 = 50\,\text{mM}$	$[\text{Protein}^{-z}] = 0$
$[\text{K}^+] = x_0 + x = 114.3\,\text{mM}$	$[\text{K}^+] = x_0 - x = 85.7\,\text{mM}$
$[\text{Cl}^-] = x_0 - zy_0 + x = 64.3\,\text{mM}$	$[\text{Cl}^-] = x_0 - x = 85.7\,\text{mM}$

- Let x represent the concentration of K^+ that crosses the membrane:

$$\{[\text{K}^+][\text{Cl}^-]\}_{\text{other side}} = \{[\text{K}^+][\text{Cl}^-]\}_{\text{protein side}}$$

$$(x_0 - x)^2 = (x_0 + x)(x_0 - zy_0 + x)$$

$$x = \frac{x_0 y_0 z}{4x_0 - y_0 z}$$

$$= \frac{(100)(50)(1)}{(4)(100) - (50)(1)} = 14.3\,\text{mM}$$

Then the state in Table 2-3 will be achieved.

- At the same time, a Nernstian concentration cell develops:

$$E = E^0 - \frac{RT}{zF} \ln \frac{[\text{K}_i^+]}{[\text{K}_o^+]}$$

- The cell can be thought of as $\text{K}|\text{K}^+||\text{K}^+|\text{K}$, so

$$E^0 = 0 \quad |E_{\text{membrane}}| = 0.059 \log \frac{114}{86} = 7\,\text{mV}$$

- This potential is generated by the electrolyte gradient.
- Cells commonly have channels for K^+ and Na^+, so in living cells, these ions move, current due to ion flow is observed, and a nonzero electrical potential is formed.
- Common condition: K^+ channels are opened, but Na^+ channels are closed.
- *Colicin-A*, a protein that makes a membrane "leaky" to Na^+, destroys the $[\text{Na}^+]$ gradient, dissipates ΔG_{Na^+} (stored), and dissipates E_{Na^+} and thus also E_{K^+}.

Active Transport Concentration Gradients

Give an example of active transport, and discuss the requirements and the consequences of this mechanism.

- *Na$^+$/K$^+$ ion Pump ATPase* (a Mg^{2+} enzyme) (Fig. 2-1) is a membrane enzyme and M^+-dependent reaction. This enzyme catalyzes the translocation of Na^+

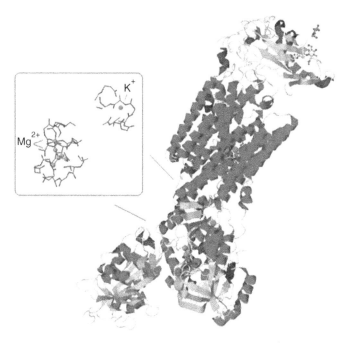

FIGURE 2-1   Crystal structure of sodium–potassium pump at 2.4 Å resolution, PDB 2ZXE (Shinoda et al., 2009).

and K^+ and requires the hydrolysis of ATP as cofactor. Mg^{2+} is also needed for maximum activity. It should be noted that the enzyme catalyzes other reactions, which do not result in translocation of cations.

- Several of the residues forming the cavity in the Na^+/K^+ ATPase are homologous to those binding calcium in the Ca^{2+} ATPase of sarco(endo) plasmic reticulum.
- Solutes are transported actively across a membrane into a region of higher free energy because of exergonic processes ($\Delta G < 0$) occurring within the membrane.
- The principal moving components are solutes.
- The K^+ concentration within the cell is maintained at ~0.1 M so it is available for:
  ○ Glucose combustion
  ○ Protein synthesis
- The Na^+/K^+ gradients are particularly important to the operation of nerve cells.
- This enzyme plays a critical role in cell homeostasis since it keeps Na^+ and K^+ gradients between the intra- and extracellular milieu, which is essential for maintenance of the cell volume.
- Additionally, Na^+/K^+ ATPase offers energy for transport activity of many secondary transporters in order to provide the cell with nutrients or regulate intracellular concentrations of ions, which are implicated in specialized cellular roles such as transmission of nerve impulses or muscle contraction.

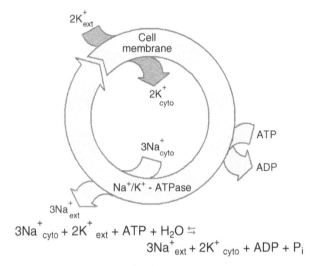

$$3Na^+_{cyto} + 2K^+_{ext} + ATP + H_2O \leftrightarrows$$
$$3Na^+_{ext} + 2K^+_{cyto} + ADP + P_i$$

FIGURE 2-2   Action of Na^+/K^+ ATPases (Forgac and Chin, 1985).

- The basic operating unit of Na^+/K^+ ATPase is composed of an α and β subunit.
- The α and β subunits can assemble in different combinations (isozymes) to form the functional pump.
- All of the isozymes are functional but different in their turnover rates depending on the composition ratio of α and β subunits.
- The α subunit carries the functional properties of Na^+/K^+ ATPase—mainly it binds and transports the cations and hydrolyzed ATP—and is intermediately phosphorylated.
- The β subunit is necessary for the structural and functional activity of the α subunit. Consequently, it influences the K^+ and Na^+ activation kinetics of the Na^+/K^+ ATPase isozyme.
- Variations in the activity of these isozymes are determined by a cooperative interaction mechanism between α and β isoforms.
- The Na^+/K^+ ATPase transports two K^+ ions into and three Na^+ ions out of the cell, using the energy of the hydrolysis of one molecule of ATP (Fig. 2-2).
- This pumping must be done against unfavorable Na^+ and K^+ concentration gradients, an endergonic process (free energy change under standard conditions, $\Delta G° > 0$) (Table 2-4).

TABLE 2-4   Pumping Against Unfavorable Na^+ and K^+ Concentration Gradients

Protein Side of Membrane	Other Side of Membrane
[Protein] = 50 mM	
$[K^+]$ = 180 mM	$[K^+]$ = 20 mM
$[Cl^-]$ = 130 mM	$[Cl^-]$ = 20 mM

- Pumps K^+ in, Na^+ out, using $\Delta G^{\circ}_{\text{hydrolysis}}$ of ATP as the energy source (Fig. 2-2):

$$\text{Adenosine triphosphate} + H_2O \rightleftharpoons \text{adenosine diphosphate}$$
$$+ HPO_4^{2-} \qquad \Delta G^{\circ} = -30\,\text{kJ/mol}$$

Adenosine triphosphate, ATP

Suggest a mechanism for translocation of Na^+ by Na^+/K^+-ATPase.

- In Scheme 2-1, species A is identified as an intermediate and is characterized by the mixed phosphate with aspartic acid from the enzyme. This intermediate is responsible for Na^+ transport outside the cell.
- The second key is the deprotonation of the phosphate group in the extracellar plasma. This favors the replacement of Na^+ by K^+.
- The K^+ complex returns to the cytoplasm. Then phosphate returns to the cytoplasm (step 5) and ATP enters (step 6).
- Protonation of the terminal phosphate of ATP (step 7), now bridging Mg^{2+} and K^+, leads to reversal of the Na^+/K^+ exchange (step 8).
- Displacement of ADP from the terminal phosphate by aspartate returns the sequence to the intermediate A.
- Therefore, if the cell is made "leaky," then the pump runs and ATP is used up.

How can charged K^+ and Na^+ cross through the nonpolar medium of the cell's wall?

- A membrane or barrier surrounds the cell.
- This barrier separates its aqueous interior from the plasma, selectively allows ions and nutrients into the cell, and allows unwanted material or material for use somewhere else.
- The membrane consists of a lipid bilayer and globular proteins (Fig. 2-3).
- The alkali metal cations must traverse this barrier passing through an unfavorable medium of low dielectric constant.
- There are two *naturally* occurring groups of compounds that have a clear impact on the permeability of membrancs to cations.
- These groups of compounds are collectively termed *ionophores*.

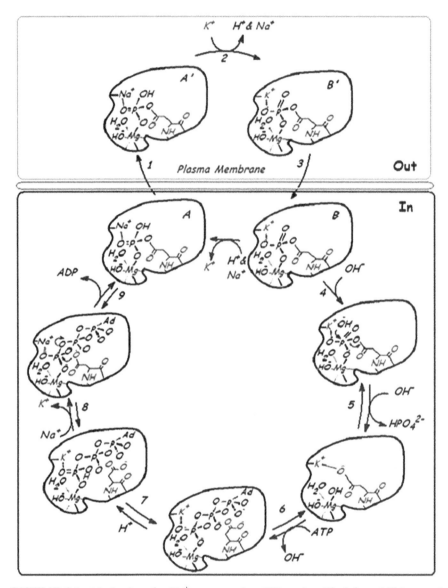

SCHEME 2-1 Mechanism for Na^+ transport by ATPase (modified from Mildvan and Grisham, 1974). (See the color version of this scheme in Color Plates section.)

- One group is *the channel-forming ionophores*. These channels:
  - Are the naturally occurring pores in the membranes
  - Are proteins that span the membrane, providing a hydrophilic channel through which cations pass
  - Have a highly specialized gating mechanism to distinguish between cations and control entry

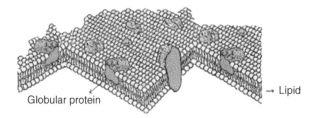

FIGURE 2-3 Lipid bilayer membrane.

Examples of channel-forming ionophores are gramicidin and F.30 alamethicin, which are linear peptides that transport cations M^+ and M^{2+}, respectively.

- The other group is the *mobile carrier ionophores*, or *antibiotic* molecules. These carriers bind and encapsulate the cations through the membrane. Mobile ionophores:
    - Can selectively transport K^+ and Na^+ across the membrane
    - Are neutral at physiological pH and act as discriminatory cation carriers by forming complexes with alkali metal cation
    - Should have an appreciable K_f value
    - Should be soluble in a lipid layer

Ionophores and Ion Transport

Design a model to demonstrate ionophores' role of ion transport through liquid membrane systems, and what is the difference between the ionophores and ion receptors?

- A model is constructed using a U-tube consisting of two aqueous layers separated by a semipermeable medium or membrane.
- $CHCl_3$ is the nonpolar medium and is chosen because its dielectric constant is similar to that of membrane.
- K–o–nitrophenolate is a colored K^+ salt introduced on one side of the barrier (Fig. 2-4a).
- A second tube is prepared with a carrier ligand (enniatin-B) in the $CHCl_3$ layer (Fig. 2-4b).
- The movement of color on transport by the ligand carrier (enniatin-B-K) can then be noted (Fig. 2-4c).
    - If the color transfers into the organic layer but no further, the ligand acts as an *ion receptor*.
    - If the color is transferred through the $CHCl_3$ layer and into the second aqueous layer, the added molecule is acting as an *ion carrier*.
    - Enniatin-B is the ionophore; the ligand carrier, enniatin-B-K^+ is the ionophore complex.

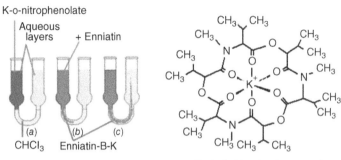

Enniatin B-potassium complex

FIGURE 2-4   Model of ionophore role (modified from Fenton, 1977).

Main Classes of Ionophores

What are the main classes of ionophores? List examples for each class.

- There are two main classes for the ionophores:
  - ○ *Synthetic ionophores* include crowns and cryptates, which are toxic.
  - ○ *Natural ionophores* are produced by fungi, lichen, as microbiological warfare agents against bacteria.
- Examples of main classes of *synthetic ionophores*:
  1. Crown ethers

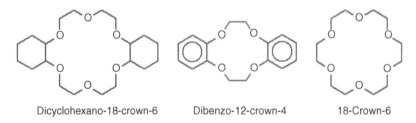

Dicyclohexano-18-crown-6   Dibenzo-12-crown-4   18-Crown-6

12-Crown (Scheme 2-2) forms $Na(12\text{-crown-4})_2^+$; Na^+ is an 8-coordinate in the "crown cube," tetragonal antiprism:

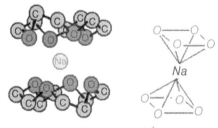

$Na(12\text{-crown-4})_2^+$

HO – CH₂– CH₂–OH

+

Br – CH₂–CH₂–Br

Br $\longrightarrow$ O O O OH

+

1,4,7,10-tetraoxo-cyclo-dodecane "12-crown-4"

SCHEME 2-2   Synthetic pathway of 12-crown-4.

2. Octopus molecule: Shows weaker binding as no macrocycle effect is present:

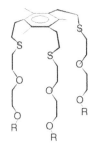

3. Lariat ether:

where R = COO⁻, NH₂, or Pyr.

4. Cryptand ligands: The complexes are formed and are called cryptates.

• These macrobicyclics have two donor atoms with three bonds:

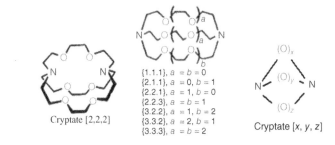

Cryptate [2,2,2]

{1.1.1}, a = b = 0
{2.1.1}, a = 0, b = 1
{2.2.1}, a = 1, b = 0
{2.2.3}, a = b = 1
{3.2.2}, a = 1, b = 2
{3.3.2}, a = 2, b = 1
{3.3.3}, a = b = 2

Cryptate [x, y, z]

- The cryptands provide three-dimensional selectivity toward cations, compared to the two-dimensional selectivity provided by the crown ethers, in terms of the size of the molecular cavity.
- Cryptate [2,2,2] totally engirds the metal ion, gives very large K_f values, and consequently has unusual chemistry. Examples:
    - (a) [2,2,2] allows insoluble $BaSO_4$ to dissolve in H_2O to 50-g/L level.
    - (b) Adding Na^0 to NH_3 or ethylamine or THF, then adding [2,2,2], yields gold crystals, an electride salt:

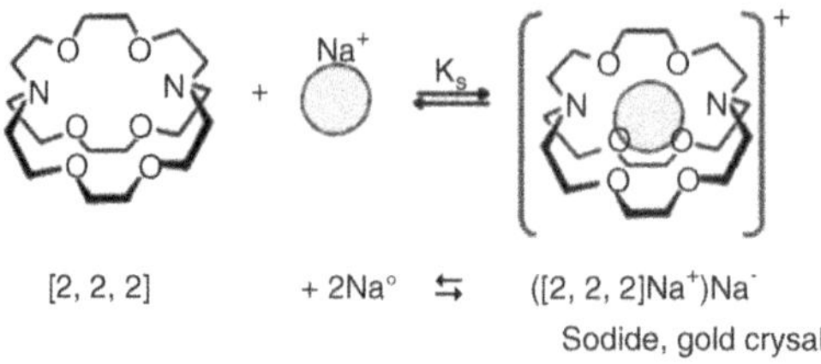

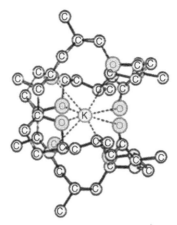

[2, 2, 2] $+ 2Na^\circ$ ⇌ $([2, 2, 2]Na^+)Na^-$

Sodide, gold crysal

- Examples of the main classes of the natural ionophores:
    1. Macrocyclic esters, macrotetrolides or actins:

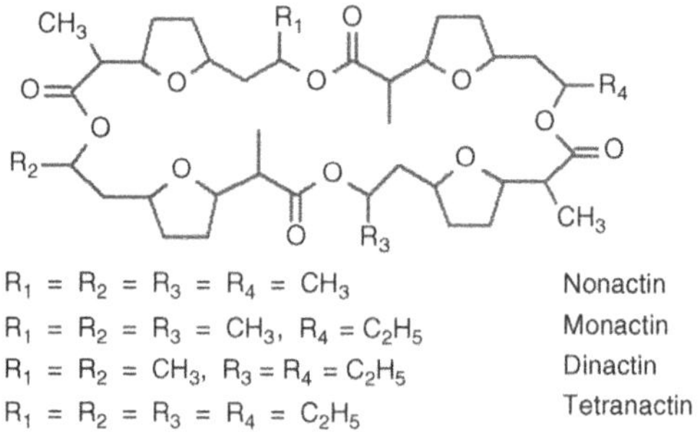

$R_1 = R_2 = R_3 = R_4 = CH_3$ Nonactin
$R_1 = R_2 = R_3 = CH_3, R_4 = C_2H_5$ Monactin
$R_1 = R_2 = CH_3, R_3 = R_4 = C_2H_5$ Dinactin
$R_1 = R_2 = R_3 = R_4 = C_2H_5$ Tetranactin

- o Actins bind K^+ by ether and carbonyl oxygen, forming a tennis ball's seam conformation complex and then has a hydrophobic exterior.
- o Most are K^+ selective, $K_{nonactin-K^+} \approx 100\, K_{nonactin-Na^+}$.

Diagrammatic representation of conformation of K^+–nonactin complex

2. Depsipeptide macrocycle, valinomycin
   - 12-Depsipeptide macrocycle uses six ester C=O for K$^+$; others H bond other peptides:

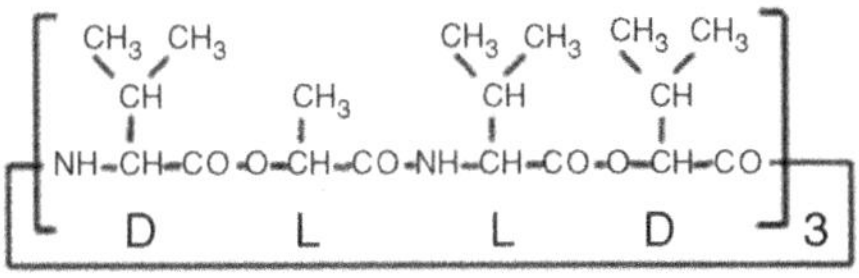

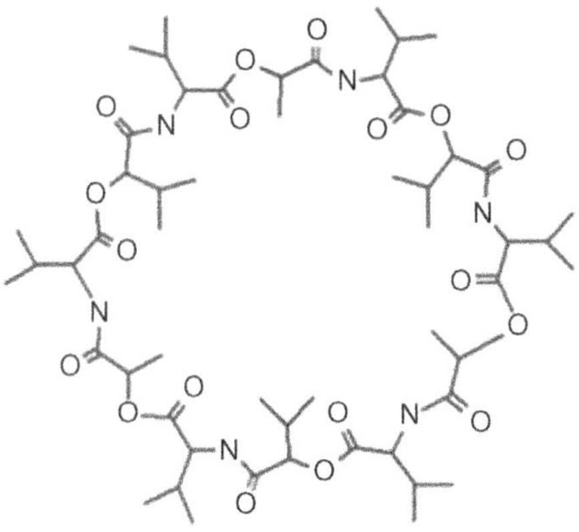

Valinomycin

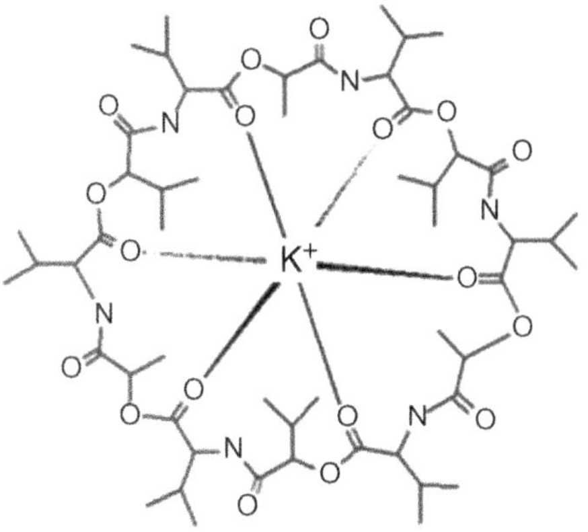

K$^+$–valinomycin complex

   - More examples: monamycin, tyrocidins, enniatins (cyclohexadepsipeptide), and gramicidins:

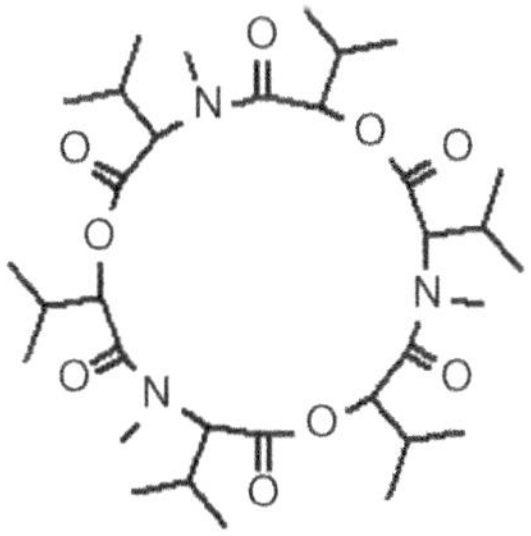

Enniatin

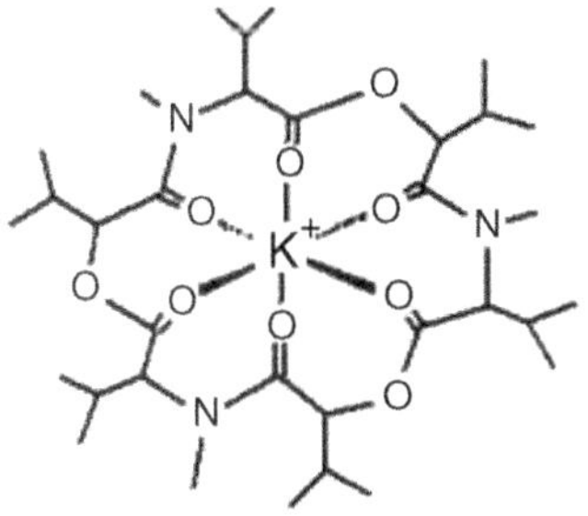

K$^+$–Enniatin B complex

3. Carboxylate ionophores

 (i) Macrocycles

Example: alamethicin (hexadecapeptide), COO⁻ in the middle:

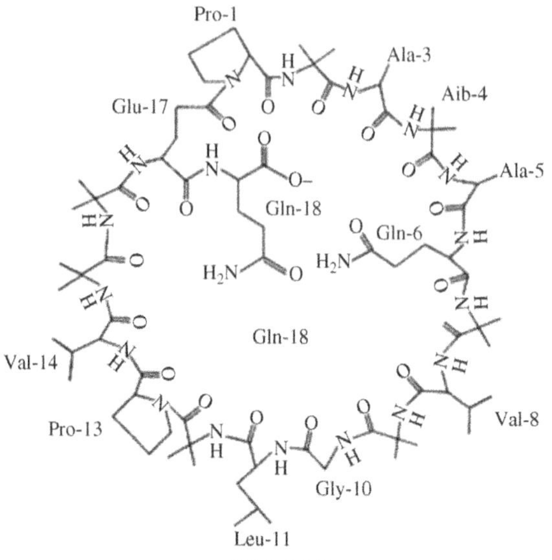

Alamethicin

 (ii) Nonmacrocycles

 (a) Nigericin

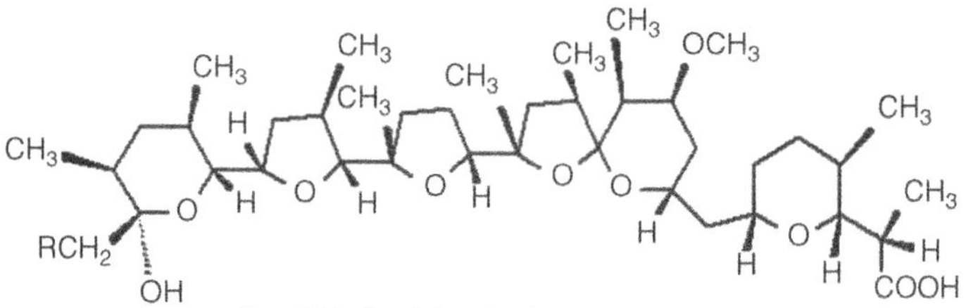

R = OH nigericin, R = H grisorixin

 (b) Monensin

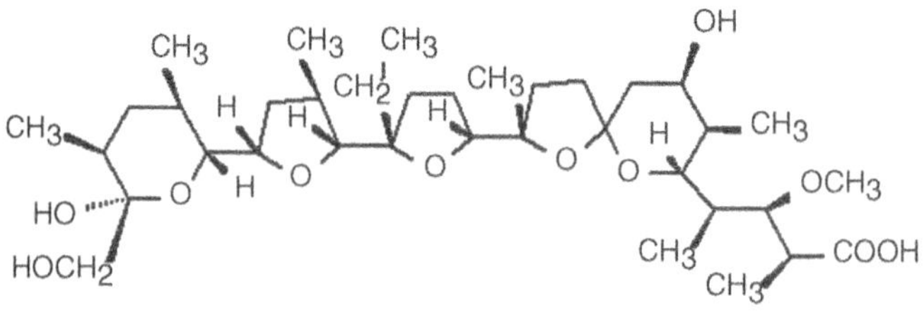

(c) Dianemycin

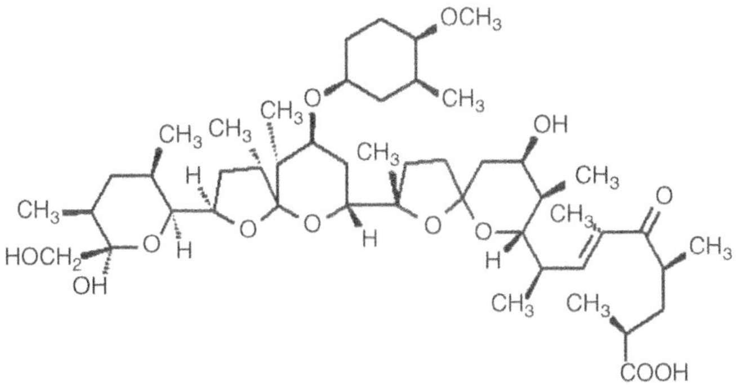

(d) Antibiotic X-206

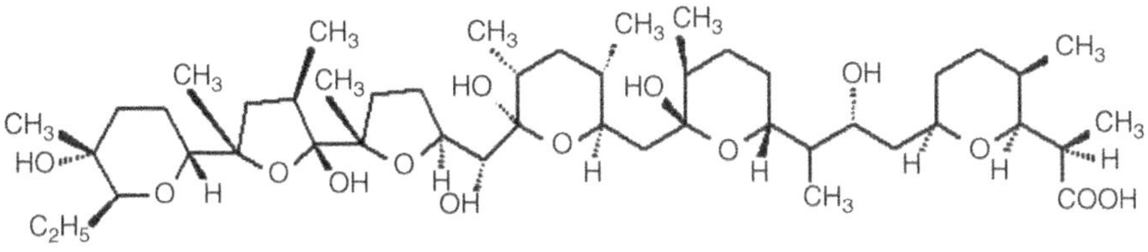

(e) Antibiotic X-537A

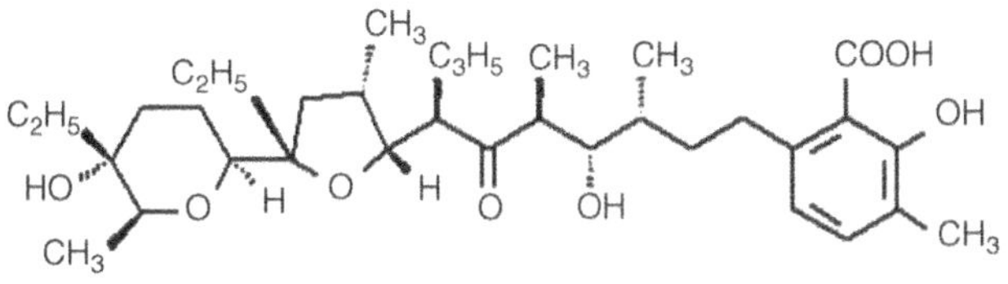

Ionophores and Selective Recognitions

How do ionophores select the metal ions? What are the factors that decide the stability of the complex formation of ionophores?

(a) Hard and Soft Acids and Bases (HSAB)

- *General rule*: The formation of stable complexes results from interactions between hard acids and hard bases.
- *Cations* are Lewis acids (electron acceptors).
- *Ligands* are Lewis bases (electron donors).
- *Hard ions* are those which:
  (i) Retain their valence electrons very strongly
  (ii) Are not readily polarized

(iii) Are of small size
- *Soft ions* do not retain their valance electrons firmly.
- Cations can be classified as:
  (i) *Hard acids*: H^+, Li^+, Na^+, K^+, Mg^{2+}, Ca^{2+}, Mn^{2+}, Cr^{3+}, Fe^{3+}, Co^{3+}
  (ii) *Borderline*: Zn^{2+}, Cu^{2+}, Ni^{2+}, Fe^{2+}, Co^{2+}, Sn^{2+}, Pb^{2+}
  (iii) *Soft acids*: Cu^+, Ag^+, Au^+, Ti^+, Pd^{2+}, Pt^{2+}, Cd^{2+}
- Ligands can also be classified as:
  (i) *Hard bases*: H_2O, OH^-, ROH, OR^-, R_2O, NH_3, NCS^-, Cl^-, PO_4^{3-}, SO_4^{2-}, F^-, NO_3^-, CO_3^{2-}
  (ii) *Borderline*: pyridine, RNH_2, N_2, N_3^-, NO_2^-, Br^-
  (iii) *Soft bases*: RSH, RS^-, R_2S, R_3P, R_3As, CO, CN^-, SCN^-, $S_2O_3^{2-}$, H^-, I^-
- Groups 1 and 2 (hard acids) prefer O donors (hard bases).
(b) Chelate and macrocyclic effect
- *Chelate effect*: Stands for the increased stability of a complex system containing chelate rings as compared to a system that is similar but contains fewer or no rings.
  (i) To understand this effect, let

$$\Delta G^\circ = -RT \ln \beta$$

$$\Delta G^\circ = \Delta H^\circ - T\,\Delta S^\circ$$

Therefore, β increases as ΔG° becomes more negative. A more negative ΔG° can result from making ΔH° more negative or from making ΔS° more positive.
  (ii) Example: The following two reactions illustrate a purely entropy-based chelate effect:

$$Cd^{2+}(aq) + 4CH_3NH_2 \rightleftharpoons \left[Cd(CH_3NH_2)_4\right]^{2+}(aq) \quad \log \beta = 6.52$$

$$Cd^{2+}(aq) + 2H_2NCH_2CH_2NH_2 \rightleftharpoons \left[Cd(en)_2\right]^{2+}(aq) \quad \log \beta = 10.6$$

Ligand	ΔH° (kJ/mol)	ΔS° (J//mol deg)	$T\,\Delta S^\circ$ (J/mol)	ΔG° (kJ/mol)
$4CH_3NH_2$	−57.3	−67.3	20.1	−37.2
$2en$	−56.5	+14.1	−4.2	−60.7

In this case the enthalpy changes are very close, and the chelate effect can be attributed to the entropy difference.
- *Macrocyclic effect*: A chelating *n*-dentate ligand gives a more stable complex (more negative ΔG° of formation) than *n*-unidentate ligands of similar type. In the same way, an *n*-dentate macrocyclic ligand gives even more stable complexes than the most similar *n*-dentate open-chain ligand.

Consider the following case:

$Cu\,(NH_3)_4^{2+}$ $\log \beta_4 = 14.3$

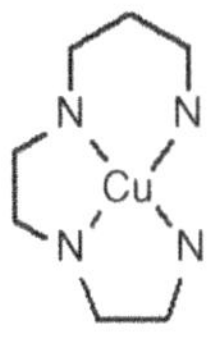

$\text{Log } K = 21.7$
Chelate effect

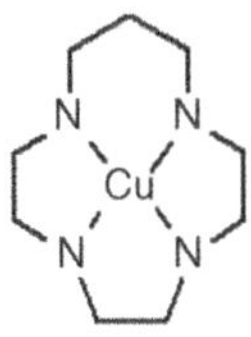

$\text{Log } K = 27.2$
Macrocycle effect

Similar to the chelate effect, the macrocyclic effect results from favorable entropy change. However, the macrocyclic effect is usually associated with a favorable enthalpy change.

(c) Solvent competition

- Complex formation involves stepwise substitution of the cation's salvation shell by the donor atoms of the ligand. Such reactions will be sensitive to the nature of both solvent and cation.

  For example, consider the reaction of dicyclohexano-18-crown-6 with Na^+ and K^+ in different solvents, e.g., H_2O, and CH_3OH:

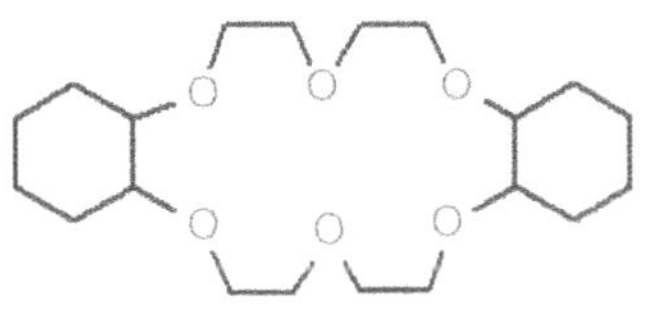

Dicyclohexano-18-crown-6

	Ionic Radii (Å)	$\log K_f$	
		H_2O	CH_3OH
Dicyclohexano-18-crown-6 + Na$^+$(solvent)$_n$	0.8	4.0	0.95
Dicyclohexano-18-crown-6 + K$^+$(solvent)$_n$	1.9	5.9	1.33

Note that H_2O and CH_3OH and dicyclohexano-18-crown-6 are oxygen donors.

- For these solvent molecules (H_2O or CH_3OH), binding to cations is determined by the polarizing power of the cation and decreases with increases in the radius of the cation. Consequently, cations of smaller size bind more strongly to H_2O molecules than to CH_3OH. If:

$$r_{Na^+} < r_{K^+}, \text{ then}$$

$$K_{Na^+-H_2O} > K_{K^+-H_2O}, \qquad K_{Na^+-CH_3OH} > K_{K^+-CH_3OH},$$

$$K_{Na^+-H_2O} > K_{Na^+-CH_3OH}, \qquad K_{K^+-H_2O} > K_{K^+-CH_3OH}$$

- The oxygen donor atoms of the dicyclohexano-18-crown-6 molecule displace coordinated oxygen–solvent molecules from the cation of larger size rather than the smaller one, and therefore, H_2O competes very effectively. This explains the sequence of the observed formation constants:

$$K_{K^+-(18-crown-6)_{H_2O,CH_3OH}} > K_{Na^+-(18-crown-6)_{H_2O,CH_3OH}}$$

$$K_{Na^+-(18-crown-6)_{CH_3OH}} < K_{Na^+-(18-crown-6)_{H_2O}}$$

$$K_{K^+-(18-crown-6)_{CH_3OH}} < K_{K^+-(18-crown-6)_{H_2O}}$$

When the donor atoms of the solvent or the ligand are different, other factors that affect the displacement of the solvent molecules must be considered. These factors include the hard/soft acid–base concept (HSAB) and the entropy change associated with the randomization of released solute/solvent molecules.

(d) Cavity size

- The formation constants for the interaction of certain synthetic and natural ionophores with cations differ significantly from cation to cation, depending upon the relative sizes of cation and the cavity:

		$\log K_f (CH_3 OH)/\log K_f (H_2O)$	Ionic Radii (Å)
[2, 1, 1]	$+Li^+$	7.7	0.60
	$+Na^+$	6.2	0.95
[2, 2, 2]	$+Li^+$	1.8	0.60
	$+Na^+$	7.2	0.95
	$+K^+$	9.9	1.33
	$+Rb^+$	8.5	1.48
	$+Cs^+$	3.6	1.69
[3, 3, 3]	$+K^+$	5.2	1.33
	$+Rb^+$	5.5	1.48
	$+Cs^+$	5.9	1.69

- These crown ethers discriminate between cations directly on a size basis by possessing a cavity that is a better match for one cation than another. *Note*: The order of cavity size increase, [2, 1, 1] < [2, 2, 2] < [3, 3, 3], matches the order of ionic radii increase, $Li^+ < K^+ < Cs^+$.

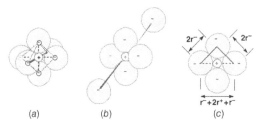

FIGURE 2-5   Small cation (gray) in octoheral hole formed by six anions.

Stereochemistry of Metal Ionophores

What are the factors that decide the stereochemistry of metal ionophores?

(a) Radius ratio r^+/r^-

- The radius ratio is a limiting factor in determining the coordination number (CN) of the cation as well as the stereochemistry of the ionophores. For example, based upon the hard-sphere model for ionic structure, $r^+ + r^-$ is equal to the equilibrium internuclear distance (Fig. 2-5).

  (i) The diagonal of the square $= 2r^- + 2r^+$.

  (ii) The angle formed by the diagonal in the corner must be 45°. Thus

$$\frac{2r^-}{2r^- + 2r^+} = \cos 45° = 0.707$$

$$r^- = 0.707r^- + 0.707r^+$$

$$0.293r^- = 0.707r^+$$

$$\frac{r^+}{r^-} = 0.414$$

  (iii) The cation will be stable in an octahedral hole only if it is at least large enough to keep the anions from touching, i.e., $r^+/r^- = 0.414$.

  (iv) Smaller cations will preferentially fit into tetrahedral holes.

  (v) By similar calculations, the ratio r^+/r^- for other geometry can be found (Table 2-5).

TABLE 2-5   Radius Ratio and Geometry

Maximum CN	Geometry	Limiting Radius Ratio	
		r^+/r^-	r^-/r^+
4	Tetrahedral	0.225	4.44
6	Octahedral	0.414	2.42
8	Cubic	0.732	1.37
12	Dodecahedral	1	1

- In 1 : 1 or 1 : 2 salts, the appropriate radius ratio is that of the smaller ion to the larger to determine how many of the latter will fit around the smaller ion.
- Examples:
  (i) NaCl

$$r^+/r^- = 95/181 = 0.52 \quad CN = 6$$

Geometry is O_h:

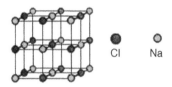

(ii) In SrF_2

$$r^+/r^- = 113/136 = 0.83 \quad \text{Maximum CN} = 8$$

$$r^-/r^+ = 1.2 \quad\quad\quad\quad\quad \text{Maximum CN} = 8$$

But $Sr^{+2} : F^- = 1 : 2$. Therefore, $Sr^{+2}_{CN} = 8$, and $F^-_{CN} = 4$ and is similar to CaF_2:

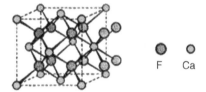

(iii) In Rb_2O

$$r^+/r^- = 148/140 = 1.06 \quad \text{maximum } Rb^+_{CN} = 8$$

$$r^-/r^+ = 0.95 \quad\quad\quad\quad \text{maximum } O^{-2}_{CN} = 8$$

But $Rb^+ : O^{-2} = 2 : 1$. Therefore, $Rb^+_{CN} = 4$, and $O^{-2}_{CN} = 8$ and is similar to CaF_2.

- For group I, the radii of the ions increase down the group:

	O^{-2}	Li^+	Na^+	K^+	Rb^+	Cs^+
R	1.20	0.6	0.95	1.33	1.48	1.74
r^+/r^-		0.5	0.79	1.11	1.23	1.45
CN		6	8	10	12	

(b) Number of donor atoms/ligands
  - 12-Crown forms NaL_2^+; Na^+ is 8-coordinate in the crown cube, tetragonal antiprism:

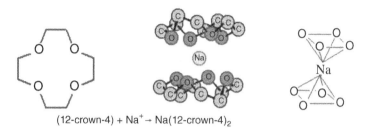

(12-crown-4) + Na$^+$ → Na(12-crown-4)$_2$

- For a larger metal M$^+$, the observed coordination number increases. For example, the 2 : 1 complex of benzo-15-crown-5 with K$^+$ gives the 10-coordinate K:

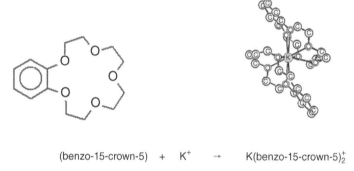

(benzo-15-crown-5) + K$^+$ → K(benzo-15-crown-5)$_2^+$

- Obviously, doubling the ligand to dibenzo-30-crown-10 gives a similar result (10-coordinate K$^+$, but 1 : 1). Note the folding of the crown ether about K$^+$:

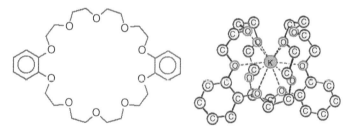

(Dibenzo-30-crown-10) + K$^+$ → K(dibenzo-30-crown-10)$^+$

What is the formula for finding the maximum coordination number that can be accommodated by a central metal ion?

Formula for finding maximum coordination number that can be accommodated by central metal ion:

$$N = \frac{2\pi}{\sqrt{3}} \left(\frac{d}{r}\right)^2 \left(\frac{1}{1 - r^2/(8d^2)}\right)$$

where d is the metal–ligand distance and r is the van der Waals radius of the ligand.

Information Transfer and Nerve Impulse

Describe the nature of nerve impulse. How does information transfer through the nerve?

- At one time, the nerve impulse was thought to be a simple electric current because it can be detected with a galvanometer.
- However, the speed of the impulse is too slow to be just electric current.
- The change in the electric potential, the nerve impulse, is related to the internal and external concentrations of $[Na^+]$ and $[K^+]$.
- In the resting stage, the membrane of the neuron is polarized, and the inside is negative with respect to the outside.
  - Outside: $K^+ < Na^+$
  - Inside: $Na^+ < K^+$

$$E_{inside} < E_{outside} \; (\sim -80 \, mV)$$

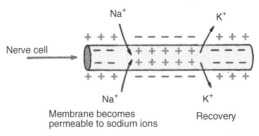

Na^+ channels close

K^+ channels open

- Stimulation (e.g., touch) of the membrane increases the permeability (along the membrane) and $[Na^+]$ rapidly enters the cell.

Na^+ channels open

K^+ channels close

E_i is directly proportional to $\ln[Na^+]$

As a result, the interior of the neuron becomes positive:

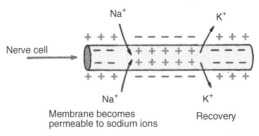

Na^+ K^+

Nerve cell

Na^+ K^+

Membrane becomes   Recovery
permeable to sodium ions

- When a maximum potential of $\sim +35 \, mV$ is reached, K^+ channels reopen, and Na^+ is pumped out (Fig. 2-6).
- Slightly later, the membrane becomes more permeable to $[K^+]$, which moves out of the cell.

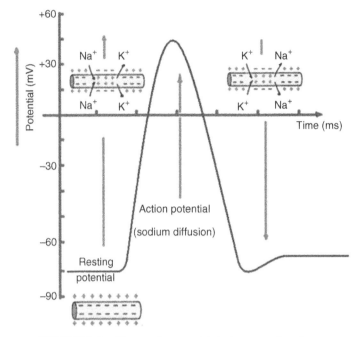

FIGURE 2-6 Changes in potential during nerve impulses.

- The change in permeability continues down the length of the neuron, that change being the nerve impulse (Fig. 2-7).
- Recovery of the neuron involves the return of K^+ to the inside of the membrane and Na^+ to the outside, which is accomplished by the active transport mechanism. So the resting state is restored.
- The result is that an E spike travels along the nerve as a function of time.
- Note that in this electrolytic conduction system, current and potential are at 90° to one another, unlike electronic conductors made by people.
- Development of this circuitry is used in the construction of an electrolytic computer.
- There can be many parallel $E–i$ events due to movement of K^+, Na^+, HPO_4^{2-}, Cl^-, Mg^{2+}, Ca^{2+}, etc., as well as acetylchlorine, glutamate, etc.

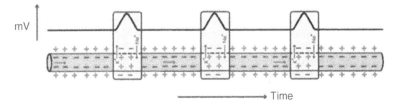

FIGURE 2-7 Passage of impulse along nerve fiber.

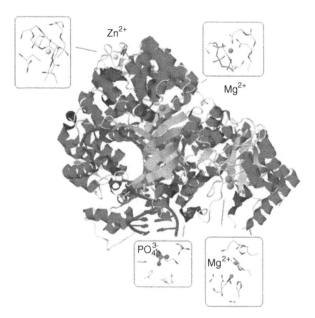

FIGURE 2-8 DNA Polymerase from *Geobacillus kaustophilus* complex at 2.4 Å resolution, PDB 3F2B: Mg^{2+} = violet, Zn^{2+} = yellow (Evans et al., 2008). (See the color version of this figure in Color Plates section.)

- H^+ also sets up potentials. For example, in Donnan equilibrium, as K^+ on the protein side is higher than on the other side, it tends to back-diffuse and is replaced by H^+ from H_2O, so pH decreases, while on the other side pH increases.

Metal Ions (Mg^{2+} and Zn^{2+}) in DNA or RNA Polymerase

Discuss the role of the metal ions (Mg^{2+}, and Zn^{2+}) in DNA or RNA polymerase.

- The crystal structure of RNA polymerase II in the act of transcription was determined at 3.3 Å resolution (Fig. 2-8). Duplex DNA is observed entering the main cleft of the enzyme and unwinding before the active site.
- The structure reveals growing base-pair interactions that lead to highly accurate nucleotide integration.
- The mechanism of DNA or RNA polymerization involves:
  - DNA molecule as template
  - A nucleoside triphosphate
  - Catalysts
    - (a) DNA polymerase (or RNA polymerase)
    - (b) Zn^{2+} is coordinated by DNA polymerase, the developing DNA strand, and probably H_2O molecules. It has been found that the coordination of *o-*

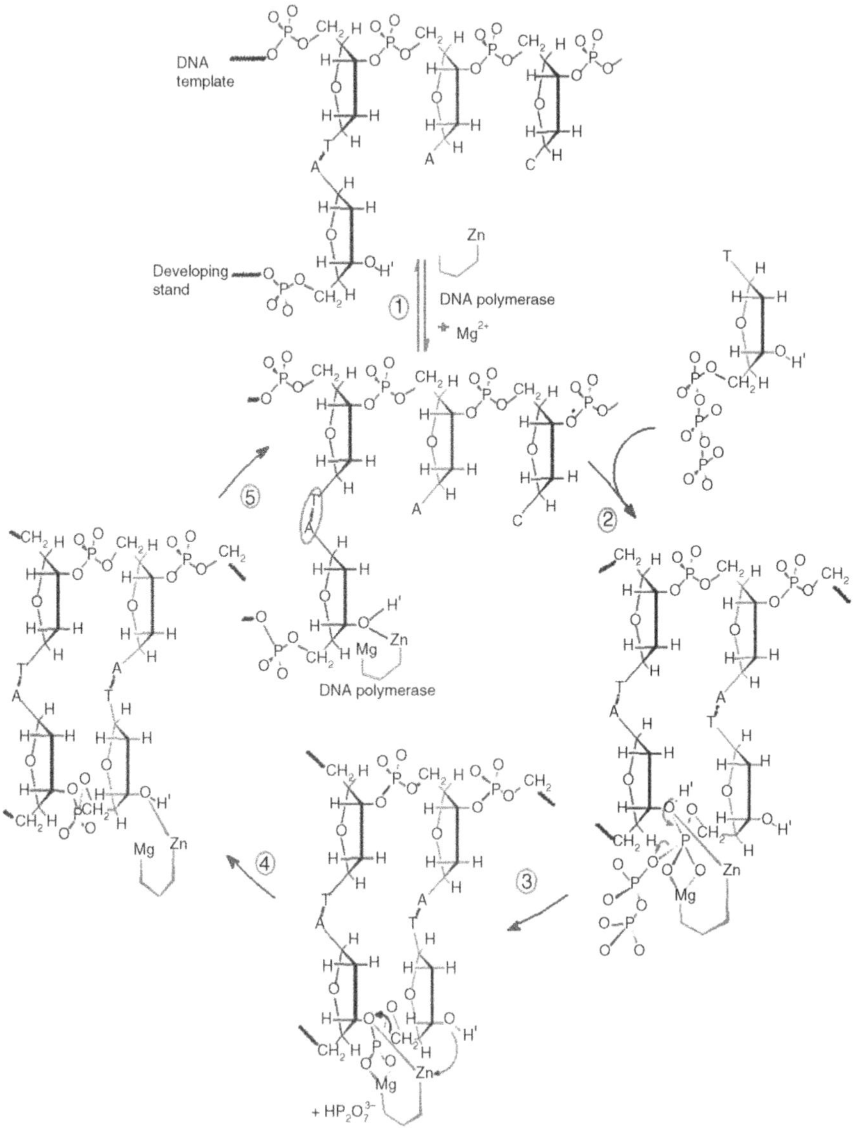

SCHEME 2-3   DNA polymerase (modified from Mildvan and Grisham, 1974). (See the color version of this scheme in Color Plates section.)

phenanthroline to the Zn^{2+} of the Zn^{2+} (oligomer)(enzyme) complex inhibits the replication. Consequently, the coordination sites about Zn^{2+} are not all occupied by DNA and enzyme donor atoms but are probably occupied by replaceable H_2O.

○ Mg^{2+} binds the entering nucleoside triphosphate to the oligomer–enzyme complex.

- The proposed replication is shown in Scheme 2-3.
- Notes:
    - The enzyme DNA or RNA polymerase contains both Zn^{2+} and Mg^{2+}.
    - In step 1, the Zn–enzyme molecule is coordinated to the sugar $3'$-OH group of the developing DNA strand.
    - In step 2, Mg^{2+} is coordinated at the α-phosphate unit of the incoming nucleoside triphosphate.
    - In step 3, the $3'$-OH group of the developing strand acts as a nucleophile upon the α-phosphate and displaces $P_2O_7^{4-}$.
    - In step 4, the Zn–enzyme complex migrates to the $3'$-OH group of the added nucleoside and starts the next nucleoside addition.

The following scheme represents the sequence of converting glucose to pyruvate in the cell cytoplasm. In the mitochondria, the pyruvate is oxidized by O_2 to CO_2, and H_2O, and energy is released (Embden-Meyerhof pathway):

(a) **Write the chemical structures for each of the involved compounds, and fill in the blanks.**

(b) **In the last step, pyruvate kinase catalyzes the breakdown of phosphenolpyruvate to pyruvate. What are the proposed mechanisms and the role of Mg^{2+} and K^+ ions?**

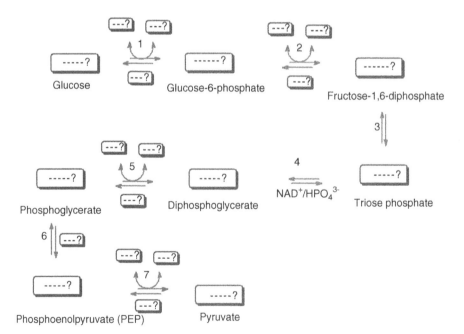

Solution

(a) The sequence for conversion of the glucose to pyruvate is shown in Scheme 2-4.

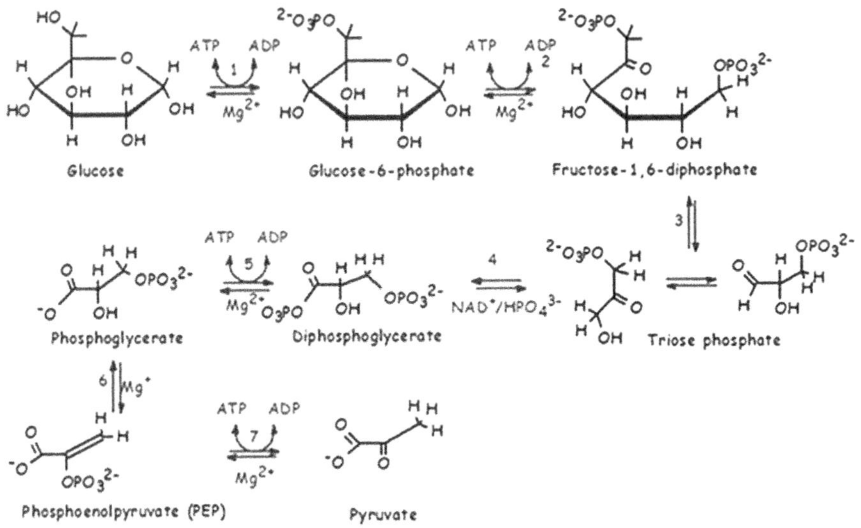

SCHEME 2-4 The sequence of converting glucose to pyruvate in the cell cytoplasm.

(b) The proposed mechanism for phosphate transfer from phosphoenolpyruvate (PEP) to ADP to form pyruvate and the regeneration of PEP from pyruvate and phosphate are shown in Scheme 2-5.

- *Notes:*
  - Mg^{2+} is coordinated to (Fig. 2-9):
    (a) Donor atoms of the enzyme

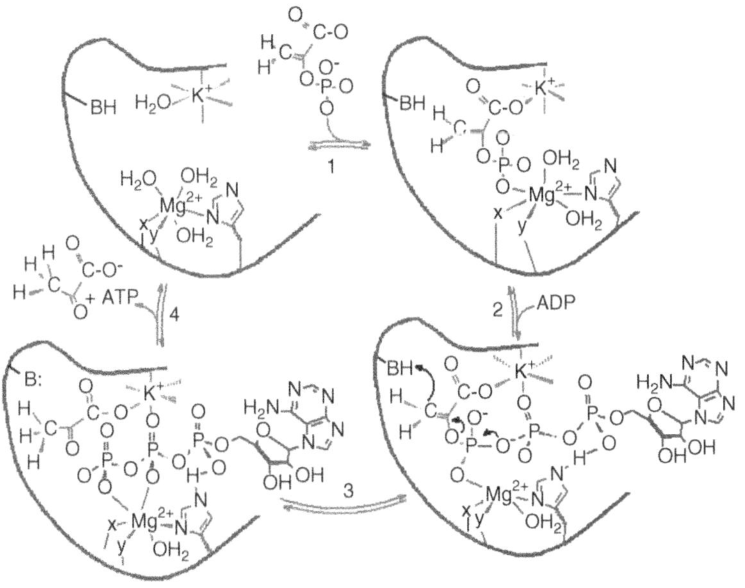

SCHEME 2-5 Pyruvate kinase, (modified from Mildvan and Grisham, 1974).

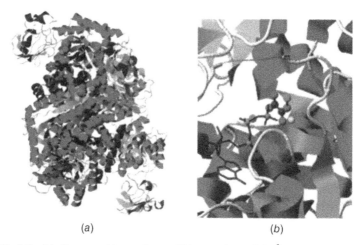

(a)   (b)

FIGURE 2-9 (a) Pyruvate kinase from rabbit muscle at 2.7 Å resolution, PDB 1AQF (Larsen et al., 1997). Mg^{2+} = green, K^+ = violet, and P = orange (b) Crystal structures of *Leishmania mexicana* pyruvate kinase in complex with ATP and oxalate, PDB 3HQO: (Morgan et al., 2010). Mg^{2+} = green, K^+ = yellow, P = orange. (See the color version of this figure in Color Plates section.)

    (b) Terminal phosphate of ADP

    (c) Phosphate of PEP.

- Coordination of K^+ to the PEP carboxylate group initiates the transfer of the terminal phosphate from PEP to ADP. K^+ also is coordinated to the enzyme.

- The oxygen of the terminal phosphate of ADP acts as nucleophile upon the PEP phosphate.

- Simultaneously, H^+ moves from an adjacent enzyme site (H-B) to the vinyl carbon of PEP to generate the pyruvate.

- The structure of rabbit muscle pyruvate kinase crystallized as a complex with Mg^{2+}, K^+, and L-phospholactate has been solved and refined to 2.7 Å resolution (Fig. 2-9).

- The crystal structure reveals an asymmetric unit that contains two tetramers. The eight subunits adopt several different conformations.

- In all of the subunits, Mg^{2+} coordinates to the protein through the carboxylate side chains of Glu 271 and Asp 295. In the subunit having the most closed conformation, Mg^{2+} also coordinates to the carboxylate oxygen, the bridging ester oxygen, and a nonbridging phosphoryl oxygen of L-phospholactate. Mg^{2+} to L-phospholactate coordination is missing in subunits exhibiting a more open conformation.

- K^+ coordinates to four protein ligands and to phosphoryl oxygen of the L-phospholactate. The position and ligation of K^+ are unaffected within the different conformations of the subunits.

When glucose-6-phosphate is not required for immediate combustion in the Embden-Meyerhof pathway, it may be converted to glycogen in a three-step process as shown in the following scheme.

(a) **Write the chemical structure of the involved compounds and fill in the spaces:**

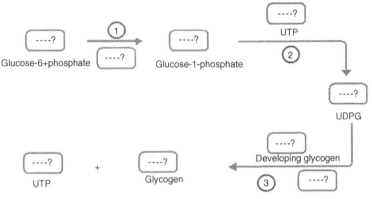

(b) **In the first step, glucose-6-phosphate is rearranged to glucose-1-phosphate by Mg^{2+} and phosphoglucomutase. Define the role of the magnesium ion.**

- The three-step mechanism is presented in Scheme 2-6.
- Phosphoglucomutase (Fig. 2-10) provides enormous rate enhancement unmatched by any other type of catalyst.
- Crystallization of phosphorylated β-phosphoglucomutase in the presence of the Mg^{2+} cofactor and either of the substrates glucose-1-phosphate or glucose-6-phosphate produced crystals of the enzyme–Mg^{2+}–glucose-1,6-diphosphate complex, which diffracted X-rays to 1.2 and 1.4 angstroms, respectively.

SCHEME 2-6 Sequence of converting glucose-6-phosphate to glycogen.

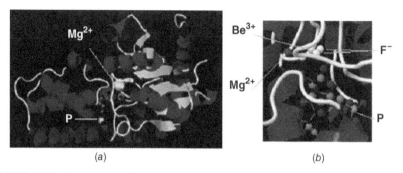

FIGURE 2-10 (*a*) Pentacovalent phosphorus intermediate of phosphoryl transfer reaction, PDB 1003 (Lahiri et al., 2003). (*b*) Structure of β-phosphoglucomutase inhibited with glucose-6-phosphate, glucose-1-phosphate, and beryllium, PDB 2WF8 (Griffin et al., 2012).

- The structure reveals a stabilized pentacovalent phosphorane formed in the phosphoryl transfer from the C(1)O of glucose-1,6-diphosphate to the nucleophilic Asp^8 carboxylate.
- In the first step of the glycogen formation, the rearrangement of glucose-6-phosphate to glucose-1-phosphate proceeds through the formation of glucose-1, 6-diphosphate. This formation required:
  - Phosphoglucomutase enzyme
  - Serine unit
  - Phosphorylation of enzyme serine unit
  - Mg^{2+}
  The serine OH group is likely coordinated to Mg^{2+} and glucose-6-phosphate.
- A proposed sequence of events is described in Scheme 2-7:
  step 1: Glucose-6-phosphate coordinates to Mg^{2+} at the 3-OH group, while the 6-phosphate binds the other site of the enzyme.
  step 2: Phosphorylation of the glucose 1-OH by serine phosphate leads to the formation of glucose-1,6-diphosphate. The formed diphosphate rotates about the Mg–OH bond.
  step 3: The serine OH displaces the 6-phosphate.
  step 4: Glucose-1-phosphate is released from the enzyme, leaving behind a serine phosphate ready for the next glucose-6-phosphate.

Back-up Store of Energy in Vertebrate Muscle and ATP

What are the back-up store of energy in vertebrate muscle, the requiring enzyme, and the suggested mechanism for the interconversion of ADP and ATP?

- The hydrolysis of ATP to ADP is the immediate source of energy for muscular contraction. The amount of ATP in the muscle is extremely small and is rapidly depleted by muscle contraction.

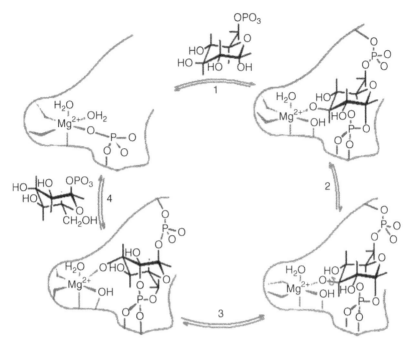

SCHEME 2-7 Conversion of glucose-6-phosphate to glucose-1-phosphate by phosphoglucomutase (modified from Mildvan and Grisham, 1974).

- The back-up store of high-energy phosphate is phosphocreatine, which has a higher phosphate group transfer potential than ATP:

- The requiring enzyme is *creatine kinase* in conjunction with Mg^{2+}.
- Creatine kinase catalyzes the reversible conversion of creatine and ATP to phosphocreatine and ADP. This conversion maintains energy homeostasis in the cell. (Homeostasis is the physiological process by which the internal systems of the body [e.g., blood pressure, body temperature, and acid–base balance] are maintained at equilibrium despite variations in the external conditions.)
- Figure 2-11 represents the X-ray structure of creatine kinase bound to a transition-state analogue complex at 2.1 Å resolution.
- The creatine is placed with the guanidino nitrogen cis to the methyl group located to promote in-line attack at the γ-phosphate of ATP–Mg^{2+}, while the

FIGURE 2-11 Structure of *Torpedo californica* creatine kinase complexed with the ADP–Mg^{2+}–NO_3^{-}–creatine transition-state analogue complex, PDB 1n16 (Lahiri et al., 2002).

ADP-Mg^{2+} is in a conformation similar to that found in the transition-state analogue complex-bound structure of the homologue arginine kinase.

- Three ligands to Mg^{2+} are contributed by ADP and nitrate and three by water molecules.
- The mechanism in Scheme 2-8 indicates the formation of phosphocreatine from ATP, while the reverse pathway is used when ADP concentration is built up during muscle contraction.

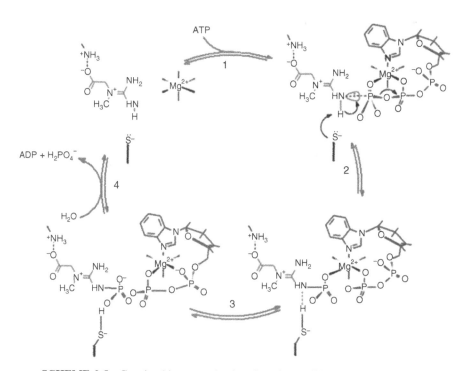

SCHEME 2-8 Creatine kinase mechanism (based on Mildvan and Cohn, 1970).

Biological Role of Calcium Ion

What was the first observation that indicates the biological role of calcium ion?

The biological role of the Ca^{2+} was realized when Ringer observed that isolated turtle hearts would continue beating for many hours if 1 mM Ca^{2+} was present in the isotonic bathing medium.

Examples of Calcium Proteins and Their Functions

List some examples of calcium proteins and specify their functions.

- *Calsequestrin* is found in sarcoplasmic reticulum (SR), the cytoplasm of the muscle fibers.

Function	Stores Ca^{2+}, has a high affinity for Ca^{2+}, binds about 43 mole Ca^{2+} per mole of protein
Molecular weight	33,000

- *Parvalbumin* is found in the white muscle of fish and amphibians:

Function	Ca^{2+}-binding regulatory protein
Molecular weight	11,500, water soluble
Tertiary structure	Six α-helices denoted A–F
Number Ca^{2+}/ molecule	Two Ca^{2+} ions/molecule, with $pK_d = 6.7$
Binding sites	One Ca^{2+} bound in the loop between C and D helices is 6-coordinate with oxygen donors from Asp[51], Asp[53], Ser[55], peptide oxygen of Phe[57], Gul[59], and Gul[62]. A second Ca^{2+} bound in the E and F regions is 8-coordinate with oxygen donors from Asp[90], Asp[94], peptide oxygen of Lys[96], both oxygens of carboxylate of Asp[92], Glu[101], and by H_2O.

- *Troponin complex* is found in vertebrate skeletal muscle:

Quaternary structure	Trimer of three separate proteins, troponin-C has been well characterized

- *Troponin–C*

Function	Ca^{2+}-binding regulatory protein
Number of Ca^{2+}/ molecule	Four Ca^{2+}/molecule, with pK_d in the range 5.5–7.5
Binding sites	Ca^{2+} binds to oxygen atoms of carboxylate group of Glu and Asp two hydroxyl oxygens of Ser[35] and Ser[67], and amide oxygen of Ar[104] Two high-affinity and two low-affinity sites
Conformational states	Three conformational states corresponding to binding of 0, 2, and 3 mol Ca^{2+}.

- *Staphylococcal nuclease*

Function	Hydrolyzes both DNA and RNA to $3'$-mononucleotides Ca^{2+} is essential for substrate binding and enzymatic catalysis
Binding sites	Approximately octahedral, coordinates to carboxyl groups of Glu^{43}, Asp^{19}, Asp^{40}, and the peptide oxygen of Thr^{41}; the sixth ligand may be H_2O or both oxygens of Asp^{21} may be involved in coordination

- *Thermolysin* is a zinc endopeptidase isolated from *Bacillus thermoproteolyticus*, which is able to function at high temperatures for a short time. The enzyme is heat stable:

Molecular weight	37,500
Metal ion/ molecule	Four Ca^{2+} and one Zn^{2+} per molecule
Function	Maintainins quaternary structure. The heat stability is due to the presence of four calcium ions.
Binding sites	Zn: His^{142}, His^{146}, Glu^{166}, H_2O
	$Ca^{2+}(1)$: Asp^{138}, Glu^{177}, Asp^{185}, Glu^{187}, Glu^{190}, H_2O
	$Ca^{2+}(2)$: Glu^{177}, Asn^{183}, Asp^{185}, Glu^{190}, H_2O, H_2O
	$Ca^{2+}(3)$: Asp^{57}, Asp^{59}, Glu^{61}, H_2O, H_2O, H_2O.
	$Ca^{2+}(4)$: Tyr^{193}, Thr^{194}, Thr^{194}, Ile^{197}, Asp^{200}, H_2O

- *Concanavalin A* is a protein agglutinin (sticking together and forming visible clumps), which binds specifically with α-glucosyl and α-mannosyl residues:

Function	Binds carbohydrates, the metal ion being essential for sugar binding
Molecular weight	25,000, exists as a dimer at pH 3.5–5.6
Metal ions/ molecule	One Ca^{2+} and one Mn^{2+} per molecule

Calcium Ions and Muscle Contractions

Explain the role of calcium ions in muscle contractions.

- When the nerve impulse arrives at the junction between the nerve ending and the muscle, the outer membrane of a muscle fiber is depolarized.
- This causes Ca^{2+} to be released from the SR and its binding sites on the muscle fibers, and the concentration of Ca^{2+} in the sarcoplasm (cytoplasm) rises 100-fold in a millisecond.
- The released Ca^{2+} interacts in a specific manner with Ca^{2+}-binding regulatory proteins.
- In vertebrate skeletal muscle, Ca^{2+} interacts with troponin, but in fish and amphibian muscle, Ca^{2+} interacts with parvalbumin.

- The binding of Ca^{2+} to troponin-C, which has been much studied, produces conformational changes that are transmitted to tropomyosin and then to actin. This permits actin to interact with myosin and results in muscle contraction.
- Muscle contraction is associated with the release of Ca^{2+}.

Magnesium versus Calcium

Compare the biological roles of magnesium and calcium.

Mg^{2+}	Ca^{2+}
• Important in early life processes	• Functions involve higher forms of life, nerve transmission, muscle contraction, and blood coagulation
• Relatively low in blood plasma and other biological fluids	• Predominates in blood plasma and other biological fluids
• Within the cell $[Mg^{2+}]$ is high	• Within cell, $[Ca^{2+}]$ is low
• Reacts with substrate rather than enzyme	• Higher coordination number (6, 7, 8)
	• Binds directly with enzyme molecule
	• Induces conformation change
• $r = 0.65\,\text{Å}$	• $r = 0.99\,\text{Å}$
• Higher charge density	• Lower charge density
• Involved in $ATP \rightarrow ADP + P_i$	• Fast substitution reaction
• Complex formation involving two macromolecules, such as interaction of two nucleic acids	• Move quite rapidly in living tissues
• Interaction of nucleic acid with protein, these reactions would lead to disorder within the cell, due to release of H_2O, unless the nucleic acids were present as complexes with Mg^{2+}	• As trigger for transferring signals between different cells (muscle contraction)
• Can substitute for Mn^{2+}	

REFERENCES

R. J. Evans, D. R. Davies, J. M. Bullard, J. Christensen, L. S. Green, J. W. Guiles, J. D. Pata, W. K. Ribble, N. Janjic, and T. C. Jarvis, *Proc. Natl. Acad. Sci. USA*, 105, 20695–20700 (2008).

D. E. Fenton, "Across the living barrier," *Chem. Soc. Rev.*, 6, 325 (1977).

M. Forgac and G. Chin, in *Metalloproteins, Part 2*, P. M. Harrison ed., Macmillan Press Ltd., Houndmills (1985).

J. L. Griffin, M. W. Bowler, N. J. Baxter, K. N. Leigh, H. R. Dannatt, A. M. Hounslow, G. M. Blackburn, C. E. Webster, M. J. Cliff, and J. P. Waltho, *Proc. Natl. Acad. Sci. USA*, 109, 6910 (2012).

S. D. Lahiri, G. Zhang, D. Dunaway-Mariano, and K. N. Allen, *Science*, 299, 2067–2071 (2003).

S. D. Lahiri, P. F. Wang, P. C. Babbitt, M. J. McLeish, G. L. Kenyon, and K. N. Allen, *Biochemistry*, 41, 13861–13867 (2002).

T. M. Larsen, M. M. Benning, G. E. Wesenberg, I. Rayment, and G. H. Reed, *Arch. Biochem. Biophys.*, 345, 199–206 (1997).

A. Mildvan and M. Cohn, *Adv. Enz.* Vol. 33, 1–70 (1970).

A. S. Mildvan and C. M. Grisham, *Struct. Bonding*, 20, 1 (1974).

H. P. Morgan, I. W. McNae, M. W. Nowicki, V. Hannaert, P. A. M. Michels, L. A. Fothergill-Gilmore, and M. D. Walkinshaw, *J. Biol. Chem.*, 285, 12892–12898 (2010).

T. Shinoda, H. Ogawa, H., F. Cornelius, and C. Toyoshima, *Nature*, 459, 446–450 (2009).

SUGGESTIONS FOR FURTHER READINGS

Ion Selectivity

1. E. Gouaux and R. MacKinnon, *Science*, 310, 1461–1465 (2005).

Polyethers and Their Complexes

2. C. J. Pedersen and H. K. Frensdorf, "Macrocyclic polyether and their complexes," *Angew. Chem., Inte. Edit.*, 11, 16 (1972).

Crytates

3. J. -M. Lehn, "Cryptates: The chemistry of macropolycyclic inclusion complexes," *Acc. Chem. Res.*, 11, 49 (1978).

4. J. -M. Lehn, *Struct. Bonding (Berlin)*, 16, 1 (1973).

Na^+/K^+ Transport by ATPase

5. B. T. Kilbourn, J. D. Dunitz, L. A. Piodo, and W. Simon, *J. Mol. Biol.*, 30, 559 (1967).

6. F. G. Donnan, *Z. Elektrochem.*, 17, 572 (1911).

7. R. L. Post, S. K. Kume, T. Tobin, B. Orcutt, and A. K. Sen, *J. Gen. Physiol.*, 54, 306 (1969).

8. K. O. Hakansson, *Mol. Biol.*, 332, 1175–1182 (2003).

9. W. J. Rice, H. S. Young, D. W. Martin, J. R. Sachs, and D. L. Stockes, *Biophys. J.*, 80, 2187–2197 (2001).

10. J. V. Møller, P. Nissen, T. L-M. Sørensen, and M. Le Maire, *Curr. Opin. Struct. Biol.*, 15, 387–393 (2005).

11. C. Toyoshima, M. Nakasako, H. Nomura, and H. Ogawa, *Nature*, 405, 647–655 (2000).

Mg^{2+}/DNA Polymerase/Pyruvate Kinase/Phosphoglucomutase

12. A. Alt, K. Lammens, C. Chiocchini, A. Lammens, J. C. Pieck, D. Kuch, K-P. Hopfiner, and T. Carrelle, *Science*, 318, 967–970 (2007).

Mg^{2+}/Creatine Kinase

13. T. M. Larsen, M. M. Benning, I. Raymen, and G. H. Reed, *Biochemistry*, 37, 6247–6255 (1998).

Ca^{2+}/Biochemistry

14. M. J. Berridge, P. Lipp, and M. D. Bootman, *Nature Rev. Mol. Cell Biol.*, 1, 11–21 (2000).
15. M. Brini, *Cell Calcium*, 34, 399–405 (2003).
16. M. Zayzafoom, *J. Cell. Biochem.*, 97, 56–70 (2005).
17. E. R. Chapman, *Nature Rev. Mol. Cell Biol.*, 3, 498–504 (2002).
18. S. Martens, M. M. Kozlov, and H. T. Mcmahon, *Science*, 316, 1205–1208 (2007).
19. E-I. Ochiai, in *General Principles of Biochemistry of the Elements*, pp. 227–234, Plenum, New York (1987).
20. E-I. Ochiai, *J. Chem. Ed.*, 68, 10–12 (1991).
21. S. Forsen and J. Kordel, in "Calcium in biological systems," in *Bioinorganic Chemistry*, I. Bertini, H. B. Gray, S. J. Lippard, and J. S. Valentine, Eds., pp. 107–66, University Science Books, Mill Valley, California (1994).

General

22. D. Midgley, "Alkali–metal complexes in aqueous solution", *Chem. Soc. Rev.*, 4, 549 (1975).
23. M. Dobler, in *Ionophores and Their Structures*, Wiley, Chichester (1981).
24. R. Harrison and G. C. Lunt, in *Biological Membranes*, Blackie, Glasgow and London (1975).
25. R. J. P. Williams, "The biochemistry of sodium, magnesium and calcium", *Quart. Rev.*, 24, 331 (1970).
26. Yu. A. Ovchinnikov, V. T. Ivanov, and A. M. Shkrob, in *Membrane Active Complexones*, Elsevier, Amsterdam (1974).
27. S. Eshaghi, D. Niegowski, A. Kohl, D. M. Molina, S. A. Lesley, and P. Nordlund, *Science* 313, 354–357 (2006).
28. K. F. Purcell and J. C. Kotz, in *Inorganic Chemistry*, Saunders Company, Philadelphia (1977).
29. V. R. William, W. L. Mattice, and H. B. Williams, in *Basic Physical Chemistry for the Life Sciences*, W. H. Freeman and Company, San Francisco (1978).
30. R. W. Hay, in *Bio-Inorganic Chemistry*, Ellis Horwood Limited, Chichester (1984).
31. F. A. Cotton and G. Wilkinson, in *Advanced Inorganic Chemistry*, 4th Ed., John Wiley & Sons, New York (1980).

3

NONREDOX METALLOENZYMES

Zinc is a necessary component of a number of enzymes, such as alcohol dehydrogenase, alkaline phosphatase, carbonic anhydrase, carboxypeptidase, and DNA/RNA polymerase. In most of these enzymes, the nonredox active Zn^{2+} ion acts as a Lewis acidic center at which substrates are coordinated, polarized, and thus activated. Additional roles of zinc include acting as a template and engaging in a structural or regulatory role as in superoxide dismutase and zinc fingers. Zinc is also necessary for the activity of many enzymes, a second metal usually required, as in thermolysin. Zinc is needed to maintain normal concentrations of vitamin A in plasma. Additionally, zinc forms complexes with insulin, and crystalline zinc–insulin is required during insulin purification. Zinc–insulin complexes are also present in B-cells of the pancreas, and there is evidence suggesting that zinc is used in these cells to store and release insulin as required. Zinc increases the duration of insulin action when given by injection. Diseases arising from deficiency of zinc are dwarfism and hypogonadism, while "metal fume fever" is associated with an excess of this element.

The zinc ion is similar to the magnesium ion, and many of their salts are isomorphous, e.g., $Zn(Mg)SO_4 \cdot 7H_2O$. The aqua ions of zinc are quite strong acids, and aqueous solutions of salts are hydrolyzed to $ZnOH^+$ ions below 0.1 M. In the presence of complexing anions, compounds such as $Zn(OH)Cl$ and $ZnNO_3^+$ may be obtained. Zinc ions tend to form coordinate bonds to F, O, S, and N donor atoms. The formed complexes commonly have coordination numbers 4, 5 (commonest), and 6.

Chemistry of Metalloproteins: Problems and Solutions in Bioinorganic Chemistry, First Edition.
Joseph J. Stephanos and Anthony W. Addison.
© 2014 John Wiley & Sons, Inc. Published 2014 by John Wiley & Sons, Inc.

In this chapter, first some model studies on metal ion catalysis of hydrolysis are considered and then three well-studied zinc metalloenzymes, carboxypeptidase, carbonic anhydrase, and alcohol dehydrogenase, are investigated in detail.

Cationic Lewis Acids

Give examples of metal-ions acting as Lewis acids in organic reactions and in biological processes. Then show how.

- Aqua ions catalyze a variety of organic reactions in solution:
  - Ester hydrolysis
  - Amide hydrolysis
  - Peptide hydrolysis
  - Phosphate ester hydrolysis
  - Acetal hydrolysis
  - Sulfate ester hydrolysis
  - Phosphate hydrolysis
  - Thiol ester hydrolysis
  - Glycoside hydrolysis
  - Schiff base hydrolysis
  - Carbonyl hydration
  - Peptide bond formation
  - Transamination
  - Carboxylation
  - Hydrogen exchange
  - Schiff base formation
  - Decarboxylation
- Metal ions acting as Lewis acids in a biological process are summarized in Table 3-1.

TABLE 3-1   Cationic Lewis Acids in Biological Systems

Enzyme	Reaction Catalyzed	Metal Ion
Carboxypeptidase	Hydrolysis of C-terminal peptide residues	Zn^{2+}
Leucine aminopeptidase	Hydrolysis of leucine N-terminal peptide residues	Zn^{2+}
Dipeptidase	Hydrolysis of dipeptides	Zn^{2+}
Neutral protease	Hydrolysis of peptides	Zn^{2+}, Ca^{2+}
Collagenase	Hydrolysis of collagen	Zn^{2+}
Phospholipase C	Hydrolysis of phospholipids	Zn^{2+}
β-Lactamase II	Hydrolysis of β-lactam ring	Zn^{2+}
Thermolysin	Hydrolysis of peptides	Zn^{2+}, Ca^{2+}
Alkaline phosphatase	Hydrolysis of phosphate esters	Zn^{2+}

Table 3-1 (*Continued*)

Enzyme	Reaction Catalyzed	Metal Ion
Carbonic anhydrase	Hydrolysis of CO_2	Zn^{2+}
α-Amylase	Hydrolysis of glucosides	Zn^{2+}, Ca^{2+}
Phospholipase A_2	Hydrolysis of phospholipids	Ca^{2+}
ATPase	Hydrolysis of ATP to ADP	Mg^{2+}

- Examples illustrating the influence of the metal ion:
  - Hydrolysis of ethyl glycinate

$$NH_2-CH_2-COOEt + H_2O \quad \rightarrow \quad NH_2-CH_2-COOH + EtOH$$

Without metal ion: $k_{OH} = 0.63\,M^{-1}\,s^{-1}$ at 25°C
$M = Cu^{2+}$: $k_{OH} = 1.4 \times 10^5\,M^{-1}\,s^{-1}$ at 25°C

Cu^{2+} ion acts as a Lewis acidic center, polarizing the carbonyl ester, and activating the hydrolysis with an acceleration of 2×10^5.

  - Decarboxylation of oxaloacetic acid, A^{2-}, to give pyruvic acid and CO_2

$$HO_2-CCOCH_2-COOH \quad \rightarrow \quad HO_2-CCOCH_3 + CO_2$$

A^{2-}: $k = 1.7 \times 10^{-5}\,s^{-1}$ at 25°C
$(ZnA)_{ketonic}$: $k = 7.42 \times 10^{-3}\,s^{-1}$ at 25°C
$(CuA)_{ketonic}$: $k = 0.17\,s^{-1}$ at 25°C (Scheme 3-1)

  - The Zn^{2+} complex decomposes 436 times faster than A^{2-}, and the Cu^{2+} complex is 23 times faster than the Zn^{2+} complex.
  - The Lewis acidity depends on the ratio of the charge to the size of the metal ion.
  - Cu^{2+} is a more effective Lewis acid than Zn^{2+} in assisting the transfer of the electrons from the C—C bond undergoing cleavage.
  - The metal ion–catalyzed reaction shows a normal $^{13}C-^{12}C$ carbon isotope effect of 6%, indicating the cleavage of the C—C is rate-determining.
  - Enzymatic reaction does not display a carbon isotope effect; therefore, the rate-determining step (RDS) is not the C—C cleavage.

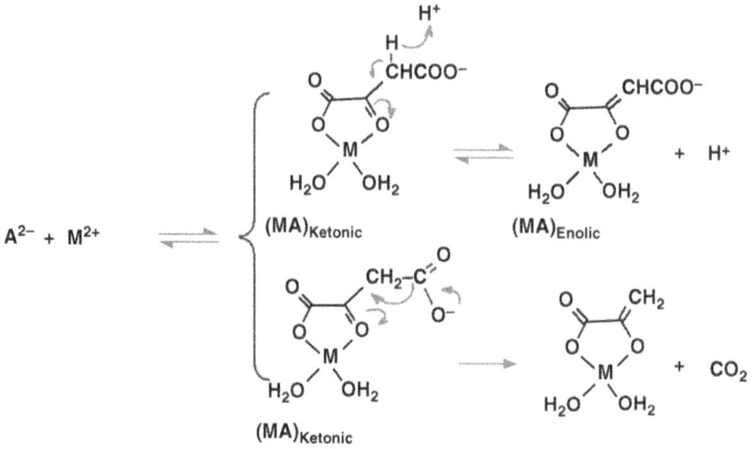

SCHEME 3-1 Decarboxylation of oxaloacetic acid. (Modified from Steinberger and Westheime, 1951.)

- o However, the enzymatic reaction is slower in D_2O. Therefore, the RDS involves a proton transfer in ketonization of the enol intermediate to release pyruvic acid.
- o Enzymatic reaction occurs by a similar general mechanism to a simple metal ion–promoted reaction but may involve change in RDS.
- o The ligand groups from the protein fine tune the Lewis acidity of the metal ion (Scheme 3-2).

Biological Role of Zinc Ions

What are the roles of zinc ions in biological processes?

- Zinc complexes:
  - o Are present to the extent of 1.4–2.3 g in the human body
  - o Are essential for normal growth
  - o Act as Lewis acid
  - o Are good buffers and are used in pH control in vivo

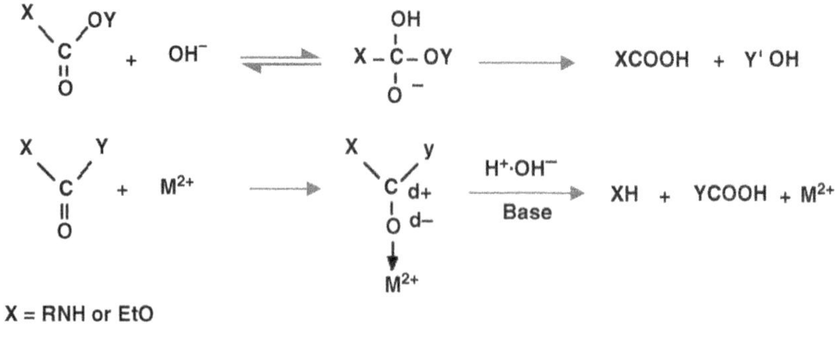

X = RNH or EtO

SCHEME 3-2 Metal ion acts as Lewis acid and promotes OH^- addition.

- Zinc(II) is a symmetrical d^{10}, borderline HSAB, and interacts strongly with O and N donor ligands.
- There are currently 18 zinc metalloenzymes and 14 zinc ion–activated enzymes known.
- Zinc in proteins can either participate directly in chemical catalysis as in carboxypeptidase, carbonic anhydrase, and alcohol dehydrogenase or be important for maintaining protein structure and stability as in insulin and superoxide dismutase.

CARBOXYPEPTIDASES

What are the main functions of carboxypeptidases?

- Carboxypeptidases aid protein digestion.
- Carboxypeptidases catalyze the hydrolysis of the C-terminal (exopeptidase) amino acid residue from a peptide or protein chain.

- Carboxypeptidases are released in the pancreatic juice of animals.

What are the main types of carboxypeptidases?

- Carboxypeptidase-A (CPA) and carboxypeptidase-B (CPB):
  - Are extracellular
  - Have maximum activity at alkaline pH
  - Are metalloenzymes
- Yeast carboxypeptidase-C:
  - Are intracellular
  - Have maximum activity at acidic pH
  - Are not metalloenzymes

What are the requirements to stimulate carboxypeptidase-A and carboxypeptidase-B?

- The carboxylate group of the C-terminal must be *free*.
- Substrates in which the amino acid side chain, R'', is aromatic and hydrophobic are favored.

○ However, carboxypeptidase-A has somewhat broad specificity in this respect.

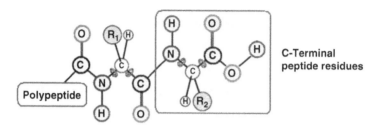

C-Terminal
peptide residues

Polypeptide

○ Carboxypeptidase-B requires the presence of a positively charged side chain.
○ Proline residues will not be hydrolyzed:

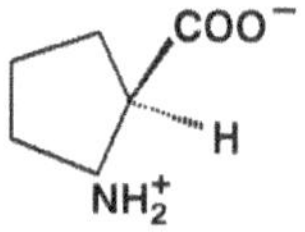

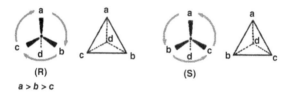

- The C-terminal residue must have S-configuration.

(R)   (S)

$a > b > c$

○ Where the four atoms immediately adjacent to the chiral center are considered, atoms are arranged with decreasing atomic number:

$$I > Br > Cl > F > O > N > C > H$$

○ When similar atoms are directly attached to the central atom, exploration of the comparison is continued. The comparison should suspend at the first difference. For example,

$$I > Br > Cl > F > O > N > C(O, H, H) > C(C, C, H) > H$$

What are the main structural features of carboxypeptidase-A and carboxypeptidase-B?

- CPA and CPB have been sequenced.
- They consist of one single peptide chain and one atom of zinc.

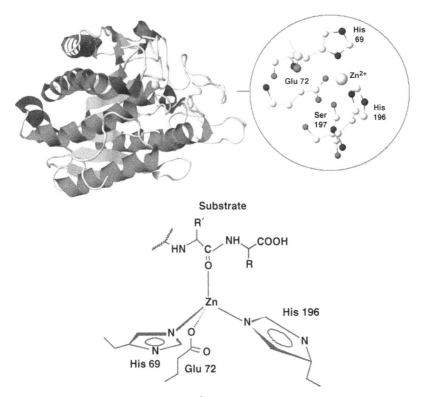

FIGURE 3-1 Structure of CPA at 1.25 Å resolution, PDB 1M4L (Kilshtain-Vardi et al., 2003).

- Their molecular weight is ~34,300.
- CPA has 307 amino acid residues; CPB has 308 residues.
- The three-dimensional structures of the two enzymes are similar.
- The active sites of CPA and CPB are very similar.
- Zn^{2+} has a distorted tetrahedral stereochemistry and is coordinated to two histidine residues (His[69] and His[196]), a glutamic (Glu[72]) acid residue, and an H_2O molecule (or OH^-) (Fig. 3-1).
- Their structures were established by X-ray diffraction. The crystal structure of bovine zinc metalloproteinase CPA has been refined to 1.25 Å resolution.
- The crystal structure studies confirm that:
  ○ The distance between the zinc-bound solvent molecule and the metal ion is strongly suggestive of a neutral water molecule and not a hydroxide ion in the resting state of the enzyme
  ○ Glu72 and Glu270 are negatively charged in the resting state of the enzyme at pH 7.5

How can Zn^{2+} be extracted from carboxypeptidase-A and carboxypeptidase-B? What are the consequences of metal ion removal?

- Zn^{2+} in CPA can be removed by dialysis either
  - at low pH or
  - at neutral pH against buffer containing a zinc-chelating agent such as 1,10-phenanthroline.
- **What is dialysis?**
  - In dialysis, solvent molecules and small ions migrate through membranes.
  - It is a laboratory technique.
  - A solution of the protein is placed in a cellophane bag.
  - The bag is securely closed.
  - The bag is immersed in distilled water or buffer.
  - The large molecules are retained within the membrane.
  - The small ion and molecules migrate through the membrane until equilibrium is attained.

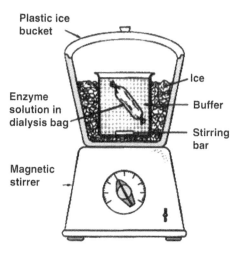

- The removal of the Zn^{2+} causes:
  - A loss in the activity of the enzyme
  - Little effect on the overall structure
- Addition of Zn^{2+} to the apoenzyme restores the catalytic activity.
- Fe^{2+}, Mn^{2+}, Co^{2+}, and Ni^{2+} replace Zn^{2+} and regenerate peptidase activity.

What are the spectral consequences when Zn^{2+} is replaced by Co^{2+} in carboxypeptidases?

- Zinc ion can be replaced by Co^{2+} in carboxypeptidase and in several other Zn enzymes.

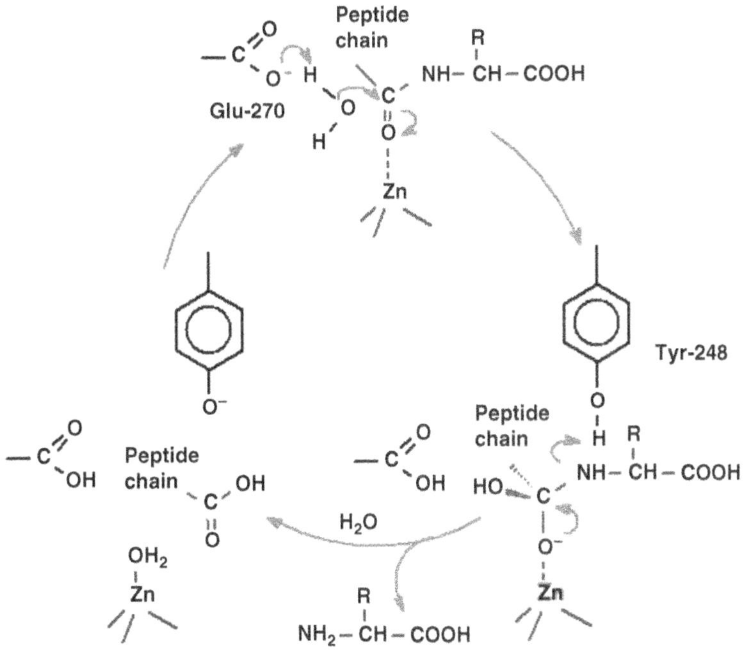

SCHEME 3-3 Carbonyl group is polarized and attacked by a water molecule, which is promoted by Glu-270. (Modified from Breslow and Wernick, 1976. Reproduced from Hughes, 1981, by permission of John Wiley & Sons.)

- Replacement of Zn^{2+} by Co^{2+} has been particularly useful because Zn^{2+} (d^{10}) form colorless complexes, whereas Co^{2+} (d^{7}) complexes absorb in the visible region.
- Thus, spectral studies (absorption, circular dichroism, and magnetic circular dichroism) and low-temperature electron spin resonance (ESR) spectra of Co-substituted enzymes indicate a distorted tetrahedral environment around the cobalt atom.
- NMR studies on manganese carboxypeptidase have shown that water or hydroxide is coordinated to the manganese, and this water molecule (or hydroxide) is displaced on formation of the enzyme–inhibitor complex with β-phenylpropionate.

Describe the role of Zn^{2+} in carboxypeptidase-A and carboxypeptidase-B.

- The amide carbonyl of the substrate molecule replaces the water molecule (or OH^-), and ligates to Zn^{2+}.
- The carbonyl group is polarized and attacked by a water molecule, which is activated by Glu-270 (Scheme 3-3).
- The resulting intermediate is protonated on nitrogen by Tyr-248.

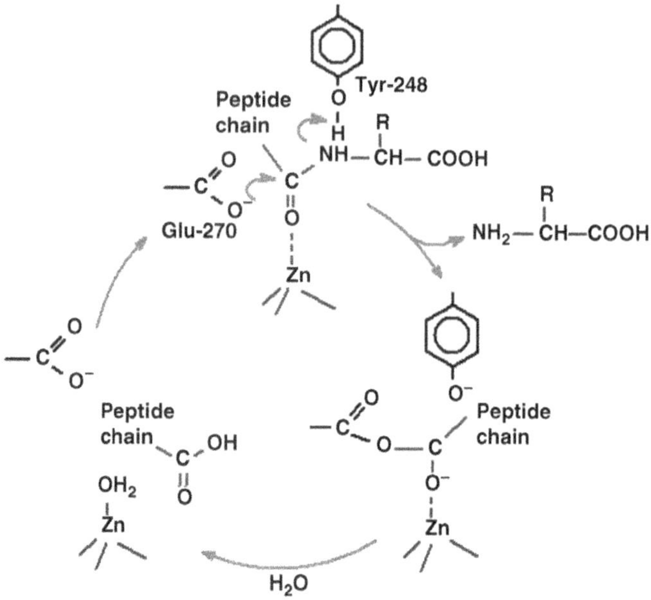

SCHEME 3-4 Carbonyl group is polarized and attacked by Glu-270 acting as a nucleophile. (Modified from Breslow and Wernick, 1976.)

- Then it is decomposed to amine and carboxylic acid.
- An alternative mechanism involves Glu-270 acting as a nucleophile rather than a general base (Scheme 3-4).
- A mixed anhydride intermediate is formed and subsequently hydrolyzed.

Design models to mimic the role of the metal ion in carboxypeptidase-A and carboxypeptidase-B.

(a) **Inert Bis(ethylenediamine)Co(III) Complex, $Co^{3+}en_2$**

- Results obtained with labile complexes are hard to explain and other mechanisms cannot be avoided.
- Kinetically inert Co^{3+} complexes, in particular bis(ethylenediamine) complexes with coordinated glycine esters, cis-[Co(en)$_2$(NH$_2$CH$_2$COOR)Cl]Cl$_2$, where R = CH$_3$, C$_2$H$_5$, and iso-C$_3$H$_7$, have a great significance in demonstrating the role of the metal ion.
  - In these complexes, the ester is coordinated through the amino group, and the complexes are stable in an aqueous solution for several hours.
  - In acid solution, Hg^{2+} reacts with the complex and removes the coordinated halide. In the absence of other nucleophiles, water molecules attack the positive center of the polarized carbonyl and hydrolysis of the ester occurs.

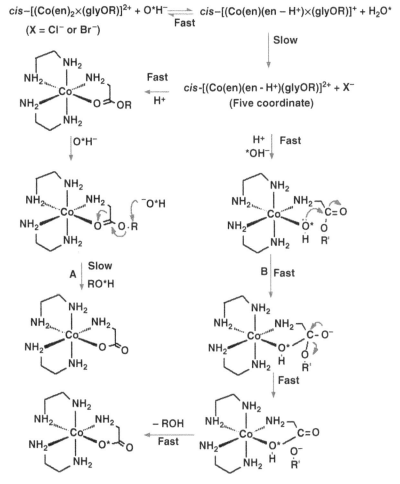

o Oxygen-18 tracer studies have been performed on the base hydrolysis of glycine esters coordinated to Co^{3+}, in cis-$[Co(en)_2X(glyOR)]^{2+}$, where $X = Cl^-$, Br^- (Scheme 3-5).

• These studies show two competing pathways, *intermolecular* hydrolysis of chelated ester (pathway A) and *intramolecular* attack of coordinated OH^- (pathway B). Both reaction pathways result from competition for a

SCHEME 3-5 Intermolecular and intramolecular base hydrolysis of glycine esters coordinated to Co^{3+}. (Modified from Buckingham et al., 1970.)

5-coordinated deprotonated intermediate formed by loss of the halide ion. Similarly, studies of these kinetically inert Co^{3+} complexes can also define the mechanistic features for hydrolysis of the amid, ester, and peptide ligands.

- An indication for the formation of an intermediate of reduced coordination number was supported by the demonstration that azide ion could be incorporated into the complex during hydrolysis of cis-$[Co(en)_2Br(glyNH_2)]^{2+}$ forming cis-$[Co(en)_2N_3(glyNH_2)]^{2+}$.

- Two pathways (A and B) for hydrolysis are also observed:
  - o In mechanism A, Co^{3+} polarizes the carbonyl group:

(X = OR or NH₂)

  - o The carbonyl is then attacked by an external hydroxide.
  - o In mechanism B, free noncoordinated carbonyl is attacked by cobalt-bound hydroxide:

$$pK_a \text{ of } Co-H_2O \sim 6 \quad \text{at pH 7} \quad Co-H_2O \rightarrow Co-OH$$

  - o For $[Co(NH_3)_5(NH_2CH_2CONH_2)]^{3+}$
    - (a) At pH 9 the rate enhancement for hydrolysis of glycine via pathway B is $\geq 10^7$ over the rate of pathway A, which is $\geq 10^{11}$ than that for the base hydrolysis of uncoordinated glycinamide.
    - (b) The pH dependence of the observed rate constant of these reactions suggests that the catalytically active form of the metal complex of the ester contains M−OH.
  - o Therefore, the hydrolysis of a coordinated amide or ester by intramolecular attack of the metal hydroxide is far more efficient than intermolecular attack by solvent OH^- on the metal-bonded carbonyl.

(b) **Labile Metal Complexes**

- The anhydride-A hydrolyzes at a rate independent of pH, $k_{obs} = 2.7 \times 10^{-3} \, s^{-1}$:

A   AZn

  - o The Zn(II) complex AZn hydrolyzes at pH 7.5 in a reaction that is first order in the complex and first order in the hydroxide with $k_{obs} = 3 \, s^{-1}$

at 25°C. The *coordinated hydroxide* may be involved in hydrolyzing the anhydride.

- Hydrolysis of 8-acetoxyquinoline-2-carboxylic acid (HA):

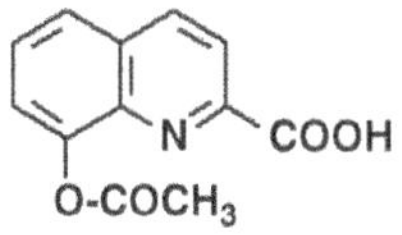

  o The rate law is given as

$$\text{Rate} = k_0[\text{A}^-] + k_{\text{OH}}[\text{A}^-][\text{OH}^-]$$

  with $k_0 = 1.69 \times 10^{-4} \, \text{s}^{-1}$. Then

$$k_{\text{OH}} = 0.84 \, \text{M}^{-1} \, \text{s}^{-1} \qquad \text{at } 25°C$$

  o For metal complex of A, MA; $M = Zn^{2+}$ or Cu^{2+}, undergo *base hydrolysis* $(k_{\text{OH}} \gg k_o)$.

  (a) The reaction involves intermediate attacked by the coordinated hydroxide (Scheme 3-6).

  (b) The rate of the hydrolysis is comparable to the reported value of the hydrolysis of ester by CPA.

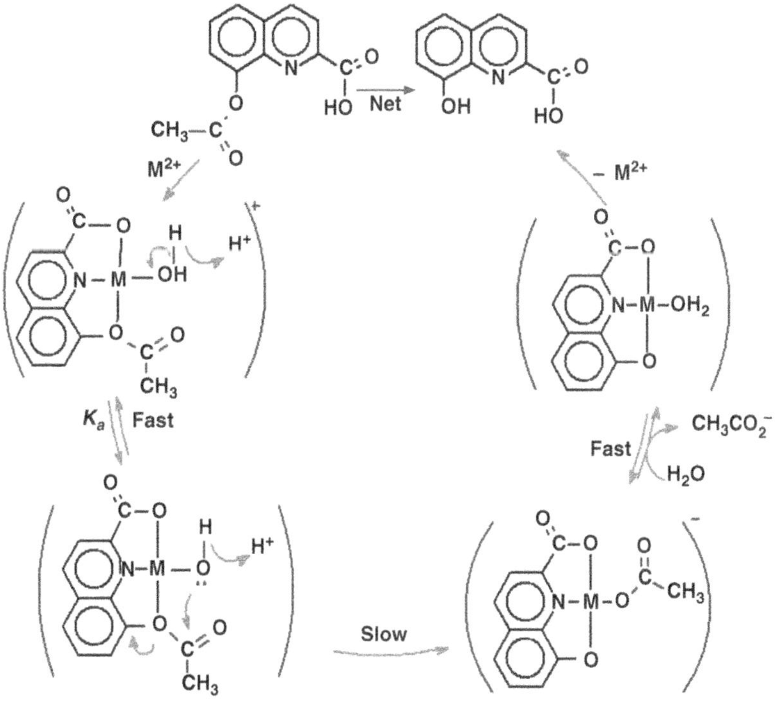

SCHEME 3-6 Intramolecular attack by coordinated hydroxide. (Modified from Hay, 1984.)

CARBONIC ANHYDRASE

What are the main functions of carbonic anhydrase?

- Carbonic anhydrase (CA) is a zinc enzyme.
- CA is present in animals, plants, and certain microorganisms.
- It catalyzes:
  - The reversible hydration of carbon dioxide:

$$CO_2 + H_2O \rightleftharpoons HCO_3^- + H^+$$

  - The hydration of many aldehydes and esters:

$$CH_3CHO + H_2O \rightleftharpoons CH_3CH(OH)_2$$

Conversion of carbonic acid to CO_2 and H_2O is a spontaneous process. Why is carbonic anhydrase needed?

- The uncatalyzed dehydration of carbonic acid or hydration of CO_2 is too slow for respiration of animals, $k_{obs} = 7 \times 10^{-4}\,s^{-1}$.
- CA can hydrate 10^6 molecules of CO_2 per second at 37°C, 10^7 times the uncatalyzed hydration rate; the rate acceleration with CA is of the order of 10^9.

What are the main structural and chemical features of carbonic anhydrase?

- Carbonic anhydrases A, B, and C are slightly different and occur in different organisms.
- The most well characterized enzyme is bovine and human carbonic anhydrase B (CA-B).
- CA-B is monomeric and contains one atom of Zn.
- Its molecular weight is 30,000.
- The turnover number for CO_2 hydration is $10^6\,s^{-1}$. That is, 1 mol of the enzyme hydrates 10^6 mol of CO_2 per second.
- The Zn ion is ligated by three histidine residues, His^{93}, His^{95}, and His^{117}, in distorted T_d geometry. The fourth coordination site is occupied by H_2O.

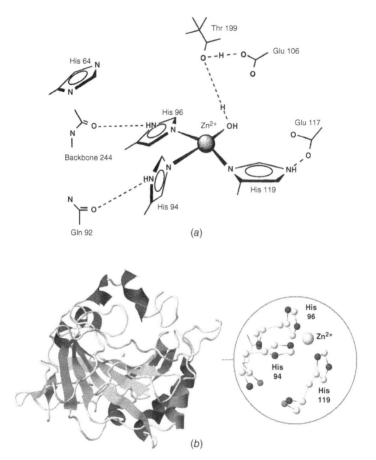

FIGURE 3-2 (*a*) Environment of zinc in CA and (*b*) active site of human CA II at 2 Å resolution, PDB 1CA2 (Eriksson et al., 1988).

- The refined structure of human CA at 2.0 Å resolution (Fig. 3-2) confirms the following:
  - ○ The zinc ion is ligated to three histidyl residues and one water molecule in a nearly tetrahedral geometry.
  - ○ In addition to the zinc-bound water, seven more water molecules are identified in the active site.
  - ○ The OH group of Thr[199] can function as a hydrogen bond acceptor only when it donates its proton to the COO^- of Glu[106] (Fig. 3-2).

Describe the role of the metal ion in carbonic anhydrase.

- The Zn ion of CA can be complexed by *o*-phenanthroline and the resulting complex can be removed by dialysis.

- The apoenzyme is completely inactive.
- The activity can be restored by the addition of Zn^{2+} in the molar ratio $1:1$.
- Optical rotatory dispersion studies have shown that the native enzyme and the apoenzyme have the same tertiary structure.
- These confirm that the function of Zn is not to stabilize the tertiary structure but it is directly involved in the catalytic activity of the enzyme.
- The marine diatom *Thalassiosira weissflogii* produces a Cd carbonic anhydrase when it is grown in a Zn-deficient environment; this Cd protein is different from the Zn protein.

What are the possible mechanisms that describe the action of carbonic anhydrase?

- A hydrophobic pocket made of valine, leucine, and tryptophan residues about 3–4 Å from the Zn in CA has been suggested as a binding site for CO_2.
- Three mechanism were proposed:
  - Zinc carbonyl mechanism: Ionization of imidazole (of a histidine in the proximity of Zn), then nucleophilic attack on CO_2 (Scheme 3-7)
  - Zinc hydroxide mechanism: Alternatively, ionization of coordinated H_2O, then nucleophilic attack by OH^- on CO_2 (Scheme 3-8)
  - Hydration–dehydration mechanism:
    - (a) Glutamic acid[106] acts as a general base, and the 5-coordinate Zn is a Lewis acid. The acidity is catalyzed by CO_2 coordination to Zn in the fifth position (Scheme 3-9).
    - (b) This mechanism incorporates both the Zn hydroxide and Zn carbonyl mechanisms.

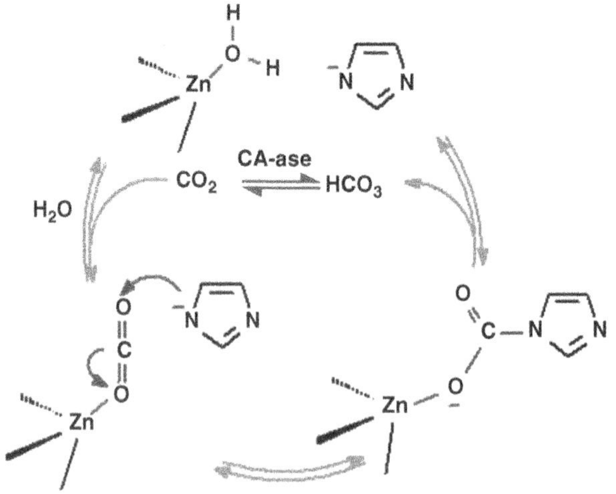

SCHEME 3-7 Ionization of imidazole and nucleophilic attack on CO_2. (Modified from Campbell et al., 1977, and Gupta and Pesando, 1975.)

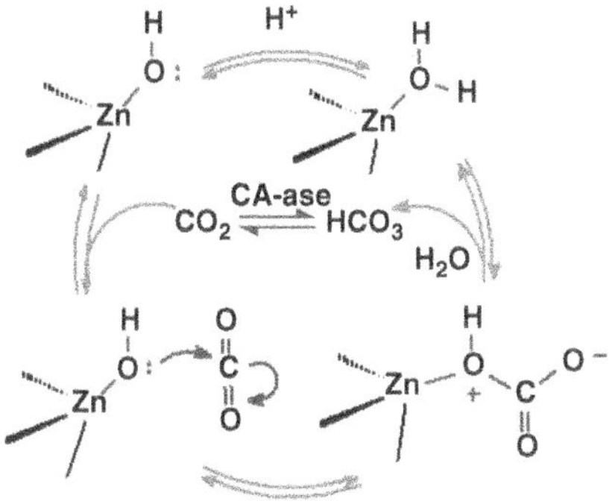

SCHEME 3-8 Ionization of coordinated H_2O and nucleophilic attack by OH^- on CO_2. (Modified from Coleman, 1967, and Lindskog and Colman, 1973. Reproduced from Hughes, 1981, by permission of John Wiley & Sons.)

Design models for the carbonic anhydrase, and what do they reveal?

- Varieties of metal complexes of coordinated hydroxide have been studied as a model for the Zn hydroxide mechanism in CA.
- ZnCROH is 5-coordinate and an effective catalyst for hydration of acetaldehyde:

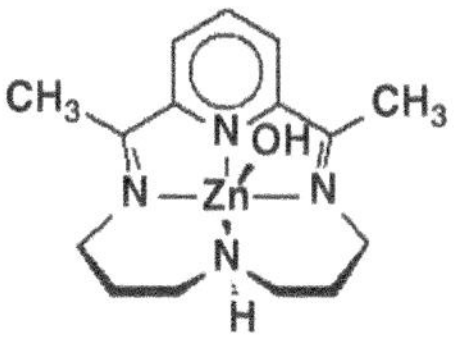

ZnCROH

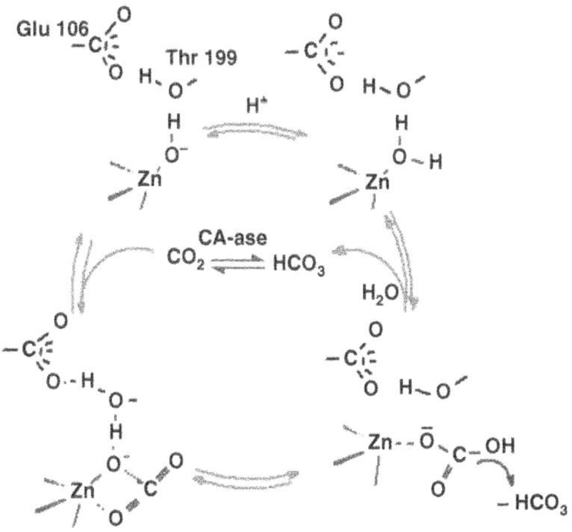

SCHEME 3-9 Glu[106] acting as general base and Zn^{2+} as Lewis acid, catalyzed by coordinated CO_2. (Modified from Kannan et al., 1977. Reproduced from Hughes, 1981, by permission of John Wiley & Sons.)

Rate Constants

The following data are the rate constants for the hydration of CO_2 and hydrolysis of *p*-nitropheny acetate (Hay, 1984). What do you conclude?

| | k (M^{-1} s^{-1}), 25°C | | |
	CO$_2$ Hydration	*p*-Nitrophenyl Acetate Hydrolysis	pK_a
Carbonic anhydrase	10^7-10^8	—	8
	—	460	7.5
H_2O	6.7×10^{-4}	—	−1.7
OH$^-$	8500	9.5	15.5
$[(NH_3)_5CrOH]^{2+}$	10	—	5.2
$[(NH_3)_5CoOH]^{2+}$	220	1.52×10^{-3}	6.4
$[(NH_3)_5CoIm]^{2+}$	—	9	10.02
$[(NH_3)_5IrOH]^{2+}$	590	—	6.7
$[(NH_3)_5RhOH]^{2+}$	470	—	6.78

- In the simple hydroxo complexes, the catalytic activity parallels the acidity, pK_a (Table 3-2).
- It is clear that the hydroxide ion bound to the Zn(II) is capable of catalyzing the hydration of CO_2 and hydrolysis of esters.

TABLE 3-2   Catalytic Activity and pK_a

| | k (M^{-1} s^{-1}), first order with respect to OH$^-$ | | |
	CO$_2$ Hydration	*p*-Nitrophenyl Acetate Hydrolysis	pK_a
H_2O	6.7×10^{-4}	—	−1.7
$[(NH_3)_5CrOH]^{2+}$	10	—	5.2
$[(NH_3)_5CoOH]^{2+}$	220	1.52×10^{-3}	6.4
$[(NH_3)_5CoIm]^{2+}$	—	9	10.02
$[(NH_3)_5RhOH]^{2+}$	470	—	6.78
$[(NH_3)_5IrOH]^{2+}$	590	—	6.7
OH$^-$	8500	9.5	15.5
Carbonic anhydrase	10^7-10^8	—	8
		460	7.5

ALCOHOL DEHYDROGENASE

What is the catalytic role and the structural features of the alcohol dehydrogenases?

- These enzymes catalyze the reversible oxidation of alcohols to aldehydes and ketones and remove alcohol both produced internally and consumed externally.

- The overall reaction can be written as:

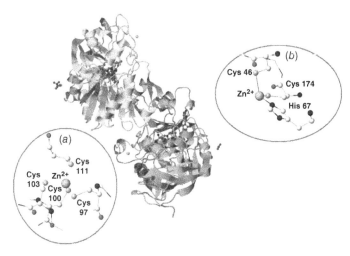

- ○ R_1R_2CHOH is the substrate.
- ○ NAD^+ is the coenzyme.
- ○ Alcohol dehydrogenase is the metalloenzyme.
- The enzyme consists of two identical subunits of molecular weight ~40,000 and 2 Zn/subunit.
- X-ray studies indicate two types of Zn ions.
- The active site of Zn is shown in Fig. 3-3.
  - ○ The zinc's ligands are:
    - (a) Cys^{46}, Cys^{174}
    - (b) Imidazole group of His^{67}
    - (c) In distorted T_d, which is completed by H_2O
  - ○ This site acts as an electrophilic center. It attaches and lowers the pK_a of the alcoholic OH, promoting hydride transfer to NAD^+, via formation of a Zn(II) alkoxide complex.

FIGURE 3-3 (*a*) Active site and (*b*) secondary site of human alcohol dehydrogenase, PDB 1AGN (Xie et al., 1997).

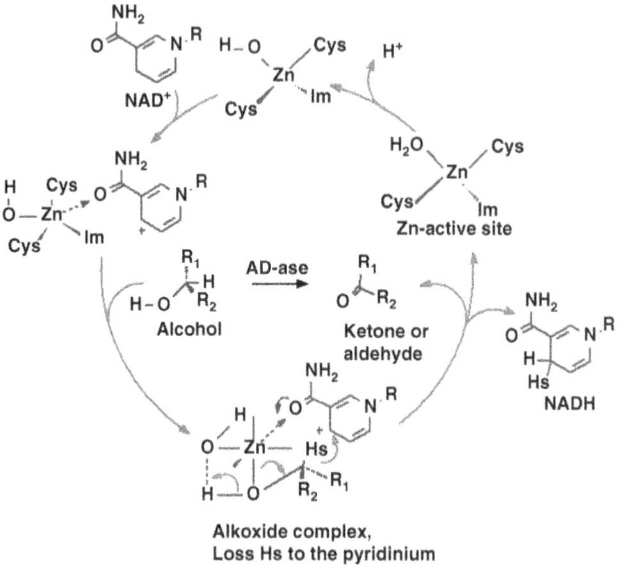

SCHEME 3-10 Possible reaction mechanism of alcohol dehydrogenase. (H_s is hydride substrate).

o The ligands of the second type of Zn (Fig. 3-3) are: Cys^{97}, Cys^{100}, Cys^{103}, and Cys^{111}.

o This site is inaccessible to solvent and substrate and the function may be structural stabilization.

What is the sequence of events during the reaction of alcohol dehydrogenase?

• The enzyme binds NAD^+, which induces a conformational change and releases a proton, $Zn–H_2O \rightarrow Zn–OH$ (Scheme 3-10).

• The alcohol substrate $R_1R_2(CH_s)OH$ binds the active zinc and converts to an alkoxide complex:

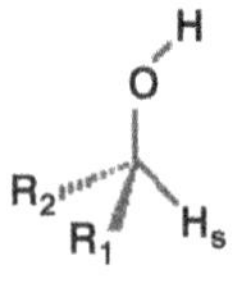

• Then transfer of H_s occurs from C_1 of the alkoxide to C_4 of the pyridinium ring of NAD^+.

• This produces aldehyde or ketone, which dissociates from the enzyme.

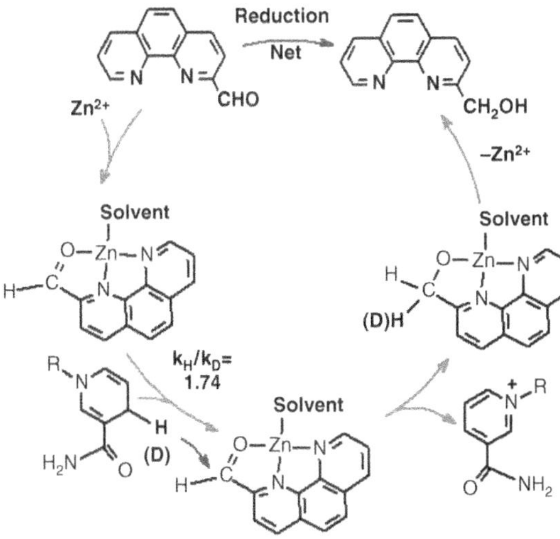

SCHEME 3-11 Reduction of 1,10-phenanthroline-2-carboxylaldehyde. (From Creighton et al., 1976.)

A number of model systems have been investigated to study the influence of Zn^{2+} on the reactivity of the carbonyl; give an example.

- Zn^{2+} catalyzes the reduction of 1,10-phenanthroline-2-carboxylaldehyde to the corresponding carbinol by 1,4-dihydronicotinamide in acetonitrile:

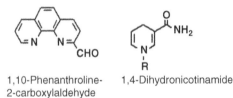

1,10-Phenanthroline-   1,4-Dihydronicotinamide
2-carboxylaldehyde

- The reduction does not take place in the absence of Zn^{2+} over a period of four days.
- It was assumed that the reaction involves the formation of Zn(II) complex (Scheme 3-11).
- Reduction is hindered if the two hydrogen atoms at C_4 are substituted by deuterium ($k_H/k_D = 1.74$), showing that the cleavage of the C_4—H bond occurs in the rate-determining step.

REFERENCES

R. Breslow and D. Wernick, *J. Am. Chem. Soc.*, 98, 259 (1976).

D. A. Buckingham, C. E. Davis, D. M. Foster, and A. M. Sargeson, *J. Am. Chem. Soc.*, 92, 5571 (1970).

D. A. Buckingham, C. E. Davis, D. M. Foster, L. G. Marzilli, and A. M. Sargeson, *Inorg. Chem.*, 9, 11 (1970).

D. A. Buckingham, C. E. Davis, D. M. Foster, and A. M. Sargeson, *J. Am. Chem. Soc.*, 92, 6151 (1970).

I. D. Campbell, S. Lindskog, and A. I. White, *Biochim. Biophys. Acta*, 484, 443 (1977).

J. E. Colman, *J. Biol. Chem.*, 242, 5212 (1967).

J. E. Colman, *Nature*, 214, 193 (1967).

D. J. Creighton, J. Hajdu, and D. S. Sigman, *J. Am. Chem. Soc.*, 98, 4619 (1976).

A. E. Eriksson, T. A. Jones, and A. Liljas, *Proteins*, 4, 274–282 (1988).

R. K. Gupta and J. M. Pesando, *J. Biol. Chem.*, 250, 2630 (1975).

R. W. Hay, in *Bio-Inorganic Chemistry*, Ellis Horwood Series Chemical Science, Chichester (1984).

M. N. Hughes, in *Inorganic Chemistry of Biological Process*, 2nd ed., Wiley, New York (1981).

A. Kilshtain-Vardi, M. Glick, H. M. Greenblatt, A. Goldblum, and G. Shoham, *Acta Crystallogr. Sect. D*, 59, 323–333 (2003).

S. Lindskog and J. E. Colman, *Proc. Natl. Acad. Sci. USA*, 70, 2505 (1973).

R. Steinberger and F. H. Westheime, *J. Chem. Soc.*, 73(1), 429 (1951).

P. Xie, S. H. Parsons, D. C. Speckhard, W. F. Bosron, and T. D. Hurley, *J. Biol. Chem.*, 272, 18558–18563 (1997).

M. N. Hughes, in *Inorganic Chemistry of Biological Process*, 2nd ed., Wiley, New York (1981).

SUGGESTIONS FOR FURTHER READING

Metalloenzymes

1. C. F. Mills, Ed., *Zinc in Human Biology*, Springer-Verlag, New York (1989).

2. A. S. Prasad, in *Biochemistry of Zinc*, Plenum, New York (1993).

3. I. Bertini, H. B. Gray, S. J. Lippard, and J. S. Valentine, in *Bioinorganic Chemistry*, University Science Books, Mill Valley, CA (1994).

4. B. L. Vallee and D. S. Auld, *Faraday Discuss.*, 93, 47 (1992).

5. B. L. Vallee and D. S. Auld, *Proc. Natl. Acad. Sci. USA*, 88, 999 (1991).

6. B. L. Vallee and D. S. Auld, *Proc. Natl. Acad. Sci. USA*, 87, 220 (1990).

7. A. R. Fersht, Ed., *Enzyme Structure and Mechanism*, Freeman, Reading, MA (1977).

8. J. E. Coleman, "Metal ion in enzymatic catalysis", in *Progress in Bioinorganic Chemistry*, E. T. Kaiser and F. J. Kezdy, Eds., Vol. 1, Wiley Interscience, New York (1971).

9. M. L. Bender, Ed., *Mechanisms of Homogeneous Catalysis from Protons to Proteins*, Wiley Interscience, New York (1971).

10. R. J. P. Williams and A. E. Dennard, "The transition metal as reagents in metalloenzymes", in *Transition Metal Chemistry*, R. L. Carlin, Ed., Vol. 2, Edward Arnold, London (1966).

Metal Ion Catalysis

11. D. P. N. Satchell, "Metal–ion–promoted reactions of organo–sulphur compounds", *Chem. Soc. Rev.*, 6, 345 (1977).

12. M. M. Jons, Ed., *Ligand Reactivity and Catalysis*, Academic, New York (1968).

13. N. E. Dixon and A. M. Sargeson, "Roles for the metal ion in reactions of coordinated substrates and in some metalloenzymes", in *Zinc Enzymes*, T. G. Spiro, Ed., Wiley, New York (1983).

14. R. W. Hay, "Metal ion catalysis and metalloenzyme", in *An Introduction to Bioinorganic Chemistry*, D. R. Williams, Ed., Chapter 4, C. C. Thomas, Springfield, IL (1976).

15. R. W. Hay, in *Metal Ions in Biological Systems*, H. Sigel, Ed., Vol. 5, Marcel Dekker, New York (1976).

16. W. N. Lipscomb and N. Sträter, *Chem. Rev.*, 96, 2375–2433 (1996).

17. D. A. Buckingham, "Metal–OH and its ability to hydrolyse (or hydrate) substrates of biological interest", in *Biological Aspects of Inorganic Chemistry*, A. W. Addison, W. R. Cullen, D. Dolphin, and B. R. James, Eds., Wiley-Interscience, New York (1977).

18. E. T. Kaiser and B. L. Kaiser, *Acc. Chem. Res.*, 5, 219 (1972).

19. R. W. Hay and P. J. Morris, "Metal ion promoted hydrolysis of amino acid esters and peptides", in *Metal Ions in Biological Systems*, H. Sigel, Ed., Vol. 5, Marcel Dekker, New York (1976).

20. W. N. Lipscombe, *Chem. Soc. Rev.*, 1, 319 (1972).

21. W. N. Lipscombe, *Tetrahedron*, 30, 1725 (1974).

Carbonic Anhydrase

22. E-I. Ochia, in *Bioinorganic Chemistry–An Introduction*, Chapter 11, Allyn and Bacon, Boston (1977).

23. W. N. Lipscomb and N. Sträter, "Recent advance in zinc enzymology", *Chem. Rev.*, 96, 2375–2433 (1996).

24. D. W. Christianson and J. D. Cox, *Biochemistry*, 68, 33–57 (1999).

25. A. Liljas, K. K. Kannen, P-C. Bergsten, I. Waara, K. Fridborg, B. Strandberg, U. Carlborn, L. Jarup, S. Lovgren, and M. Petef. *Nature New Biology, Lond.*, 235, 131 (1972).

26. J. E. Colman, *Prog. Bioorg. Chem.*, 1, 159 (1971).

27. K. K. Kannen, B. Nostrand, K. Fridborg, S. Lovgren, and M. Petef, *Proc. Natl. Acad. Sci. USA*, 72, 51 (1975).

28. K. K. Kannen, M. Petef, K. Fridborg, H. Cid-Dresdner, and S. Lovgren, *FEBS Lett.*, 73, 115 (1977).

29. R. W. Hay, *Inorg. Chim. Acta*, 46, L115 (1980).

30. S. Lindskog, L. E. Henderson, K. K. Kannen, A. Liljas, P. O. Nyman, and B. Strandberg, in *The Enzymes*, P. D. Boyer, Ed., 3rd ed., p. 587, Academic, New York (1971).

31. T. W. W. Lane and F. M. M. Morel, "A biological function for cadmium in marine diatoms", *Proc. Natl. Acad. Sci. USA*, 97, 4627–4631 (2000).

Alcohol Dehydrogenase

32. Z-X. Liang and J. P. Klinman, *Curr. Opin. Struct. Biol.*, 14, 648–655 (2004).

33. C. I. Bronden, H. Jornvall, H. Eklund, and B. Furugren, *The Enzymes*, 11, 104 (1975).

34. R. T. Dworschack and B. V. Plapp, *Biochemistry*, 16, 2716 (1977).

35. D. J. Creighton, J. Hajdu, and D. S. Sigman, *J. Am. Chem. Soc.*, 98, 4619 (1976).

4

COPPER PROTEINS

INTRODUCTION

Copper-containing proteins have been found to be widely distributed in both plants and animals. Copper is the third most abundant transition element in the body, following iron and zinc. The adult human body contains 100–150 mg of copper; about 64 mg is found in muscles, 23 mg in the bones, and 18 mg in the liver. Animals and humans have homeostatic mechanisms for absorption, transport, utilization, and excretion of copper.

Copper is a principal component of several metalloproteins and some naturally occurring pigments. Copper-containing proteins have been related to metabolic processes such as hydroxylation, oxygen transport, electron transfer, and oxidative catalysis. They are also essential to hemoglobin synthesis and to the action of certain enzymes, including cytochrome oxidase, tyrosinase, catalase, uricase, ascorbate oxidase, and monoamine oxidase. Copper plays a role in bone formation and maintenance of myelin of the nervous system.

Both blood cells and serum contain copper. The red blood cell copper is present as superoxide dismutase. In plasma, 80–95% of the copper is tightly bound to ceruloplasmin, a copper-binding plasma protein; some plasma copper is loosely bound to albumin. Other cuproprotein enzymes present in animal tissues include

Chemistry of Metalloproteins: Problems and Solutions in Bioinorganic Chemistry, First Edition.
Joseph J. Stephanos and Anthony W. Addison.
© 2014 John Wiley & Sons, Inc. Published 2014 by John Wiley & Sons, Inc.

dopamine hyroxylase and hemocyanin (an oxygen carrier in hemoglobin of certain invertebrates).

In the presence of a deficiency of copper, the movement of iron from the tissues to plasma is halted, hypoferremia results, and hemoglobin production is retarded. Wilson's disease and Menkes' kinky hair syndrome are associated with a genetic disorder in the metabolism of copper.

Copper has a single s electron outside the filled $4d^{10}$ shell but cannot be categorized in group I of alkali metals. The filled d^{10} shell is much less effective than is a noble gas shell in shielding the s electron from the nuclear charge. Consequently, the first ionization enthalpy of Cu is higher than those of the alkalis; however, the second ionization enthalpy of Cu is very much lower than those of the alkalis. Copper complexes are generally square planar or distorted octahedral, and the coordinating atom preference is $S > N > O$, which is approximately 10-fold stronger than that of zinc. The association of copper with small peptides is characterized by a tendency to coordinate with the nitrogen atoms of the peptide bond with simultaneous displacement of the peptide bond proton.

In this chapter, it is intended to survey briefly the electronic and ESR spectra of the copper ions in order to understand the copper–ligand interaction and then use this information to understand the chemistry of copper-containing proteins such plastocyanin, azurins, superoxide dismutase, hemocyanins, and ascorbic acid oxidase. The design and use of model compounds will be inspected together with the information that may be gained from a study of their similarities and differences when compared with the in vivo system.

ELECTRONIC SPECTRA OF COPPER IONS

Identify the following:

  (i) The ground-state term symbol of Cu^{2+} and Cu^+

 (ii) The crystal field stabilization energy (CFSE) and electronic spectral selection rules of copper ions

(iii) Which spectral transition produces the blue color of blue-type copper proteins and how π-bonding can effect these transitions

 (iv) How the Jahn–Teller effect induces structural distortion

  (v) The relationship between the symmetry of Cu^{2+} complexes and the spectral electronic transitions

(i) Ground state and term symbol

- The five d orbitals of the free metal ion (no ligands) are degenerate at the *ground state*.

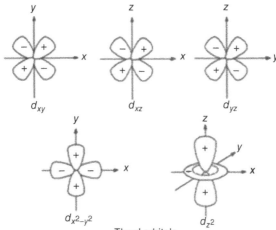

The d orbitals

- However, due to the spin–spin interactions, the following distributions of $3d^9$ electrons in Cu^{2+} will differ in energy: $(d_{xz}^1, d_{xz}^2, d_{xz}^2, d_{z^2}^2, d_{x^2-y^2}^2)$, $(d_{xz}^2, d_{xz}^1, d_{xz}^2, d_{z^2}^2, d_{x^2-y^2}^2)$, $(d_{xz}^2, d_{xz}^2, d_{xz}^1, d_{z^2}^2, d_{x^2-y^2}^2)$, $(d_{xz}^2, d_{xz}^2, d_{xz}^2, d_{z^2}^1, d_{x^2-y^2}^2)$, and $(d_{xz}^2, d_{xz}^2, d_{xz}^2, d_{z^2}^2, d_{x^2-y^2}^1)$.
- Hund's Rule defines how electrons are arranged in an orbital, providing insight into the atom's reactivity and stability.
- The ground state of the Cu^{2+} ion and its term are determined by Hund's rules. For a given atom:
  (a) The electronic configuration with maximum S (maximum spin multiplicity, $2S+1$) has the lowest energy.
  (b) The electronic configuration for a given multiplicity, maximum L (maximum M_L) has the lowest energy.
  (c) For a given term, if the outmost subshell is half-filled or less, the level with the lowest value of J ($J=L-S$) has the lowest energy. If the outermost subshell is more than half-filled, the level with the highest value of J ($J=L+S$) is lowest in energy.

Maximum $S=\frac{1}{2}$

Maximum multiplicity: $2S+1=2$

m_ℓ	2	1	0	−1	−2
	↑↓	↑↓	↑↓	↑↓	↑

Maximum $L=2\times2+2\times1+2\times0+2\times-1+1\times-2=2$

When $L=0$, 1, 2, 3, and 4, the corresponding terms are S, P, D, F, and G.

Therefore

o The ground-state term is D.

o The outmost subshell is more than half-filled, $J=L+S=2+\frac{1}{2}=\frac{5}{2}$.

o Ground-state term symbol, $^{\text{Multiplicity}}L_J$: $^2D_{5/2}$.

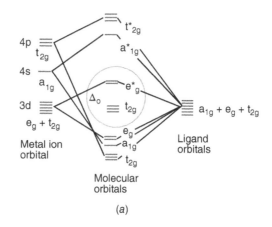

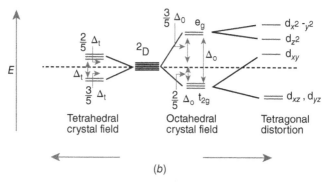

FIGURE 4-1 (*a*) Qualitative molecular orbital scheme for octahedral complex. (*b*) Splitting of d orbitals by crystal field ($\Delta_o = \Delta_{octahedral}$ and $\Delta_t = \Delta_{terahedral}$).

- The five d orbitals of Cu^{2+} are degenerate in the $^2D_{5/2}$ state.
- The five d orbitals of Cu^+ are degenerate in the $^1S_0$ state.

(ii) CFSE and electronic spectra selection rules of copper ions

- The ^2D state splits into two states in O_h and in T_d environments (Fig. 4-1).
- Note that $\Delta_t = -\frac{4}{9}\Delta_o$.
- For Cu^{2+} (d^9)
  - Octahedral: $CFSE = (6)(-\frac{2}{5}\Delta_o) + (3)(\frac{3}{5}\Delta_o) + 4\pi = -\frac{3}{5}\Delta_o + 4\pi$
  - Tetrahedral: $CFSE = (4)(-\frac{3}{5}\Delta_t) + (5)(\frac{2}{5}\Delta_t) + 4\pi = -\frac{2}{5}\Delta_t + 4\pi$
- For Cu^+ (d^{10})
  - Octahedral: $CFSE = (6)(-\frac{2}{5}\Delta_o) + (4)(\frac{3}{5}\Delta_o) + 5\pi = 5\pi$
  - Tetrahedral: $CFSE = (4)(-\frac{3}{5}\Delta_t) + (6)(\frac{2}{5}\Delta_t) + 5\pi = 5\pi$
- In the above, π is the electron pairing energy.
- Electronic spectra selection roles: If ground and excited states for a possible transition have the required characteristics, the transition is said to be *allowed*. If not, the transition is *forbidden*.

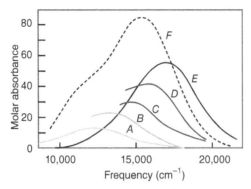

FIGURE 4-2 Absorption spectra of $[Cu(H_2O)_6]^{2+}$ (*A*), $[Cu(NH_3)(H_2O)_5]^{2+}$ (*B*), $[Cu(NH_3)_2(H_2O)_4]^{2+}$ (*C*), $[Cu(NH_3)_3(H_2O)_3]^{2+}$ (*D*), $[Cu(NH_3)_4(H_2O)_2]^{2+}$ (*E*), and $[Cu(NH_3)_5H_2O]^{2+}$ (*F*) at 25°C. (Data from Cotton and Wilkinson, 1980. Reproduced by permission of John Wiley & Sons.)

○ Transitions may occur only between energy states with the same spin multiplicity ($\Delta S = 0$).

○ In centrosymmetric environments, transitions may occur only between states of opposite parity (u $\leftrightarrows$ g, Laporte's rule). The transitions g → g and u → u can be vibronically allowed.

(iii) Spectral transition that produces blue color of blue-type copper proteins and how π-bonding can effect these transitions

- Amine complexes of copper are much more intensely blue than the aqua (Fig. 4-2).
- This is because the amines produce a stronger ligand field, which causes the absorption band to move from the far red to the middle of the red region.
- In the aqua ion, the absorption maximum is at ~800 nm, whereas in $[Cu(NH_3)_4(H_2O)_2]^{2+}$ it is at ~600 nm.
- π-Bonding must be considered (Fig. 4 3) whenever the ligand has suitable p_π or π molecular orbitals available for overlap with d_π or t_{2g} atomic orbitals of the metal ion.

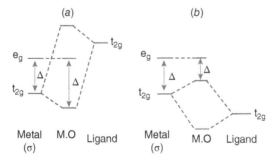

FIGURE 4-3 Interaction of ligand orbitals and metal t_{2g} orbitals.

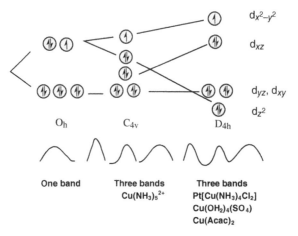

FIGURE 4-4 Jahn–Teller effect and tetragonal distortion.

- When ligands have *empty* π orbitals (unsaturated, less stable), the orbitals t_{2g} of the ligand will lie at *higher energy* than that of the corresponding metal orbital (Fig. 4-3*a*).
- Other ligands contain donor atoms such as oxygen or fluorine (saturated, more stable), and the only π orbitals available, t_{2g}, are *filled*. The p orbitals of the ligand (t_{2g}) lie at *lower energy* than the metal t_{2g} orbitals (Fig. 4-3*b*).
- It may be seen that the effect of (*a*) is to increase Δ, while in (*b*) Δ is decreased.
- Every ligand has specific and characteristic donor–acceptor influence.
- Thus, π-bonding is important in understanding the positions of ligands in the *spectrochemical series* $I^- < Br^- < Cl^- < -SCN^- < F^- < OH^- <$ oxalate$^- \leq H_2O$ $< -NCS^- \leq NH_3 < Pyr^- <$ ethylenediamine $<$ Dipyr $< o$-phenantholine $< NO_2^-$ $< CN^- < CO$. In a biological system

$$SH < CO_3^{2-} < \text{amide} < \text{imidazole}$$

(iv) Jahn–Teller effect and induced structural distortion

- The addition of fifth and sixth molecules of ammonia to $[Cu(NH_3)_4(H_2O)_2]^{2+}$ is difficult. The reason for this unusual behavior is connected with the *Jahn–Teller effect* (Fig. 4-4):
  - The Jahn–Teller effect may be seen by considering the effect of unevenly filled e_g and/or t_{2g} shells on the symmetry of the complexes.
  - In Cu^{2+}, d^9 configuration, let us assume we have one electron in the d_{z^2} orbital and two electrons in $d_{x^2-y^2}$ orbital. Repulsion between ligand electrons and metal electrons will then be greater in the x–y plane than

on the z axis, producing a tetragonally distorted complex with short bonds on the z axis.

- ○ If the opposite arrangement of electrons holds, then the bond on the z axis will be longer than the bonds on the plane.
- Stereochemistries and oxidation states
  - ○ Cu^+ (d^{10})

CN 2	Linear	
CN 3	Planar	
CN 4	Tetrahedral	(common state)

  - ○ Cu^{2+} (d^9)

CN 4	Square planar	(common state)
CN 4	Distorted tetrahedral	
CN 5	Square pyramidal	
CN 5	Trigonal bipyramidal	
CN 6	Distorted octahedral	(common state)

(v) **Relationship between symmetry of Cu^{2+} complexes and spectral electronic transitions**

- Copper (I) complexes, being of d^{10} configuration, will only show *charge transfer* bands, $Cu^+ \rightarrow L$. These bands have been observed up to 650 nm in the visible region.
- Copper (II) complexes (d^9) will also show charge transfer bands, $Cu^{2+} \leftarrow L$, beside the $d \leftarrow d$ transitions.
- In O_h there is only one transition: $^2T_g \leftarrow {}^2E_g$.

- The common stereochemistry is that of the tetragonally distorted octahedral one. The effect of a crystal field upon the degeneracy of the d orbitals can be seen in Fig. 4-4.

- In D_{4h}, as the ligands on the z axis are withdrawn, the energy of the d_{z^2} and d_{zx}, d_{yz} orbitals fall while that of the $d_{x^2-y^2}$, and d_{xy} will rise relative to these. Three electronic transitions are anticipated:
  - $^2A \leftarrow ^2B$
  - $^2B \leftarrow ^2B$
  - $^2E \leftarrow ^2B$

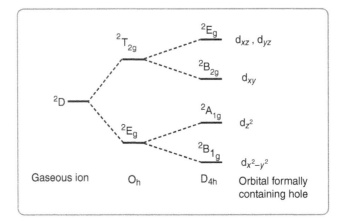

- When the distortion is slight, only one band will be seen as the energy separation between the split bands will be small, where:

  - A, B denote one-dimensional (1D) symmetry representation
  - E denotes 2D symmetry irreducible representation
  - $T(F)$ denotes 3D symmetry irreducible representation
  - A has $\chi = 1$ and is *symmetric* with respect to C_n
  - B has $\chi = -1$ and is *antisymmetric* with respect to C_n
  - Subscripts:
    - 1   Representation is symmetric with respect to $C_2 \perp C_n$, or if no $C_2 \perp C_n$, then with respect to σ_v
    - 2   Representation is antisymmetric with respect to $C_2 \perp C_n$, or if no $C_2 \perp C_n$, then with respect to σ_v
    - g   Representation is symmetric with respect to i
    - u   Representation is antisymmetric with respect to i
  - Superscripts:
    - $'$ Representation is symmetric with respect to σ_h
    - $''$ Representation is antisymmetric with respect to σ_h

- On the other hand, in T_d one transition is expected, while in D_{2d}, three bands will be observed (Fig. 4-5).
  - In Cs [$CuCl_4$] and [$Pt(NH_3)_4$][$CuCl_4$], $CuCl_4{}^{2-}$ is a distorted isomer, D_{2d}.

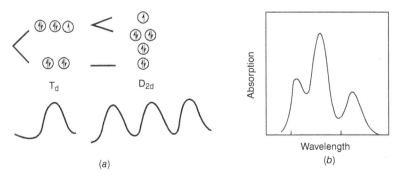

FIGURE 4-5 (*a*) Transitions in T_d and D_{2d} ($T_d \rightarrow D_{2d}$, due to Jahn–Teller effect). (*b*) Three bands are expected in the electronic spectrum of ($[CuCl_4]^{2-}$).

What are the structural and spectral features of copper(II)–peptide complexes in the visible region?

- In copper(II)–peptide complexes:
  - Cu^{2+} is coordinated to oxygen and nitrogen.
  - The geometry is usually tetragonal with four short and two long bonds (a result of Jahn–Teller distortion).
  - One or both long bonds may be removed completely, resulting in square planar and square pyramidal conformations.
  - The 6-coordinate → blue green ($\lambda_{max} \sim 730$ nm)
  - The 5-coordinate → blue ($\lambda_{max} \sim 635$ nm)
  - The 4-coordinate → violet pink ($\lambda_{max} \sim 500$ nm)

- Within compounds of the same coordination number, the band position depends on the electron donor–acceptor interactions of the attached ligands.
  - The 6-coordinate (tetragonally distorted symmetry) occurs with glycylglycylhistidine, NaCu(II)(GlyGlyGlyHis) H_2O:

and with three deprotonated amide groups, $Na_2Cu(II)(GlyGlyGlyGlyGly)\cdot$
$4H_2O$, ($\lambda_{max} \sim 510\,nm$).

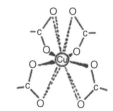

Na$_2$ Cu^{II} (GlyGlyGlyGly).4H$_2$O

o Cu (II) (6–aminohexanoic acid)$_4$(CLO$_4$)$_2$ has a CuO_4O_4' chromophore, with
   four short and four long bonds forming a distorted dodecahedron:

Cu^{II} (6-aminohexanoic acid)$_4$.(ClO$_4$)$_2$

The $\bar{\nu}_{max}$ of d–d bands can be expressed as the sum of the individual ligand field
contribution:

$\bar{\nu}_N(\text{peptide}) = 4.85\,kK$

$\bar{\nu}_N(\text{amino}) = 4.53\,kK$

$\bar{\nu}_N(\text{imidazol}) = 4.3\,kK$

$\bar{\nu}_O(\text{carboxylate}) = 3.42\,kK$

$\bar{\nu}_O(\text{peptide, } H_2O \text{ or } OH^-) = 3.01\,kK$

The effect of axial coordination of OH^-, COO^-, and NH_2 is to shift ν_{max} to lower
energy by 1 kK (where $kK = 1000\,cm^{-1}$)

Find the expected $\bar{\nu}_{obs}$ for Cu(GluGly)·H_2O:

where

$$\bar{\nu}_{obs} = \bar{\nu}_N \text{ (amino)} + \bar{\nu}_N \text{ (peptide)} + \bar{\nu}_O \text{ (carboxylate)}$$
$$= 4.53 + 4.85 + 3.42 + 3.01 = 15.81 \text{ kK}$$
$$\lambda = 632 \text{ nm}$$

ESR SPECTRA OF COPPER IONS

How can ESR spectra be used as a "spectral probe" to study copper enzymes?

- The ESR spectra only monitor ions that have an odd number of unpaired electrons in the ground state. Thus it is a promising tool to provide a great deal of insight into copper-containing proteins.
- Rotation of an electron will generate a magnetic field. For simplicity it may be considered as a small magnet.
- When this rotated electron exists in a strong magnetic field, a parallel alignment will occur (ground state), which can be excited to antiparallel alignment (Fig. 4-6).

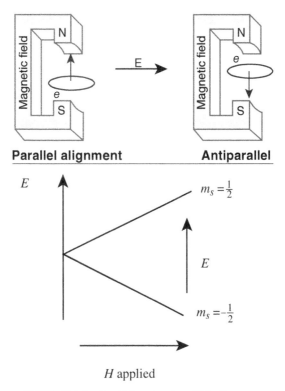

FIGURE 4-6 Parallel and antiparallel alignment.

- The energy difference between the ground and the excited states, E, is directly proportional to the applied magnetic field, H:

  Energy difference $\propto$ applied magnetic field

  $E \propto H\beta$, where $\beta =$ Bohr magneton $= 9.2741 \times 10^{-21}$ erg/G

  $$\beta = \frac{eh}{4\pi mc}$$

  $\therefore E = h\,\nu = g\,H\beta$, where $g =$ proportionality constant

  $g = 714.44\,(\nu/H)$

  $g_e = 2.0023$ for free electron

- Usually records the ESR spectrum at constant frequency, ν, and H is varied until absorption occurs.

- It is significant that ESR measurements be obtained at low temperature to enhance the occupancy of the ground state at the expense of the excited state in order to avoid the effect of line broadening.

- The change in the factor g provides valuable information on the occupation of the orbitals, the degree of hybridization, the orbital moment contribution, the oxidation state of the copper, confirms the presence of covalency, and may help in the characterization of the ligand.

- The area under the absorption curve is proportional to the amount of unpaired electrons and can be converted to $[Cu^{2+}]$ by calibrating the spectrometer with suitable standers (Fig. 4-7).

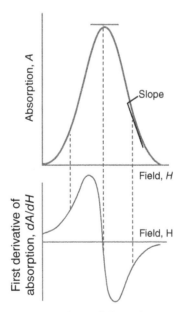

FIGURE 4-7   Absorption and signal (slope of absorption signal at each point; spectrometer display).

Define:
  (a) the anisotropic effect,
  (b) the isotropic effect, and
  how do these affect the ESR spectra?

(a) **Anisotropic effect**

- For radicals with low symmetry, the value of g changes as a function of the *orientation* of the *crystal* relative to the external magnetic field. (Why?)
- The value of g arises from *coupling* of the *spin* angular momentum with the *orbital* angular momentum.
- The spin angular momentum is oriented *with the field*.
- However, the orbital angular momentum, which is associated with the electron moving in molecular orbitals, is *locked* to the molecular wave function.
- Consider a case where there is an orbital contribution to the moment from an electron in a circular molecular orbital that can process about the z axis of the molecule.
- When the orbital angular momentum and the spin angular momentum are in the same direction, a maximum μ value is expected:

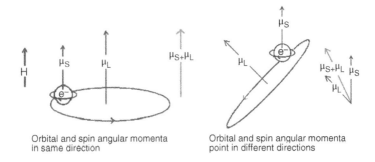

Orbital and spin angular momenta in same direction

Orbital and spin angular momenta point in different directions

- A different orientation of molecule, the results of μ_s and μ_L do not point in the same direction and become less than the sum of μ_s and μ_L.
- Therefore, the value of g changes as a function of the orientation of the crystal relative to the external magnetic field.
- When the effects of the orbital moment are small, they are integrated into the g value, and now we see that this g value will be anisotropic.
- In magnetically anisotropic crystals, if an electron in a circular molecular orbital is rapidly rotating about the z axis, a maximum g_z will be achieved, so the same result is obtained for x or y parallel to the field (Fig. 4-8). When $g_z \neq g_x$, $g_y \rightarrow$ anisotropic:

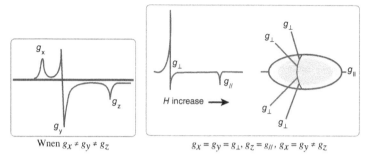

Wnen $g_x \neq g_y \neq g_z$ $g_x = g_y = g_\perp, g_z = g_{//}, g_x = g_y \neq g_z$

- The ESR spectrum of mushroom laccase shows signals at 2.05 and 2.20. ESR signals disappeared after treatment with 10 mM catechol (see Nakamura and Ogura, 1967). What information does this give us?
  - Copper (II) complexes (d^9) with a single unpaired electron
  - Easy resolved ESR signal at g value slightly higher than that of free electron
  - Reduction of Cu^{2+} to Cu^+ when catechol is added (Fig. 4-9).

(b) **Isotropic spectra**

- Isotropic spectra are obtained when the radical under consideration has spherical or cubic symmetry, $g_x = g_y = g_z$.
- For radicals with lower symmetry, anisotropic effects are manifested in the solid spectra.

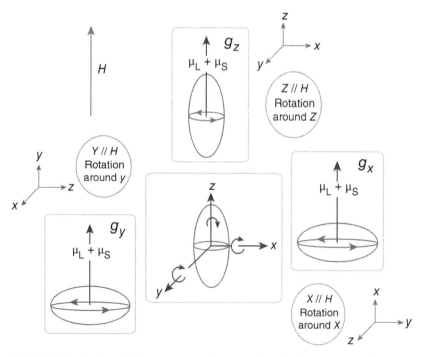

FIGURE 4-8 Both orbital and spin angular momenta point in z, x, and y directions.

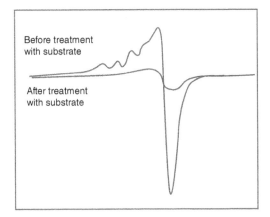

FIGURE 4-9   Example of changes in ESR signals after treatment with substrate.

- The solution spectra qualitatively appear as isotropic spectra because anisotropic effects are averaged to zero by the rapid rotation of the molecules.
- If it were not for the orbital contribution, the moment from the electron would be isotropic.
- "Typical" ESR spectra for copper complexes are represented in Fig. 4-10.

How do the nuclear–spin of Cu^{2+} ion and the attached ligands affect the ESR signal? How can the hyperfine and superhyperfine splitting be used as a "spectral probe" for studying the copper enzymes?

- *Hyperfine splitting* is as a result of coupling the *electron* magnetic moment μ_e with the *nuclear* magnetic moment μ_p:

$$^1H \;\; \rightarrow I = \tfrac{1}{2} \rightarrow \mu_p = 1$$
$$^{57}Fe \rightarrow I = \tfrac{1}{2} \rightarrow \mu_p = 0.03$$
$$^{63}Cu \rightarrow I = \tfrac{3}{2} \rightarrow \mu_p = 0.26$$

where I is the nucleus spin quantum number. For simplicity, consider Fig. 4-11.

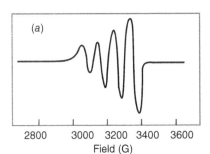

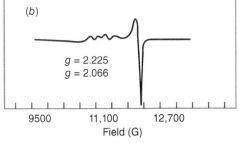

FIGURE 4-10   (*a*) Isotropic spectrum (solution) for square planar Schiff bas ligand. (*b*) Anisotropic spectrum (glass or powder sample) for axial complex.

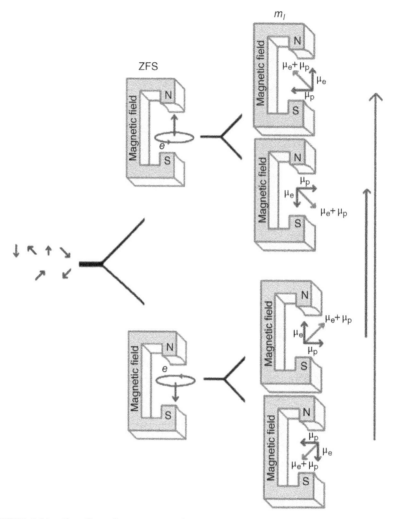

FIGURE 4-11 Coupling electron magnetic moment μ_e with nuclear magnetic moment μ_p.

- In $S > \frac{1}{2}$ systems, consider the zero field splitting (ZFS) arising from possible values of S_z. Now we deal with values of m_I (I_z) from $-I$ to $+I$ in steps of 1. For example, for ^1H, $m_I = -\frac{1}{2}, +\frac{1}{2}$ (Figs. 4-12 and 4-13).
- ESR selection rules: Give the resonance at:
  - $\Delta m_s = \pm 1$
  - $\Delta m_I = 0$
- When an unpaired electron of spin S interacts with the nuclear spin I of the metal, $2I + 1$ bands are obtained. The separation associated with these is the

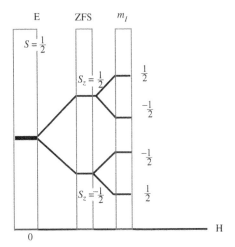

FIGURE 4-12   ZFS and influence of nuclear magnetic moment for ^{1}H.

hyperfine splitting A and will depend upon the stereochemistry of the complex:

$$\Delta E \rightarrow g\beta H + 2A/4$$
$$g\beta H - 2A/4$$

A is hyperfine coupling constant of I values, where
$$A \ (cm^{-1}) = 0.935 \times 10^{-4} \ (g/g_e) \ A(G)$$

I values $\rightarrow (2I+1)$

m_I values $\rightarrow (2I+1)$ resonances

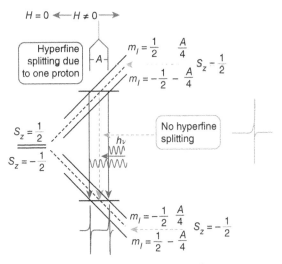

FIGURE 4-13   Hyperfine splitting A due to one proton of ^{1}H. No hyperfine splitting for free electron.

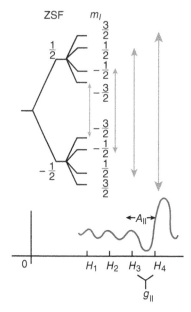

FIGURE 4-14 Four resonances separated by A.

- For Cu^{2+} $(I = \frac{3}{2})$
  - $m_I = -\frac{3}{2}, -\frac{1}{2}, +\frac{1}{2}, +\frac{3}{2}$
  - Therefore, we get four resonances separated by A (Fig. 4-14).

  Note: $\mu_{Cu} = 0.26$, while $\mu_{Fe} = 0.03$, expect A_{Fe} is lower by $\sim$10 times, so HF lines merge.
- See Fig. 4-15.
- Additional superhyperfine splitting may take place by the unpaired electron if the electron is under the effect of multiple sets of equal nuclei.
- At room temperature ESR spectrum of Cu^{2+} triglycine (see Wiersema and Windle, 1964).
- $Cu^{2+} \rightarrow I = \frac{3}{2}$. Therefore, the signal is split into 4 lines.
- $N \rightarrow I = 1$, three N donor atoms, number of lines $n = 3$

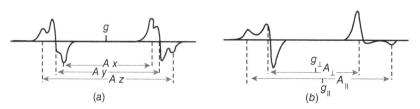

FIGURE 4-15 (*a*) Anisotropic, with $I = \frac{1}{2}$, $A_x > A_y > A_z$. (*b*) Axial symmetry, with $I = \frac{1}{2}$, $g \perp > g_{||}$, $A_{||} > A$.

- Each ESR signal of an electronic system that interacts with a group of n *equivalent* nuclei of spin I is split into $2nI + 1$ lines, and the number of lines $= (2)\,(3)\,(1) + 1 = 7$ lines:

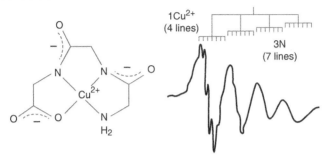

Seven superhyperfine lines are expected,

- A direct demonstration of covalent bonding in a complex can be obtained if a superhyperfine structure is observed in the ESR spectrum.

How may the spectrochemical series and the stereochemistry affect g-values of Cu^{2+} ion and what is the significance of $A_{\parallel}$–$g_{\parallel}$ trend?

g Values

- The observed g value is different from $g_e = 2.0023$.
- This difference is due to spin–orbital coupling.
- We find that the electron is not a simple free electron when it moves among different types of atomic orbitals (AOs), i.e., when it can acquire orbital angular momentum as well as spin angular momentum.
- For electron $g = 2.0023$, the change in this value (Δg) is a measure of the contribution of the spin and orbital motion to the total angular momentum.
- For example, when $d_{x^2-y^2}$ and d_{xy} have close energy ($E_{d_{x^2-y^2}} \cong E_{d_{xy}}$):

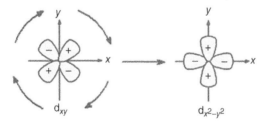

Degeneracy maximizes spin–orbital coupling as e^- indistinguishably occupies processes about z.

- Therefore, it is expected that the value of Δg will be directly proportional to the sum of the spin–orbital coupling, $n\,\lambda$, and indirectly proportional to the energy difference between the two orbitals ($\Delta E = E_{d_{x^2-y^2}} - E_{d_{xy}}$). For a d^1 system:

$$\Delta g = \pm \frac{n\lambda}{\Delta E}$$

Use a plus sign if coupling is to an empty level and a minus sign if coupling is to a filled level. The value of n for different coupling levels is given as:

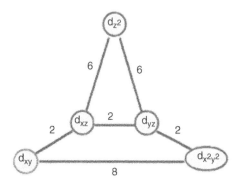

That is, for $d^1_{x^2-y^2}(d^9)$

$$\Delta g_\| = \frac{8\lambda}{\Delta E(d_{x^2-y^2} \leftrightarrow d_{xy})}$$

$$\Delta g_\perp = \frac{2\lambda}{\Delta E(d_{x^2-y^2} \leftrightarrow d_{xz}, d_{yz})}$$

Then,

$$g_\| = 2 + \frac{8\, k_\| |\lambda|}{\Delta E}$$

$2 \rightarrow g_e \qquad \Delta E \rightarrow E(d_{x^2-y^2}) - E(d_{xy})$

$k_{//} \rightarrow$ electron delocalization = fudge factor $\rightarrow (0 \leq k_{//} \leq 1)$

For a p^1_z system

$$\Delta g_\| = 0$$

$$\Delta g_\perp = \frac{\pm 2\lambda}{\Delta E}$$

where λ is the spin–orbital coupling constant.

(a) *Effect of Ligand Field Δ*

- The value of $g_{//}$ becomes more than $g_e = 2.00$ as ΔE get smaller ($g_{//} \propto 1/\Delta E$). Therefore:

  (i) Weak-field ligand $X^- < O < N < C \rightarrow$ (LF α Δ),

  (ii) Tetrahedral complexes (or D_{2d}) have higher $g_{//}$ than tetragonal (C_{4v}, D_{4h}) because $(1/\Delta_t) > (1/\Delta_o)$,

For example:

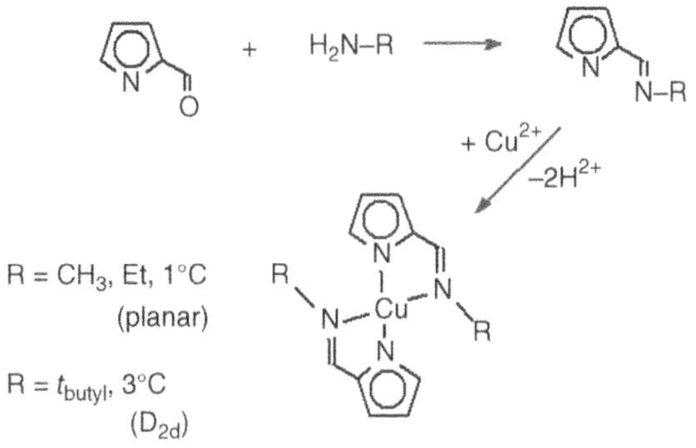

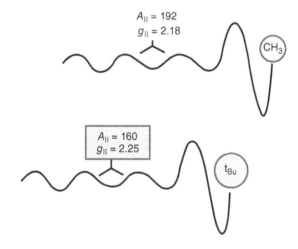

+
H2N–R
N
O
N
N–R
+ Cu2+
–2H2+
R = CH3, Et, 1°C
(planar)
R = tbutyl, 3°C
(D2d)
R
N
Cu
N
N
R

- In these types of copper complexes, the changes in the ESR spectra that follow variations of R are controlled by the change in the symmetry:

A|| = 192
g|| = 2.18
CH3
A|| = 160
g|| = 2.25
tBu

When $R = CH_3$, the copper complex is D_{4h}, while D_{2d} corresponds to $R = t$-butyl, and

$$g_{//} = 2 + \frac{8k_{//}|\lambda|}{\Delta} \qquad \Delta_{D_{2d}} < \Delta_{D_{4h}}$$

Then $g_{\|}(\Delta_{D_{2d}}) > g_{\|}(\Delta_{D_{4h}})$.

(b) *Electron Delocalization:*

- The value of $k_{\|}$ approaches unity when the electrons show less delocalization onto ligands, e.g., $k_{\|}$ (SP^3) < $k_{\|}$ (SP^2).
- The combination of the ligand fields strength (Δ) and the electron delocalization gives the *ESR spectrochemical series*:
  $k_{\|}/\Delta$: $F^- > O$, N $sp^3 \geq$ N $sp^2 > S > Cl$. For example:

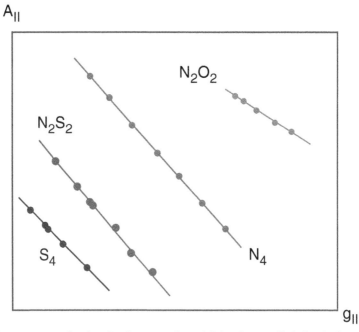

In these types of copper complexes, the changes in the ESR spectra follow the variations of the ligand interactions, and are controlled by $(k_{||}/\Delta)$.

In $Cu(NH_3)_5$, the copper is attached to N, while in $Cu(S\text{-}C(CH_3)\text{=}C(CH_3)\text{-}S)$, the copper is attached to S, and: $g\ ((k_{||}/\Delta)_N) > g\ ((k_{||}/\Delta)_S))$.

Hyperfine Trends ($A_{||}-g_{||}$ Trend):

- Values of $A_{||}$ versus $g_{||}$ of metal complexes of the same moieties are on the same line:

- ESR spectrum of spinach plastocyanin exhibits four well-defined signals at $g_| = 2.05$, $g_{||} = 2.23$, and $A_{||} = 0.005\ cm^{-1}$. The values of $A_{||}$ and $g_{||}$ lie on the same line of the N_2S_2 moiety. All the complexes falling close to this line show small values of $A_{||}$, and are distorted tetrahedral species. Note: the values $A_{||}$ and $g_{||}$ of octahedral and square planar complexes fall close to a different line.

- Values of $A_{||}$ and $g_{||}$ suggest N_2S_2 moiety.
- Low $A_{||}$ indicates pseudotetrahedral coordination.
- Low $g_{||}$ suggests S donor(s) Cys or Me.

COPPER PROTEINS

Identify the significant roles of copper in the biological process, classify, give examples, and explain their functions and main chemical properties.

- Copper-containing proteins are involved in a variety of biological functions:
  - Electron transport
  - Copper storage
  - Catalyze reactions involving molecular oxygen (O_2)
- Copper sites in proteins are classified into three categories:

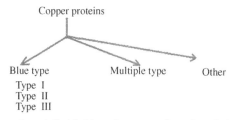

- Blue-type proteins can be subdivided into three types based on their spectral properties:
- Type I
  - Function: rapid electron transfer
  - Color: bright blue
  - ESR: unusual
  - Examples
    - (a) Plastocyanins

      Source: plants, algae

    - (b) Azurins

      Source: bacteria

    - (c) Cucumber basic blue protein

      Source: cucumber

    - (d) Amicyanins

      Source: Pseudomonas AMI (Amicyanin was first isolated from Methylobacterium extorquens AM1 [then known as Pseudomonas AMI])

- Type II
  - Function: involved in substrate binding; form adducts with N_3^-, CN^-, etc.
  - Color: less intense at normal concentrations
  - Optical and ESR: normal
  - Examples
    - (a) Galactose oxidase

Function:	alcohol oxidase
Source:	fungi

    - (b) Superoxide dismutase

Function:	superoxide (O_2^-) scavenging
	$2O_2^- + 2H^+ \rightarrow 2H_2O_2 + O_2$
Source:	yeast, mammals

    - (c) Dopamine–β–hydroxylase

Function:	catecholamine synthesis
Source:	human chromaffin granules

- Type III
  - Examples
    - (a) Tyrosinase

Function:	oxidation of monohydric phenols (cresol) to *o*–dihydric compound; oxidation of *o*-dihydric phnols (catechol) to *o*–quinine
Source:	fungi, mammals

    - (b) Hemocyanins

Function:	often involved in O_2 binding
Source:	arthropods, mollusks
Color:	gray-blue color when oxygenated, colorless when deoxygenated
ESR:	no ESR signals, antiferromagnetic

- Multiple types
  - Ascorbate oxidase

Function:	oxidation of ascorbate, $O_2 \rightarrow H_2O$
Source:	plants
Number of Cu atoms:	8
Types of Cu:	I, II, and III

  - Laccase

Function:	oxidation of catechols, $O_2 \rightarrow H_2O$
Source:	bacteria

Number of Cu atoms: 4
Types of Cu: I, II, and III

○ Cytochrome–*c*–oxidase

Function: $O_2 \rightarrow H_2O$
Source: bacteria, vertebrates

○ Nitrite reductase

Function: denitrification, $NO_2^- \rightarrow NO$
Source: denitrifying bacteria

- Other
  ○ Copper thionein

Color: colorless, Cu^I; hard to oxidize to Cu^{II}
Function: binding and transport, copper homeostasis
Source: vertebrates

○ Factor V

Function: blood coagulation
Source: human

○ Factor VIII

Function: blood coagulation
Source: human

○ Nitrous oxide reductase

Function: denitrification, $N_2O \rightarrow N_2$
Source: denitrifying bacteria

○ Uricase

Function: purine catabolism
Source: pigs and soybeans

PLASTOCYANIN

Identify the biological function of plastocyanin and the role of the cooper ion.

- See Fig. 4-16.
- Type I: Intense blue color, blue absorption maximum at 597 nm.

FIGURE 4-16   Crystal structure of spinach plastocyanin at 1.7 Å resolution, PDB 1AG6 (Xue et al., 1998).

- Molecular weight: 10,500, one Cu/molecule, 99 amino acids/molecule.
- Function: Electron transfer from cytochrome-f to P700 chlorophyll.
- Function of Cu ion in plastocyanin (or copper proteins): Copper is known in two oxidation states in solution, Cu^+ and Cu^{2+}. Copper's main role in vivo is in redox reactions.

    For:

$$Cu^+(aq) + e \rightarrow Cu^0 \qquad\qquad E^0 = 0.52 \text{ V}$$

$$Cu^+(aq) \rightarrow Cu^{+2}(aq) + e \qquad E^0 = -0.153 \text{ V}$$

$$Cu^+ \leftrightarrows Cu^{2+} + e^-$$

Find the concentration of Cu^+ ion in an aqueous solution at 298 K. Then, identify the role of the polypeptide chain in the biological function of plastocyanin.

- Redox potentials of copper ion:

$$Cu^+(aq) + e \rightarrow Cu^0 \qquad\qquad E^0 = 0.52 \text{ V}$$

$$Cu^+(aq) \quad \rightarrow Cu^{+2}(aq) + e \qquad E^0 = -0.153 \text{ V}$$

$$2\,Cu^+(aq) \quad \rightleftharpoons Cu^0(s) + Cu^{+2}(aq) \;\; E^0 = 0.37 \text{ V}$$

$$\Delta G^\circ = -zFE^0 \qquad \Delta G^\circ = -RT \ln K$$

Then

$$K = \frac{[Cu^{2+}][Cu]}{[Cu^+]^2} \cong 10^6$$

Show that Cu^+ in an aqueous solution is disproportionate to Cu^{2+} and Cu^0.

- The thermodynamic stability of Cu^+ in solution can be considerably increased if the solvent is other than water and increases in the order water < methanol < ethanol. Copper (I) is not disproportionate in acetonitrile, and Cu(II) acts as a strong oxidizing agent in this medium.

- The relative thermodynamic stabilities of Cu^+ and Cu^{2+} in aqueous solutions are also strongly dependent on the nature of the ligand.

- Simulations of such surrounding requirements of the copper redox center are provided by the polypeptide chain.

- Furthermore, the polypeptide or protein component tunes the metal center to the required redox role.

- The protein enables electrons to move over considerable distances from one redox center to another.

What are the upper and lower limits of redox potential of the biological systems at pH 7?

$$H_2(g) + \tfrac{1}{2}O_2(g) \rightarrow H_2O(l) \qquad \Delta G° = -236.60 \text{ kJ/mol at } 25°C.$$

Assume:

Biological Redox System	Redox Potential (V) at pH 7, 25°C
Cytochrome oxidase	0.4
Plastocyanin	0.37
Cytochrome c	0.26
HPIP (High potential iron protein)	0.35
Flavoproteins	0.0
Rubredoxin	−0.06
NAD^+/NDAH	−0.3
Ferredoxins	−0.4

Arrange these potentials in a diagram showing possible electron pathway.

All biological redox systems have redox potentials between that of the hydrogen electrode and that of the oxygen electrode.

- Hydrogen electrode:

$$H^+(aq) + e^- \rightarrow \tfrac{1}{2}H_2 \qquad E^0 = 0.00 \text{ V at pH } 0$$

- Oxygen electrode:

$$H_2(g) + \tfrac{1}{2}O_2 \rightarrow H_2O(l)$$

$$2H^+ + 2e^- + \tfrac{1}{2}O_2 \rightarrow H_2O$$

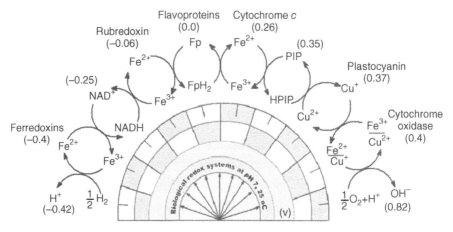

FIGURE 4-17 Redox potentials. (See the color version of this figure in Color Plates section.)

or

$$H^+(aq) + \tfrac{1}{2}O_2(g) + 2\,e^- \quad \rightarrow \quad OH^-(l) \qquad \Delta G^\circ = -236.60\ \text{kJ/mol at }25^\circ C$$

$$\Delta G^\circ = -nFE^o$$

$$E^0 = \frac{\Delta G^\circ}{-nF} = \frac{236.6 \times 1000}{2 \times 96,500} = +1.2259\ \text{V}$$

Two electrons are involved. This value is related to pH 0, and the value of E^0 is readjusted to pH 7 using

$$E^{0'} = E^0 - 0.059 \times pH \quad (25^\circ C) \qquad or \qquad E^{0'} = E^{0'} - 0.061 \times pH \quad (37^\circ C)$$
$$E^{0'} = +0.82\ V \quad \text{at pH 7} \quad (25^\circ C)$$

Figure 4-17 represents the redox potentials for biological redox systems showing the possible electron pathway.

o The reduced NADH is oxidized by flavorprotein $\rightarrow NAD^+$.

What is the common mechanism of electron transfer for simple copper complexes and give a model example? Explain the factors that may account for the unusual rapid electron transfer in plastocyanin.

- For simple Cu^{2+} complexes:
  o $Cu^{2+} + e^- \leftrightarrows Cu^+$ and the rate of reaction is given by

$$k = \frac{\kappa T}{h}\exp\left(\frac{-\Delta G^{\#}}{RT}\right)$$

where k-rate constant, κ-Boltzmann constant, h-Planck constant, R-gas law constant, and $\Delta G^{\#}$ - free energy of activation.

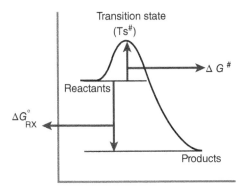

o Rate of the reaction, k, is inversely proportional to the free energy of activation, $\Delta G^{\#}$.

o All experimental data for Cu^{2+}/Cu^{+} are consistent with the two-step reaction involving a change in the coordination number $\Delta(CN)$ 1 or 2. This major structural change in interconverting Cu^{2+} and Cu^{+} significantly halts the redox rate.

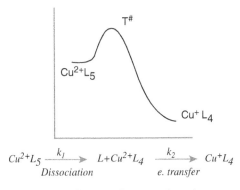

$$Cu^{2+}L_5 \xrightarrow[\text{Dissociation}]{k_1} L+Cu^{2+}L_4 \xrightarrow[\text{e. transfer}]{k_2} Cu^+L_4$$

Assume k_2 is the electron transfer step: fast step, $k_1 \ll k_2$; most of $\Delta G^{\neq}$ involved in k_1, as $Cu^{2+}-$ L bond dissociation energy.

• Model thiolates

$$\underset{\diagup}{\diagdown} Cu^{2+} = [Cu\,(N_{Im})_2(S\,\,met)]^+$$

$$\underset{\diagup}{\diagdown} Cu^{2+} + {}^-SR \xrightarrow{k_1} \underset{\diagup}{\diagdown} Cu^{2+}\!\!- SR \xrightarrow{k_2} \underset{\diagup}{\diagdown} Cu^+ \quad +{}^1\!/_2\ RSSR$$
$$\textit{Deeply colored}$$

The observed rate is give as

$$\text{Rate} = k[Cu]^2[^-SR]^2$$

Therefore, the rate-determining step, (RDS, the slowest step in the reaction) is k_1,

$$2 \underset{\diagup}{\diagdown} Cu^{2+} + 2^- SR \xrightarrow[\substack{slow \\ RDS}]{k_1} \left(\underset{\diagup}{\diagdown} Cu^{2+}\!\!- SR\right)_2 \xrightarrow[Fast]{k_2} 2 \underset{\diagup}{\diagdown} Cu^+ + RSSR$$

$$\begin{array}{c} R \\ | \\ S \end{array}$$

Cu   Cu

Mechanism involves association,
change in CN

- In plastocyanin,

$$\mathrm{Cu^{2+} + e^- \; \leftrightarrows \; Cu^+}$$

that is, 10^3–10^6 faster for plastocyanin than for simple $\mathrm{Cu^{2+}}$ complexes.

- One essential factor that affects electron transfer reactions is the *Frank–Condon principle*, which states that there must be no movement of nuclei during the period of the electronic transition, requesting the symmetry requirements of the allowed electronic transition. This requests the same molecular geometry of the reactants, transition state, and products during the period of the electron transfer ($\Delta CN = 0$).

- Consequently, as the geometries of the initial, transition, and final states of plastocyanin approach being identical, the rate of the electron transfer process increases ($\Delta CN = 0$).

- The importance of the protein in establishing this structural feature is confirmed by the observation of little structural differences during the period of electron transition.

- The following suggests that the active metal site of plastocyanin is in a geometry approaching that of the transition state of the electron transfer reaction. This state has been termed the "entatic state":

1. Copper ligands

$$N - \text{Im of His-37, His-87}$$

$$S \text{ of Cys-84, Cys(Met)-92}$$

- *Hard N* and *soft S* represent a *compromise* between the requirements of $\mathrm{Cu^{2+}}$ and those of $\mathrm{Cu^+}$ (Fig. 4-18).

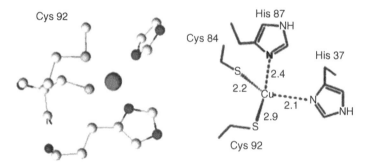

FIGURE 4-18   Active site of spinach plastocyanin at 1.7 Å resolution, PDB 1AG6 (Xue *et al.*, 1998).

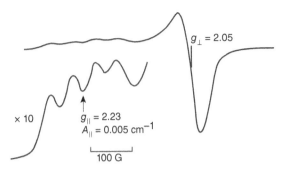

FIGURE 4-19   ESR spectrum of spinach plastocyanin. The lower portion of the spectrum is recorded at higher sensitivity. (From Malkin, 1973. Reproduced by permission of Elsevier.)

2. Sterochemistry
   - *Intermediate* between planar and tetrahedral geometries, favored by Cu^{2+} and Cu^+, so rapid electron transfer is expected.
   - This intermediate structure is supported by:
     (a) *ESR spectrum* of spinach plastocyanin exhibits well-defined signals at $g_\perp = 2.05$, $g_\parallel = 2.23$, and $A_\parallel = 0.005 \text{ cm}^{-1}$ (Fig. 4-19).
       (i) The four copper hyperfine lines in the parallel direction are clearly resolved.
       (ii) Only one set of ESR parameters is resolved. Therefore, Cu^{2+} has an identical environment.
       (iii) Lower value of $A_\parallel$ (0.005 cm^{-1}) indicates that the Cu^{2+} site is one of lower symmetry than the tetragonal, intermediate between square planar and tetrahedral.
       (iv) No direct demonstration of covalency as evidenced by superhyperfine structure has been observed.
     (b) *Biological redox reactions*: Calculating the expected redox potential of plastocyanin,

$$Cu^{2+} + e^- \quad \rightarrow Cu^+ \qquad\qquad E^0 = 0.15 \text{ V}$$
$$Cu^{2+} \, Im_4 + e^- \rightarrow Cu^+ \, Im_4 \qquad E^0 = 0.40 \text{ V}$$

Then,

$$E^0_{Rx} = E^0_{(Cu^{2+} \rightarrow Cu^+)} + \Sigma(\Delta E_L) \qquad E^0_{Rx} = 350 \sim 0.4 \text{ V}$$
$$E^0_{Rx} = 150 \text{ mV} + 4 \times 50 = 350 \sim 0.4 \text{ V}$$

where, Im is imidazole, ΔE_L is the change in the redox potential due to the ligand attachment, then

$$\Delta E_{Im} = +50 \text{ mV}.$$

Similarly, the contribution of any donor ligand to the redox potential of the copper center can be estimated:

Donor	ΔE_L (mV)
$N_{Aromatic}$	+50
COO^-	-25
$S_{thioether}$	+140
$S^-_{Cys/thiolate}$	-300
$O^-_{phenolate}$	-380

Compute $E^{0\prime}$ for type I $Cu^{2+/+}$:

$$E^0_{Rx} = E^0_{Cu^{2+}/Cu^+} + 2 \times \Delta E_{N\text{-}Im} + \Delta E_{S\text{-}thioether} + \Delta E_{S\text{-}thiolate}$$

$$E^0_{Rx} = 150 + 2 \times (+50) + (+140) + (-300) = 90 \, mV$$

Plastocyanins have $E^{0\prime} \sim 360 \, mV$. (Why there is a difference between observed and predicted values of E^0_{Rx}.)

(i) Tetrahedral structures have more $+E$ than tetragonal. This supports the intermediate structure between T_d/C_{4v}.

(ii) Nonpolar environment destabilizes $[Cu^{2+}]$ relative to $[Cu^+]$.

3. The central Cu^{2+} is coordinated to equatorial $Cu-S_{thiolate}$ coordination. (Why?)

(a) To tune central E^0

(b) Tyr–phenolate may not coordinate to Cu^+

(c) $Cu-S_{equatorial}$ has faster kinetics

- Equatorial S–thiolate coordination is supported by the deep blue color due to LMCT, ligand to metal charge transfer band, which is only observed in the S-equatorial attachment to Cu^{2+}.

Unsaturated axial ligands

$d_{z^2} \rightarrow$ is full

No Cu $\rightarrow$ S CT

(No MLCT)

equatorial S–thiolate (pseudoaxial H_2O)
Cu – S = 2.9 Å equatorial S–thiolate

$d_{x^2-y^2} \rightarrow$ is unsaturated

$S_\pi \rightarrow Cu^{2+}$ $(d_{x^2.y^2})$

LMCT, 330 – 350 nm

- This agrees with the distorted square planar geometry for the plastocyanin type 1 blue type.

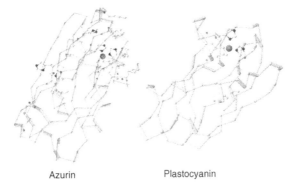

FIGURE 4-20 Crystal structure of azurin II, PDB 2CCW (Paraskevopoulos et al., 2006), and plastocyanin (Xue et al., 1998). Notice the common structural features between azurin and plastocyanin.

AZURIN AND STELLACYANIN

Show the resemblances among plastocyanin, azurin, and stellacyanin

- Azurin: Type I
- Source: isolated from several bacterial genera: *Pseudomonas*, *Bordatella*, and *Achromobacter*
- Function: respiratory, electron transfer
- Very close to plastocyanin. See (Figs. 4-20 and 4-21 and Table 4-1).
- Stereochemistry: short Cu – S 2.1 Å, 2 Cu – N_{Im}, Cu – S $_{Met}$ (Fig. 4-21).

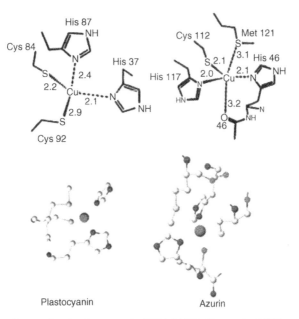

FIGURE 4-21 Active sites in plastocyanin (PDB 1AG6, Xue et al., 1998) and azurin (PDB 2CCW, Paraskevopoulos et al., 2006).

TABLE 4-1 Chemical Properties of Cupredixins

Common Name	Source	MW	His, Cys, Met	Optical Spectra (nm: mM^{-1} cm^{-1})	ESR	A (cm^{-1})	Redox Potential (mV)
Bacterial Cupredixins							
Azurin	*Pseudomonas aeruginosa*	14,600	4,3,6	625,3.5	2.052,2.26	0.006	330
Azurin	*Alcaligines denitrificans*	14,000	5,3,6	—	—	—	—
Azurin	*Alcaligines sp.* NCIB11015	14,000	4,3,4	620,10.5	2.055,2.26	0.006	230
Azurin	*Achromobacter cyclociastes*	12,000	3,1,5	600,2.0	—	—	245
Azurin	*Paracoccus denitrificans*	13,790	4,1,4	595,1.5	2.052,2.29	0.077	230
Blue protein	*Alcaligines faecalis S-6*	12,000	3,1,5	593,2.9	—	—	—
Amicyanin	*Pseudomonas AMI*	11,723	3,1,2	596,4.5	—	—	180
Rusticyanin	*Thiobacillus ferroxidans*	16,500	5,1,3	597,1.95	2.019,2.064,2.229	0.0065,0.002,0.0045	680
Plant Cupredixins							
Plastocyanin	*Spinach*	10,800	2,1,2	597,4.9	2.053,2.26	0.005	340-370
Umecyanin	*Horseradish root*	14,600	3,3,4	610,3.4	2.05,2.317	0.0035	283
Stellacyanin	*Lacquer tree*	20,000	4,3,0	604,4.0	2.03,2.08,2.29	0.0057,0.0029,0.0035	184
Plantacyanin	*Cucumber seedlings*	10,100	2,3,2	597,3.4	2.02,2.08,2.207	0.006,0.001,0.0055	317
Mavicyanin	*Green squash fruit*	18,000	5-6,5,1-2	600,5.0	2.03,2.08,2.29	0.0057,0.0029,0.0035	285

Source: Data from Adman, 1985. Reproduced by permission of John Wiley & Sons.

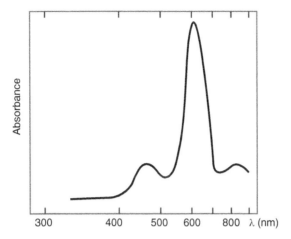

FIGURE 4-22 Typical absorption spectrum of blue type-I Cu-protein.

- The common structural and chemical features of
  - bacterial cupredoxins (azurins, blue protein, amicyanin, rusticyanin) and
  - plant cupredoxins (plasocyanin, umecyanin, stellacyanin, plantacyanin, mavicyanin, Table 4-1) are responsible for sharing common functions.
- Spectroscopy
  - Charge transfer bands
    - (a) The most salient property is the blue color.
    - (b) The absorption at about 600 nm, due to LMCT between $\pi S \rightarrow Cu\,(d_{x^2-y^2})$, is in fact usually accompanied by two less intense peaks on either side, one at 440 nm and one at 780 nm, due to $\sigma S \rightarrow Cu\,(d_{x^2-y^2})$ (Table 4-1 and Fig. 4-22).
- Raman spectra
  - (a) Blue-type proteins show a peak near $400\,\mathrm{cm}^{-1}$ that is due to mixing contribution for both Cu–S and Cu–N stretching frequencies.
  - (b) The peak at $260\,\mathrm{cm}^{-1}$ suggests the trigonal CuN_2S arrangement, where in plastocyanin $Cu–S_{Cys92} = 2.9\,\text{Å}$ and in azurin $Cu–S_{Met121} = 3.1\,\text{Å}$.
- ESR spectra of blue-type proteins also show that the characteristically small hyperfine splitting constants (Table 4-1), could be best explained by the intermediate between square planar and tetrahedral due to the thermal vibronic movement of the ligands.
- Stellacyanins
  - Stellacyanins are blue (type I) copper glycoproteins that slightly differ from other members of the cupredoxin family in their spectroscopic and electron transfer properties (Fig. 4-23, Table 4-1).
  - The ligands to copper are two histidines, one cysteine, and one glutamine, the latter replacing the methionine typically found in mononuclear blue copper proteins. The Cu–Gln bond distance is one of the shortest axial ligand distances observed to date in structurally characterized type I copper proteins.

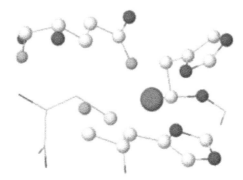

FIGURE 4-23   Active site of cucumber stellacyanin at 1.6 Å resolution, PDB 1JER (Hart et al., 1996).

○ Molecular weight 20,000

○ $E^{O'} = 184\,mV$ (Table 4-1).

○ The primary structure has no Met (no Cu–S ligation) (Table 4-1). This variation may be related to the change in the redox potential. Removing S_{met} from the calculation results in 220 mV.

○ The characteristic spectroscopic properties and electron transfer reactivity of stellacyanin, which differ significantly from those of other well-characterized plant cupredoxins (Table 4-1), can be explained by:

(a) Its more exposed copper site

(b) Its distinctive amino acid ligand composition

(c) Its nearly tetrahedral ligand geometry
    Stellacyanin has a typical ESR spectrum with an N/O donor.

Design synthetic models for plastocyanin

• The reaction of Cu(SR) or Cu(SR)ClO₄ (SR = p-nitrobenzene-thiolate or o-ethylcysteinate) with hydrotris-(3,5-dimethyl-pyrazolyl) borate gives $Cu^{+}N_3(SR)$ and $Cu^{2+}N_3(SR)$, respectively:

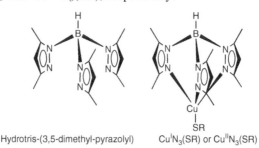

Hydrotris-(3,5-dimethyl-pyrazolyl) $Cu^{I}N_3(SR)$ or $Cu^{II}N_3(SR)$

- The coordination geometry about the copper atom for such complexes indicates the structural similarity.

Cu $\longmapsto$ 2.19 Å $\longmapsto$ S

Cu $\longmapsto$ 2.01 Å $\longmapsto$ N

∠ S–Cu–N :130°

∠ N–Cu–N :90°

- The visible spectra of [Cu-HB(3,5-Me$_2$pz)$_3$(p-NO$_2$C$_6$H$_4$S)] at −78°C in tetrahydrofuran [Cu-HB(3,5-Me$_2$pz)$_3$(o-ethylcysteinate)] at −78°C are similar to that of *Pseudomonas aeruginosa* azurin (see Thompson et al., 1977).

SUPEROXIDE DISMUTASE

The direct reduction of O$_2{}^-$ is toxic. How do biological systems handle this product? Give an example of Cu type II proteins; what is the significant role of the metal ion?

- Superoxide dismutase (SOD) is present in all aerotolerant organisms for the purpose of minimizing the concentration of superoxide, O$_2{}^-$, and thus providing protection against oxygen toxicity.
- There are three types of SOD: Ni SOD, either Fe or Mn SOD, which seem to be the same protein, and Cu/Zn SOD.
- Fe or Mn SOD is found in the prokaryotes or the mitochondria of eukaryotic cells, while Cu/Zn SOD is found in the cytoplasm of eukaryotic cells.
- Biological role:

$$2\,O_2{}^- \quad \text{SOD} \quad \xrightarrow{H^+} H_2O_2 + O_2$$

- Cu/Zn SOD (Scheme 4-1).
  - Molecular weight: 31,400
  - Two identical subunits
  - Amino acids/subunit: 151
  - Metals/subunit: 1 Cu, 1 Zn
  - Absorption maximum

λ_{max} (nm)	ε_{max} (M^{-1} cm^{-1})
258	10,300
270, 282, 289	Shoulders
680	300

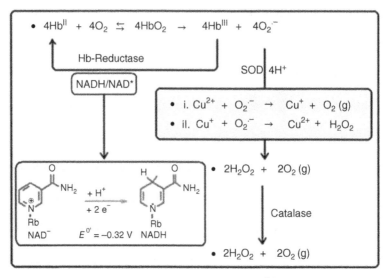

SCHEME 4-1 Biological role of superoxide dismutase. Cu^{2+} is fully oxidized and Cu^+ is reduced.

- ○ ESR: $g_\perp = 2.08$, $g_{||} = 2.265$, $A_{||} = 0.015$
- ○ E^0: 0.42 V
- ○ Color: blue green
- • Catalysis of SOD is not restricted to protein-pound metals, the aqua complexes of Cu^{2+} and Mn^{2+}, but also occurs in a variety of small complexes of Cu^{2+} and Fe^{2+}. In fact, Cu^{2+}_{aq} is roughly four times more effective than the Cu of Zn/Cu SOD.

What are the structural features necessary for a metal ion to have SOD activity? Propose a mechanism for the action of SOD.

- • In this type of catalyst:
  - ○ The metal ion must have at least one coordination position accessible for binding O_2^- in two adjacent valence states.
  - ○ The redox potential of the metal should be between ~−0.1 and +0.8 V.
  - ○ The metal ion must be able to switch between two valence states faster than the process of spontaneous dismutation.
- • The suggested mechanism is shown in Scheme 4-2.
- • Hypothesis: Protein is needed for (step ii). The E^0 of Cu drops to 300 mV so: it is easier to do (step ii)

Explain the in vitro toxicity of superoxide "O_2^-"

- • $O_2^{\bullet -} + H_2O_2 + H^+ \rightarrow OH^\bullet + H_2O + O_2$
- • $OH^\bullet$ causes:

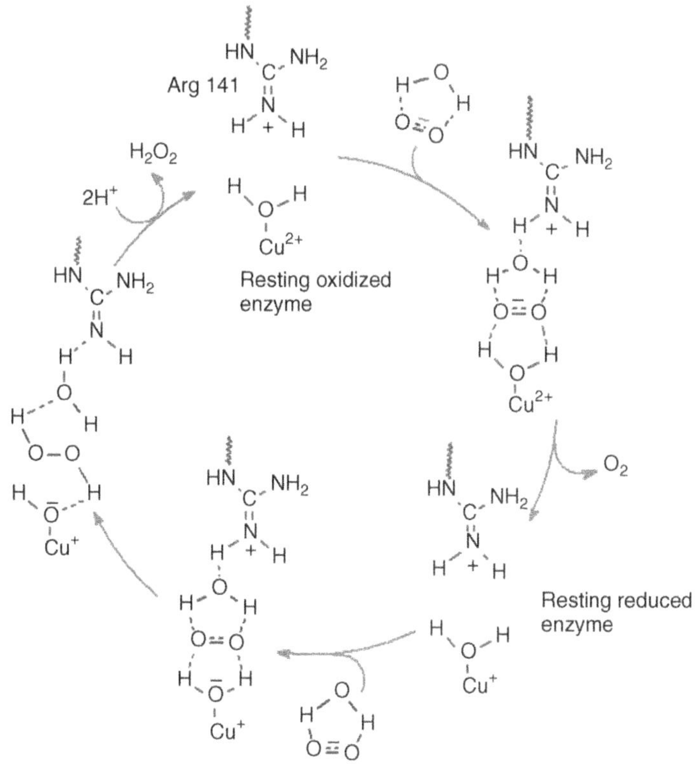

SCHEME 4-2 Proposed mechanism for action of SOD. (Modified based on Cass, 1985. Reproduced by permission of John Wiley & Sons.)

- o DNA degradation/depolymerization
- o Lipid peroxidation
- o Depolymerization of polysaccharides
- o Hydroxylation of aromatic substances
- o Ethylene formation from methional, methionine: $CH_3-SCH_2-CH_2(NH_2)-COOH$
- o Cell killing

- Toxicity can be explained by the following sequence reactions, superoxide–mediated Fenton chemistry:

$$M^{n+} + O^{\bullet-} \longrightarrow M^{(n-1)+} + O_2$$

$$M^{(n-1)+} + H_2O_2 \xrightarrow{H^+} MO \text{ or } (OH^{\bullet} + M^{n+})$$

$$\left.\begin{array}{c} MO \\ \text{or} \\ OH^{\bullet} \end{array}\right\} + RH \longrightarrow \left.\begin{array}{c} R^{\bullet} \\ \text{or} \\ ROH \end{array}\right\} \rightarrow \text{degradation product}$$

Where M^{n+} is a metal such as Cu^{2+} or Fe^{3+} and MO is a metal–oxy compound having properties similar to the hydroxyl radical $OH^\bullet$.

- Fenton reaction

$$O_2{}^{\bullet-} + M^{n+} \longrightarrow M^{(n-1)+} + O_2$$

$$M^{n+} + H_2O_2 \longrightarrow M^{(n-1)+} + 2H^+ + O_2{}^{\bullet-}$$

$$M^{(n-1)+} + H_2O_2 \xrightarrow{H^+} OH^\bullet + M^{n+} + H_2O$$

$$OH^\bullet + H_2O_2 \longrightarrow O_2{}^{\bullet-} + H_2O + H^+$$

$$2\,O_2{}^{\bullet-} + 2H^+ \longrightarrow H_2O_2 + O_2$$

Or:

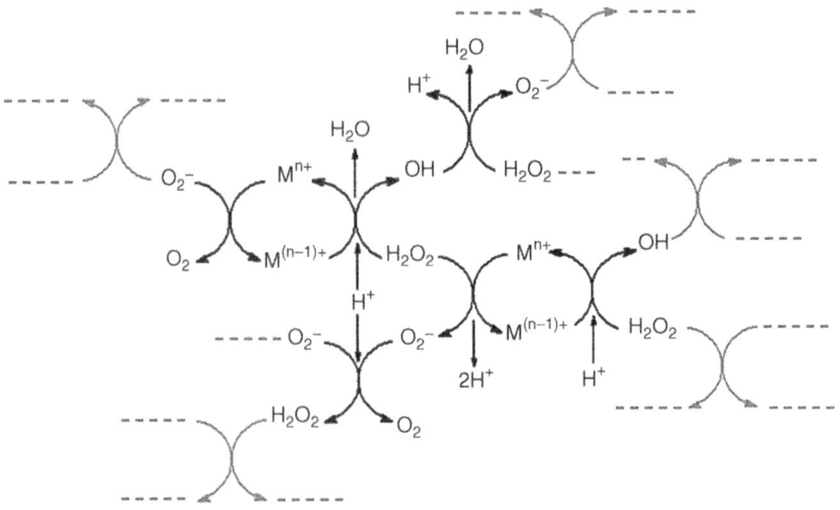

- If there is an organic substance RH, an additional reaction takes place:

$$OH^\bullet + RH \rightarrow R^\bullet + H_2O$$

What are the structural features of superoxide dismutase?

- Geometry of SOD is presented in Fig. 4-24 (distorted square planar).
- When Cu and Zn are removed by dialysis, the inactive apoenzyme is obtained.
- If Zn is replaced by Co^{2+}, an *antiferromagnetic* interaction between the two centers is observed supporting the imidazolate bridge between Cu^{2+} and Zn^{2+}.
- The Zn ion has a structural role in the stabilization of the active site.
- The copper center represents the catalytic center.
- Derivative that maintains Cu^{2+} in the native site retains almost full activity.
- If other metals replace native Cu^{2+}, this will inhibit the reactivity.

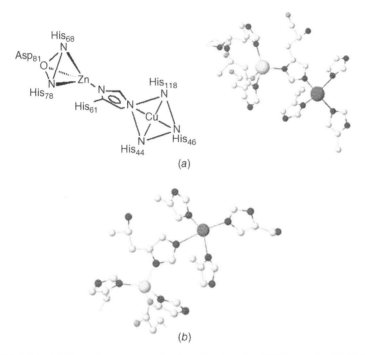

FIGURE 4-24 (*a*) Crystal structures of active site of Cu/Zn SOD treated with H_2O_2, PDB 2Z7U. (*b*) Crystal structure of active site of monomeric human Cu/Zn SOD, PDB 2XJK (Leinartaite et al., 2010).

HEMOCYANIN

Biological Function of Hemocyanin and Role of Copper Ion

The spectroscopic studies of oxyhemocyanin exhibit absorption bands at 440 and 700 nm, associated with intense bands at 570 and 347 nm ($\varepsilon \sim 8900$). The Raman spectrum of the same protein exhibits a stretching frequency at $745\,cm^{-1}$. Oxyhemocyanin indicates strong antiferromagnetic coupling, while met-hemocyanin is ESR silent. EXAFS studies confirm that (Cu^+) in deoxy and (Cu^{2+}) in oxyhemocyanin. The Cu–Cu distances in $Hc(Cu^{2+})O_2$ and $Hc(Cu^+)$ are 3.67 *and* 3.38 Å, respectively. The average distance between Cu atoms and their immediate neighbors is the same (1.95 Å) in both ligation states.

Use the above data to propose the geometric structures of the active site in oxyhemocyanin and deoxyhemocyanin and give a biomimetic model for hemocyanin.

- Hemocyanin, "blue blood," type III
- Hemocyanins are found in:
  - Horseshoe crab
  - Crawfish, lobster, crabs

- ○ Squid, cuttlefish, octopus
- ○ Scorpions, spiders
- Metal: Cu
- Metal/O_2: 2 Cu/1 O_2
- Deoxy protein: Cu^+
- Coordination of metal: protein side chains
- Quaternary structure: monomer, dimer, tetramer, or octamer
- Number of subunits: variable, (molecular weight ~400,000–9000,000)
- Color:

  Oxygenated → blue

  Deoxygenated → colorless

$$LCu^+(colorless) + O_2 \leftrightarrows LCu^{2+} O_2^-$$
$$LCu^{2+}O_2^- + LCu^+ \quad \leftrightarrows LCu^{2+} - O_2^{-2} - Cu^{2+}L(blue)$$

- Oxyhemocyanin: LCu^{2+}–O_2^{-2}–$Cu^{2+}L$ (blue)
  - ○ Electronic spectra: d–d bands at 440 and 700 nm and intense charge transfer band at 570 ($O_2^{-2} \rightarrow Cu^{2+}$) and 347 nm (Im → Cu^{2+}, $\varepsilon \sim 8900$), suggesting tetragonal Cu coordination environment.
  - ○ Raman studies: ν_{O-O} at ~745 cm^{-1} in agreement with peroxide (O_2^{-2} – ligand)-like structure, comparable to dinuclear Cu(II) complexes of $\mu - \eta^2 : \eta^2$ type peroxo binding:

  - ○ Magnetic properties: antiferromagnetic interaction between the two copper ions to give diamagnetic complex. The copper ions are magnetically coupled by an internal ligand, as indicated by the possibility of obtaining a met-hemocyanin, deoxy-Cu^{2+}, which is ESR silent.
- EXAFS studies confirm Cu^+ (Fig. 4-25) in deoxy and Cu^{2+} in oxyhemocyanin.
- Hemocyanins are multimeric oxygen transport proteins in the hemolymph of several mollusks and arthropods. Despite the diverse molecular architectures of mollusks and arthropod hemocyanin, they have a similar binuclear type III copper center to bind oxygen in a side-on conformation.
- The Cu–Cu distance in $Hc(Cu^{2+})O_2$ is different from that of $Hc(Cu^+)$, 3.67 vs. 3.38 Å, and the average distance between Cu atoms and their immediate neighbors is the same (1.95 Å) in both ligation states.
- In response to these data, the coordination number of Cu is 4 or 5 for oxy and 3 for deoxy:

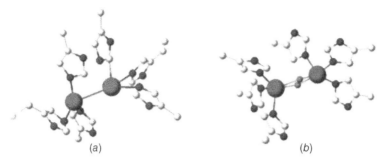

FIGURE 4-25 (*a*) Oxygen binding center of limpet hemocyanin, PDB 3QJO (Jaenicke et al., 2011) and (*b*) oxygenated *Limulus polyphemus* hemocyanin, PDB 1NOL (Hazes et al., 1993).

Deoxy HC   Oxy Hc

Proposed hemocyanin active site

- Biomimetic model for hemocyanin: Possible models for type III have been synthesized. The macrocyclic ligand (L) can bind two copper ions:

L $Me_6[22]$

- L gives the complex $Cu_2L(N_3)_4$, with a Cu–Cu separation of 5.145 Å and is completely diamagnetic between 4.2 and 390 K.
- The macrocycle $Me_6[22]$ gives a dinuclear complex with bridging methoxy groups:

- The Cu–Cu separation is 3.03 Å. The magnetic moment is 0.87 BM (Bohr Magneton) at 299 K and 0.19 BM at 81 K.
- The imidazolate-bridged Cu(II) was prepared with glycylglycine:

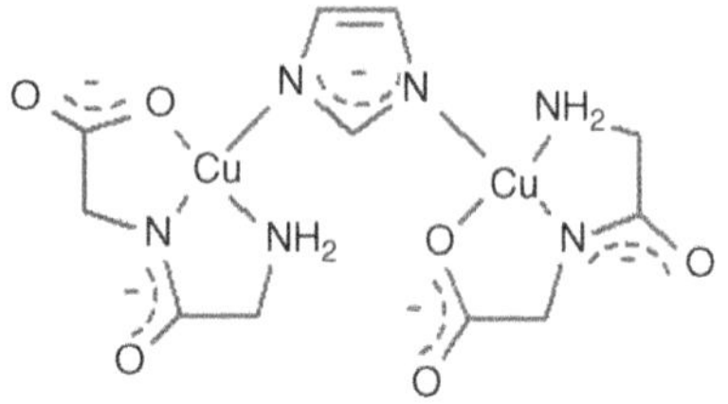
Cu
Cu
NH_2
NH_2

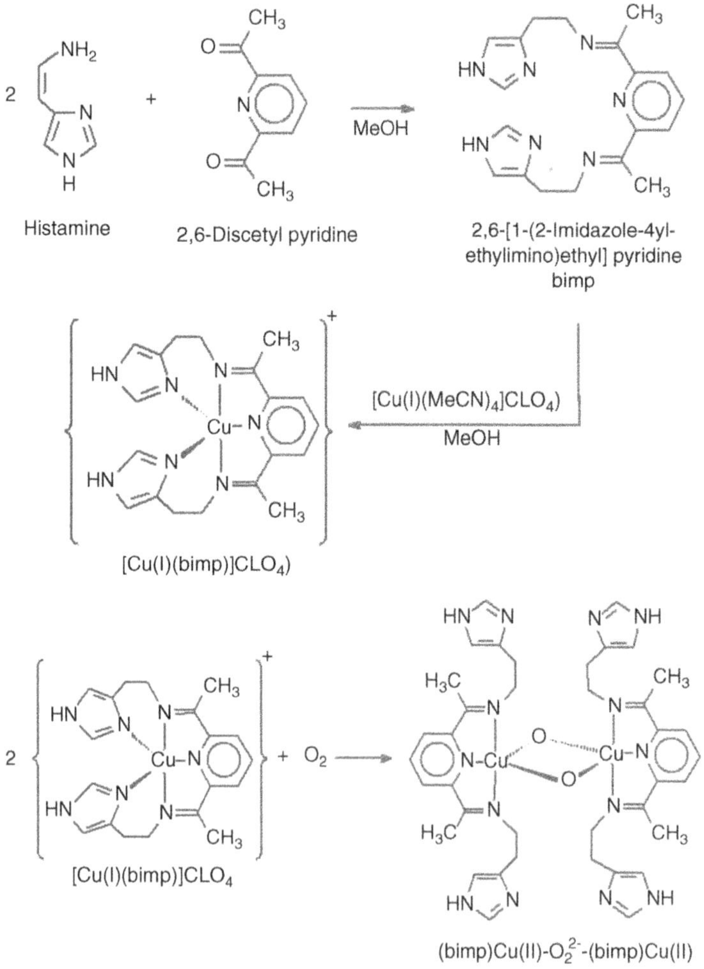
Histamine
2,6-Discetyl pyridine
MeOH
2,6-[1-(2-Imidazole-4yl-ethylimino)ethyl] pyridine
bimp
[Cu(I)(MeCN)$_4$]ClO$_4$)
MeOH
[Cu(I)(bimp)]CLO$_4$)
$+ O_2$
[Cu(I)(bimp)]CLO$_4$
(bimp)Cu(II)-O_2^{2-}-(bimp)Cu(II)

SCHEME 4-3 Synthesis of Cu(I) complex which binds dioxygen reversibly. (Postulated based on Simmons and Wilson, 1978.)

- The temperature dependence of the magnetic susceptibility reveals antiferro-magnetic coupling between the two Cu(II).
- *Note the imidazolate-bridge in:*
  - Cu–Im–Zn in superoxide dismutase
  - Fe–Im–Cu in cytochrome *c* oxidase
- The condensation of 2,6-diacetyl pyridine with 2 mol of histamine gives the ligand "bimp." Addition of $[Cu(I)(MeCN)_4](ClO_4)$ under N_2 gives the dark red $[Cu(I)(bimp)](ClO_4)$ (Scheme 4-3).
- The formed complex reacts with another $[Cu(I)(MeCN)_4](ClO_4)$ and yields a dinuclear complex:

$$(bimp)Cu(I) \quad + O_2 \quad\quad \leftrightarrows (bimp)Cu(II)O_2{}^-$$
$$\text{dark red}$$

$$(bimp)Cu(II)O_2{}^- + (bimp)Cu(I) \leftrightarrows (bimp)Cu(II)\text{-}O_2{}^{2-}\text{-}(bimp)Cu(II)$$
$$\text{Green}$$

ASCORBIC OXIDASE

What is the biological role of the ascorbic acid oxidase and the significance of copper in ascorbate oxidase? Give a biomimetic model.

- The enzyme catalyzes the oxidation of L–ascorbic acid:

$$ASC^- \rightarrow DHA^- + 2H^+ + 2e^-$$

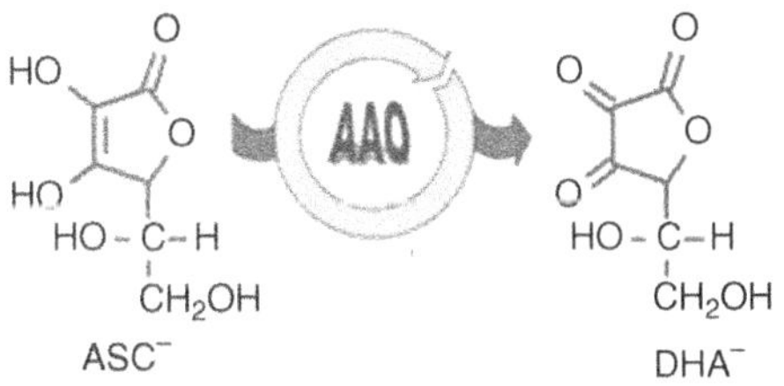

- Ascorbic oxidase obtained from summer crook neck squash
- It is blue green
- It is a polynuclear Cu oxidase
- Molecular weight 140,000
- Eight copper atoms per molecule
- The blue solution of the enzyme turns yellow when a small amount of ascorbic acid is added ($Cu^{2+} \rightarrow Cu^+$). The yellow color of enzyme returns to the blue when exposed to oxygen.
- Consists of Types I, II, and III.

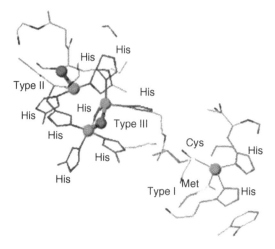

FIGURE 4-26 Active centers of ascorbate oxidase at 1.9 Å resolution, PDB 1AOZ (Messerschmidt et al., 1992).

- The crystallographic studies confirm the dimeric quaternary structure. Each subunit has four copper atoms bound as mononuclear and trinuclear species (Fig. 4-26).
- The mononuclear copper has two histidines, a cysteine, and methionine residues and replicates the type I copper. The bond lengths of the type I copper center are similar to the values for oxidized plastocyanin.
- The trinuclear species has eight histidine ligands and is divided into a pair of copper atoms with six histidine ligands whose ligating N atoms are arranged trigonal prismatic. The pair is the representation of type III copper.
- The remaining copper has two histidine ligands and represents a type II copper.
- Two oxygen atoms are bound to the trinuclear species as OH^- or O^{2-} and bridging the copper pair of type III and as OH^- or H_2O bound to the copper atom of type II.
- The bond lengths within the trinuclear copper site are similar to comparable binuclear model compounds.
- In Type III copper center O_2 binding site. The dinuclear Cu is believed to be the binding site for O_2. Laccase contains almost the same set of four Cu atoms.
- Type I: e^- buffer site
- Type II: substrate binding site, e.g.:
  - $(Cu_2^I)_{type\ III} + O_2 \rightarrow (Cu_2^{II} - O_2^{2-})_{type\ III}$
  - $(Cu^I)_{type\ I} \rightarrow (Cu^{II})_{type\ I} + e^-$
  - $(Cu_2^{II} - O_2^{2-})_{type\ III} + e^- \rightarrow O_2^{3-}$
  - $O_2^{3-} \equiv O^{\bullet -} + O^{2-}$ or $O_2^{3-} + 2H^+ \equiv OH^{\bullet} + OH^-$
  - $O^{\bullet}$ ($OH^{\bullet}$) is a powerful 1 e^- oxidant, may be the second electron remover in ASC oxidation process ($Cu^{2+} \rightarrow Cu^+$, is the first electron remover)

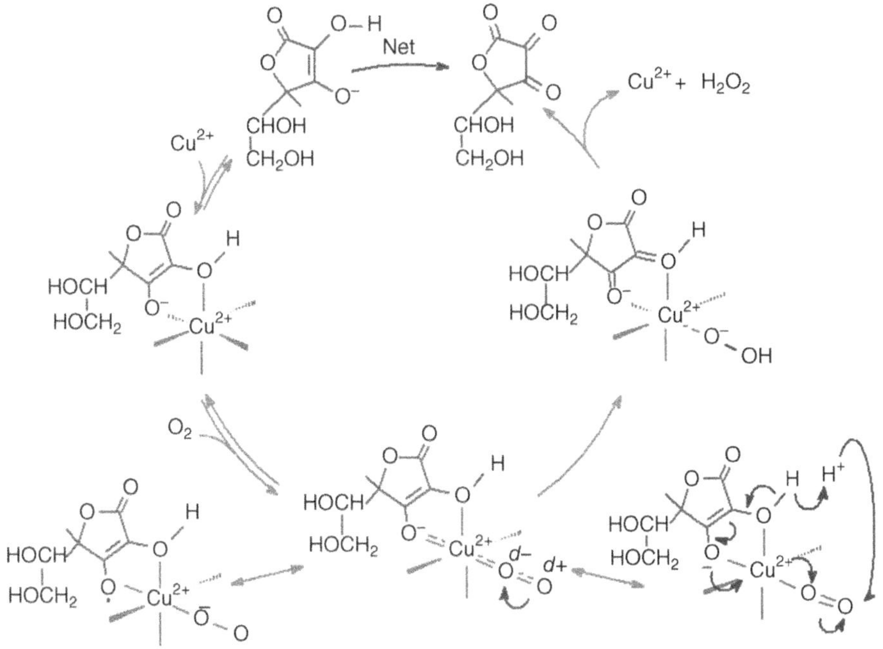

SCHEME 4-4 Mechanism of copper^{2+}-catalyzed oxidation of ascorbic acid. (Modified from Taqui Khan and Martell, 1967, 1968.)

- The absorption spectrum of ascorbic acid oxidase shows two bands at 606 nm ($\varepsilon = 770\,\mathrm{dm^3\,mol^{-1}\,cm^{-1}}$) and a shoulder at 412 nm ($\varepsilon = 500\,\mathrm{dm^3\,mol^{-1}\,cm^{-1}}$), and the charge transfer (LMCT) band at 606 nm disappeared ($Cu^{2+} \to Cu^+$) in the presence of substrate (ascorbic acid) (see spectra in Dawson, 1960).
- Model systems: The kinetic observations support the mechanism in Scheme 4-4 for Cu^{2+} ion–catalyzed oxidation of ascorbic acid:
    - (i) Initially oxygen–ascorbate mixed ligand Cu(II) complex is formed.
    - (ii) Through resonance, a free-radical intermediate is produced as a result of the first electron shift.
    - (iii) A slow proton shift forms the weak donor, dehydroascorbic acid– hydroperoxide mixed ligand complex.
    - (iv) The latter complex rapidly dissociates to regenerate the catalyst.
- In the presence of Cu^{2+} chelates, the reaction mechanism appears to be different (Scheme 4-5):
    - ○ The rate is not dependent on [O_2].
    - ○ O_2 is reduced to H_2O.
    - ○ The RDS is $Cu^{2+} \to Cu^+$, first electron transfer.
    - ○ A rapid second electron transfer produces the dehydroascorbic acid.

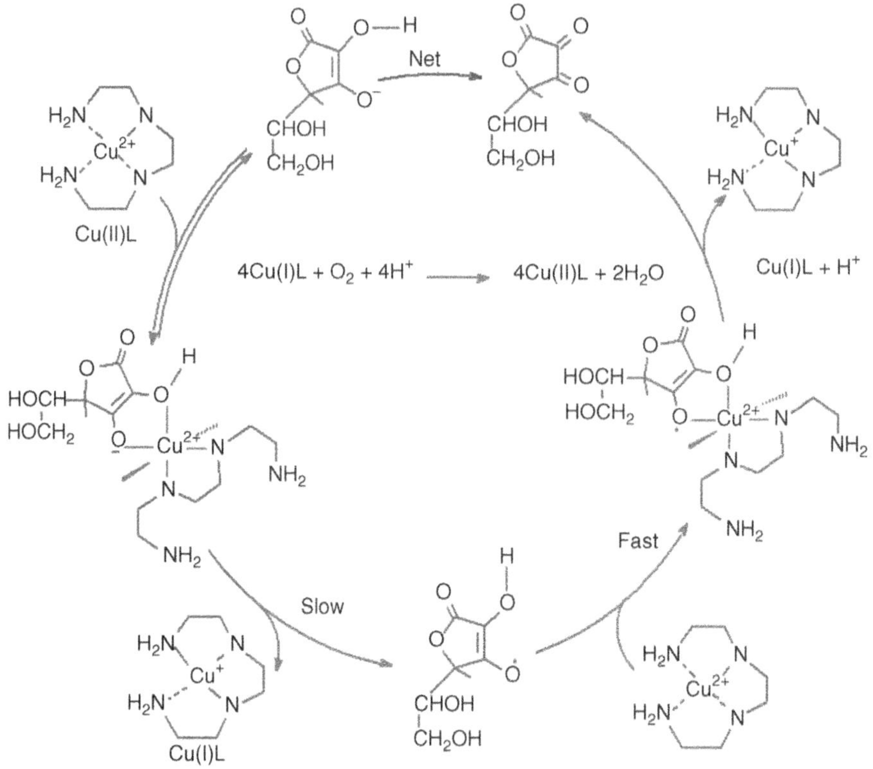

SCHEME 4-5 Metal chelate–catalyzed oxidation of ascorbic acid (free-radical intermediate ascorbic acid oxidase model). (Modified based on Taqui Khan and Martell, 1967.)

REFERENCES

E. T. Adman, in *Metalloproteins*, P. M. Harrison, Ed., Part 1, Chapter 1, Machillan, London (1985).

A. E. G. Cass, in *Metalloproteins*, P. M. Harrison, Ed., part 1, p. 121, Macmillan, Hong Kong (1985).

F. A. Cotton and G. Wilkinson, in *Advanced Inorganic Chemistry*, 4th ed., Wiley, New York, p. 816 (1980).

C. D. Dawson, *Ann. NY Acad. Sci.*, 88, Art2, 353 (1960).

P. J. Hart, A. M. Nersissian, R. G. Herrmann, R. M. Nalbandyan, J. S. Valentine, and D. Eisenberg, *Protein Sci.*, 5, 2175–2183 (1996).

B. Hazes, K. A. Magnus, C. Bonaventura, J. Bonaventura, Z. Dauter, K. H. Kalk, and W. G. Hol, *Protein Sci.*, 2, 597–619 (1993).

E. Jaenicke, K. Büchler, H. Decker, J. Markl, and G. F. Schröder, *Iubmb Life*, 63, 183–187 (2011).

L. Leinartaite, K. Saraboji, A. Nordlund, D. T. Logan, and M. Oliveberg, *J. Am. Chem. Soc.*, 132, 13495 (2010).

A. Messerschmidt, R. Ladenstein, R. Huber, M. Bolognesi, L. Avigliano, R. Petruzzelli, A. Rossi, and A. Finazzi-Agro, *J. Mol. Biol.*, 224, 179–205 (1992).

T. Nakamura and Y. Ogura, in *Magnetic Resonance in Biological Systems*, A. Ehrenberg, B. G. Malmström, and T. Vanngard, Eds., Pergamon, New York (1967).

K. Paraskevopoulos, M. Sundararajan, R. Surendran, M. A. Hough, R. R. Eady, I. H. Hillier, and S. S. Hasnain, *Dalton Trans.*, 25, 3067 (2006).

M. G. Simmons and L. J. Wilson, *J. Chem. Soc. Chem. Commun.*, 498, 634 (1978).

M. M. Taqui Khan and A. E. Martell, *J. Am. Chem. Soc.*, 89, 7014 (1967).

M. M. Taqui Khan and A. E. Martell, *J. Am. Chem. Soc.*, 90, 6011 (1968).

J. S. Thompson, T. J. Marks, and J. A. Ibers, *Proct. Natl. Acad. Sci. USA*, 74, 3114, (1977).

A. K. Wiersema and J. J. Windle, *J. Phys. Chem.*, 86, 2316 (1964).

Y. Xue, M. Okvist, O. Hansson, and S. Young, *Protein Sci.*, 7, 2099–2105 (1998).

SUGGESTIONS FOR FURTHER READING

Copper

1. R.S. Drago, in *Physical Methods in Chemistry*, W. B. Saunders, Philadelphia (1977).

2. S. D. Karlin and Z. Tyeklar, Eds., *Bioinorganic Chemistry of Copper*, Chapman and Hall, New York (1993).

3. W. Kaim and J. Rall, *Angew. Chem. Int. Ed. Engl.*, 35, 43 (1996).

4. N. Katajima and Y. Moro-oka, *Chem. Rev.*, 94, 737 (1994).

5. K. D. Karlin and J. Zubieta, Eds., *Copper Coordination Chemistry, Biochemical and Inorganic Perpectives*, Adenine, Albany, NY (1983).

6. C. A. Owen, Ed., *Copper Deficiency and Toxicity Aquired and Inhertied in Plants, Animals, and Man*, Noyes Publications, Park Ridge, NJ (1981).

7. S. K. Chapman, in *Perspectives on Bioinorganic Chemistry*, R. W. Hay, J. R. Dilworth, and K. B. Nolan, Eds., Vol. 1, JAI Press, London (1991).

8. P. Cassidy and M. A. Hitchman, *J. Chem. Soc. Chem. Commun.*, 20, 837 (1975).

Metal Complexes of Amino Acids and Proteins

9. H. C. Freeman, "Crystal structures of metal–peptide complexes," *Adv. Protein Chem.*, 22, 257 (1967).

10. R. J. Sundberg and R. B. Martin, "Interactions of histidine and other imidazole derivatives with transition metal ions in chemical and biological system," *Chem. Rev.*, 74, 471 (1974).

ESR

11. N. M. Atherton, in *Electron Spin Resonance*, Ellis Horwood, Chichester (1973).

12. R. S. Alger, in *Electron Paramagnetic Resonances*, Wiley Interscience, New York (1968).

Blue – Copper Proteins

13. R. Malkan, in *Inorganic Biochemistry*, G. L. Eichhorn, Ed., Vol. 2, Chapter 21, Elsevier, New York (1973).

14. E. L. Ulrich and J. L. Markley, "Blue–copper proteins, nuclear magnetic resonance investigations," *Coord. Chem. Rev.*, 27, 109 (1978).

15. P. M. Collman, H. C. Freeman, J. M. Guss, M. Murata, V. A. Norro, J. A. M. Ramashaw, and M. P. Venkatappa, *Nature*, 272, 319 (1978).

16. R. A. Marcus and N. Sultin, *Biochim. Biophys. Acta*, 811, 265–322 (1985).

Superoxide Dismutase and Superoxide

17. M. J. Maroney, *Curr. Opin. Chem. Biol.*, 3, 188–199 (1999).

18. J. A. Garden, L. M. Ellerby, J. A. Roe, and J. S. Valentine, *J. Am. Chem. Soc.*, 116, 9743–9744 (1994).

19. I. Fridovich, *Acc. Chem. Res.*, 5, 321 (1972).

20. I. Fridovich, *Ann. Rev. Biochem.*, 44, 147 (1975).

21. U. Weser, *Struct. Bonding*, 17, 1 (1973).

22. A. M. Michelsen and J. M. McCord, Eds., *Superoxide and Superoxide Dismutase*, Academic, New York (1977).

23. O. Hayaishi and K. Asada, Eds., *Biochemical and Medical Aspects of Active Oxygen*, University of Tokyo Press, Tokyo (1977).

24. I. Fridovich, "Oxygen radicals, hydrogen peroxide and oxygen toxicity," in *Free Radical in Biology*, W. A. Pryor, Ed., pp. 239–277, Academic, New York (1976).

25. S. J. Lippard, A. R. Burger, K. Ugurbil, J. S. Valentine, and M.W. Pantoliano, *Adv. Chem. Ser.*, 16, 251 (1977).

Hemocyanins

26. N. Kitajima, K. Fujisawa, Y. Moro-oka, and K. Toriumi, *J. Am. Chem. Soc.*, 111, 8975–8976 (1989).

27. E. I. Solomon, U. M. Sundaram, and T. E. Machonkin, *Chem. Rev.*, 96, 2563–2605 (1996).

28. R. J. M. Klein Gibbink, C. F. Martens, P. J. A. Kenis, R. J. Jansen, H-F. Nolting, V. A. Solé, M. C. Feiters, K. D. Karlin, and R. J. M. Nolte, *Inorg. Chem.*, 38, 5755–5768 (1999).

29. R. Lontie and R. Witters, in *Inorganic Biochemistry*, G. L. Eichhorn, Ed., Vol. 1, Chapter 12, Elsevier, New York (1973).

30. R. Lontie and L. Vanquickenborne, in *Metal Ions in Biological Systems*, H. Sigal, Ed., Vol. 3, Chapter 6, Marcel Dekker, New York (1974).

31. N. M. Senozan, *J. Chem. Ed.*, 53, 684 (1976).

32. M. Brunori, M. Coletta, and B. Giardina, in *Metalloproteins*, P. M. Harrison, Ed., Part 2, p. 263, Macmillan, Hong Kong (1985).

33. R. Lontie and R. Witters, in *Inorganic Biochemistry*, G. L. Eichhorn, Ed., Vol. 1, Chapter 12, Elsevier, New York (1973).

34. M. G. Simmons and L. J. Wilson, *J. Chem. Soc. Chem. Commun.*, 15, 634 (1978).

Ascorbate Oxidase

35. E. I. Solomon, U. M. Sundaram, and T. E. Machonkin, *Chem. Rev.*, 96, 2563–2605 (1996).
36. M. M. Taqui Khan and A. E. Martell, *J. Am. Chem. Soc.*, 89, 4167 (1967).

Metallothioneins

37. Y. Kojima and J. H. R. Kagi, *Trends Biochem.*, 3, 90 (1978).
38. M. G. Cherian and R. A. Goyer, *Ann. Clin. Lab. Sci.*, 8, 91 (1978).
39. P. E. Hunziker and J. H. R. kagi, in *Metalloproteins*, P. M. Harrison, Ed., Part 2, p. 149, Macmillan, Hong Kong (1985).

5

IRON PROTEINS

INTRODUCTION

Iron is the second most abundant metal, after aluminum, and the fourth element in Earth's crust. It has two readily interconverted oxidation states. Iron (II) forms salts with every anion and generally is a green, hydrated, crystalline substance, while Fe(III) occurs in salts with most anions.

Iron is also the most abundant transition metal in the human body, 4.2–6.1 g in the average human. The iron porphyrin (heme) group is an essential constituent of hemoglobin, myoglobin, cytochrome, catalase, and peroxidase. These are known as hemoproteins and are responsible for oxygen transport and electron transport. The rest of the iron in the body (nonheme iron) is almost entirely protein bound. These non-hemoproteins include the intracellular iron-containing flavoproteins (NADH dehydrogenase and succinate dehydrogenase) and iron–sulfur proteins. Iron is stored as ferritin in the liver, spleen, bone marrow, and principally in reticuloendothelial. It is transported in serum by transferrin. The approximate distribution of iron-containing compound in the normal adult is 2.6 g in hemoglobin, 0.13 g in myoglobin, 0.004 g in cytochrome c, 0.004 g in catalase, 0.007 g in transferrin, 0.4–0.8 g in ferritin, and 0.48 g in hemosiderin.

Anemia arises from iron deficiency, while hemochomatosis (increases in iron concentration) and siderosis (deposition of FeO dust in the lungs) are associated with excess of iron.

Iron is also found in the whole range of life, from humans to bacteria, as nitrogenase, various oxidases, hydrogenases, reductases, deoxygenases, and dehydrases.

Chemistry of Metalloproteins: Problems and Solutions in Bioinorganic Chemistry, First Edition.
Joseph J. Stephanos and Anthony W. Addison.
© 2014 John Wiley & Sons, Inc. Published 2014 by John Wiley & Sons, Inc.

TABLE 5-1   Iron Proteins

	Nonheme Iron Proteins				Heme Proteins
Active site	Iron phenolate proteins	Apo protein is functional	Binuclear Fe proteins	Iron–sulfur proteins	Iron porphyrin
Role	Fe Transport and O_2 catalysis	Fe storage buffer	Transport of oxygen	Transfer of electron	Transport of oxygen and electron
Examples	1. Catechole dioxygenase	1. Ferritins 2. Transferrin (Fe storage and transport)	1. Hemerythrin 2. Ribonucleotide reductase 3. Purple acid phosphate 4. Methanemono- oxygenase	1. Rubredoxin 2. Ferredoxin 3. Conjugate Fe-S proteins 4. Phthalate dioxygenase 5. Hydrogenase 6. Nitrogenase	1. Myoglobin (oxygen storage and oxygen carrier) 2. Hemoglobin (oxygen transport and oxygen carrier) 3. Cytochrome c (electron transfer) 4. Catalase (decomposition of H_2O_2) 5. Chloroperoxidase (halogenate organic substrate) 6. Lactoperoxidase (oxidation of NCS) 7. Cytochrome P 450 (hydroxylation organic substrate) 8. Cytochrome c peroxidase (reduction of H_2O_2)

In many cases little is known, whereas in others, such as hemoglobin, ferredoxins, and cytochromes, we know the molecular structures and electronic properties of the active sites in considerable detail.

In this chapter, the possible electronic ground states and the allowed electronic transition bands of iron complexes are discussed. This is followed by an introduction to Mössbauer and ESR spectroscopy and the advantages of using ESR to probe the magnetic properties of the iron proteins.

The next few sections treat the biochemistry of iron under two broad headings: nonheme proteins and heme proteins. These are subdivided into a number of topics because of the great diversity of the function and widely differing chemistries of the iron-containing species.

Physiological roles, classifications, and examples of iron proteins are summarized in Table 5-1.

ELECTRONIC SPECTRA OF IRON IONS

Discuss the possible electronic ground states and the allowed electronic transitions bands of iron (II) and iron (III) complexes.

Method used to establish ground-state term (see p. 96):

- Hund's Rules: For a given atom, the electronic configuration
    (i) With maximum S (maximum spin multiplicity, $2S + 1$) has the lowest energy

(ii) For a given multiplicity, maximum L (maximum M_L) has the lowest energy.

(iii) For given term, if the outmost subshell is half-filled or less, the level with the lowest value of J ($J=L-S$) has the lowest energy. If the outmost subshell is more than half-filled, the level with the highest value of J ($J=L+S$) is lowest in energy.

- Fe^{2+}, d^6

  o m_ℓ:

2	1	0	-1	-2

  L_{max}, S_{max}:   ↑↓   ↑   ↑   ↑   ↑

  L_{max}: $2 \times 2+$ $1 \times 1+$ $1 \times 0+$ $1 \times -1+$ $1 \times -2 = 2$

  S_{max}: $4 \times \frac{1}{2} = 2$

  o When $L =$

0	1	2	3	4

  Term   →   S   P   D   F   G

  Therefore the ground-state term is D

  Multiplicity: $2S+1 = 5$

  o Ground-state term symbol → $^5D_4$

- Fe^{3+}, d^5

  o m_ℓ:

2	1	0	-1	-2

  L_{max}, S_{max}:   ↑   ↑   ↑   ↑   ↑

  $L_{max} =$ $1 \times 2+$ $1 \times 1+$ $1 \times 0+$ $1 \times -1+$ $1 \times -2 = 0$

  S_{max}: $5 \times \frac{1}{2} = \frac{5}{2}$

  Therefore the ground-state term is S

  Multiplicity: $2S+1 = 6$

  o Ground state term symbol → $^6S_{-5/2}$

  o These ground states split into T_{2g} and E_g states in octahedral symmetry (Fig. 5-1).

- In an octahedral molecular environment, the metal valence d orbitals split into two sublevels.

- The *more energetic* sublevel is called e_g and includes the two d orbitals, $d_{x^2-y^2}$ and d_{z^2}, which interact strongly in σ-*bonding* with ligands.

- The three d orbitals in the t_{2g} sublevel (d_{xz}, d_{yz}, d_{xy}) can only participate in π-*interaction* and are always *lower in energy* than the e_g orbitals (Fig. 5-2).

- Fe^{2+} complexes, d^6: Two cases are distinguished:

  o If Δ_o is relatively *small*, as in $Fe (H_2O)_6{}^{2+}$, d^6 electrons spread over t_{2g} and e_g to give maximum spin multiplicity (Fig. 5-3):

$$\text{Multiplicity} = 2S+1,$$

$$= 2(4 \times 1/2) + 1 = 5$$

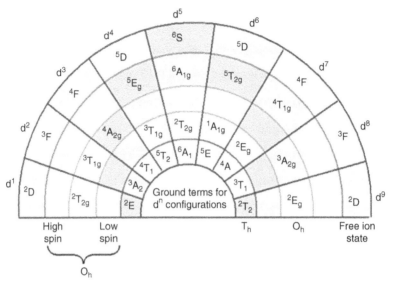

FIGURE 5-1   Ground state terms for d^n configurations and the splitting in octahedral and tetrahedral symmetries.

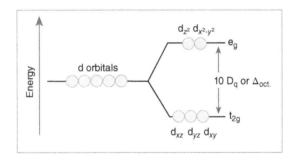

FIGURE 5-2   Splitting of d orbital in octahedral complex, $(\Delta_o = \Delta_{oct.})$.

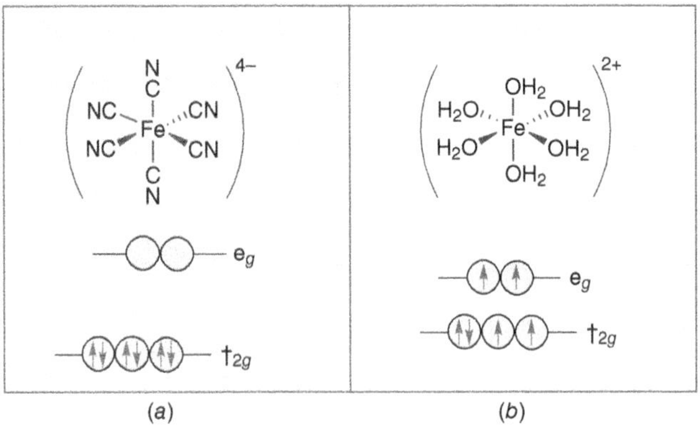

FIGURE 5-3   (a) Low-spin ground state: $(t_{2g})^6, {}^1A_{1g}$. (b) High-spin ground state: $(t_{2g})^4(e_g)^2, {}^5T_{2g}$.

- ○ If Δ_o is *larger* than the energy required to pair electrons in t_{2g} orbitals, as in $Fe(CN)_6{}^{4-}$, d^6 electrons reside in t_{2g} (Fig. 5-3):

$$\text{Multiplicity} = 2S + 1 = 1$$

Crystal Field Stabilization Energy (CFSE)

- Fe^{2+} (d^6), high spin
  - ○ Octahedral: $\text{CFSE} = (4)(-\frac{2}{5}\Delta_o) + (2)(\frac{3}{5}\Delta_o) + \pi = -\frac{2}{5}\Delta_o + \pi$
  - ○ Tetrahedral: $\text{CFSE} = (3)(-\frac{3}{5}\Delta_t) + (3)(\frac{2}{5}\Delta_t) + \pi = -\frac{3}{5}\Delta_t + \pi$
- Fe^{2+} (d^6), low spin ($\pi = $ electron pairing energy)
  - ○ Octahedral: $\text{CFSE} = (6)(-\frac{2}{5}\Delta_o) + (0)(\frac{3}{5}\Delta_o) + 3\pi = -\frac{12}{5}\Delta_o + 3\pi$
  - ○ Tetrahedral: $\text{CFSE} = (4)(-\frac{3}{5}\Delta_t) + (2)(\frac{2}{5}\Delta_t) + 2\pi = -\frac{8}{5}\Delta_t + 2\pi$
- Fe^{3+} (d^5), high spin
  - ○ Octahedral: $\text{CFSE} = (3)(-\frac{2}{5}\Delta_o) + (2)(\frac{3}{5}\Delta_o) = 0$
  - ○ Tetrahedral: $\text{CFSE} = (2)(-\frac{3}{5}\Delta_t) + (3)(\frac{2}{5}\Delta_t) = 0$
- Fe^{3+} (d^5), low spin
  - ○ Octahedral: $\text{CFSE} = (5)(-\frac{2}{5}\Delta_o) + (0)(\frac{3}{5}\Delta_o) + 2\pi = -\frac{10}{5}\Delta_o + 2\pi$
  - ○ Tetrahedral: $\text{CFSE} = (4)(-\frac{3}{5}\Delta_t) + (1)(\frac{2}{5}\Delta_t) + 2\pi = -\frac{10}{5}\Delta_t + 2\pi$

Electronic Spectra and Selection Roles

- Electronic spectra *selection roles* (see p. 98)

$$\Delta S = 0$$

$$u \leftrightarrows g \quad (\textit{Laporte's rule})$$

Absorption bands arise in the NIR, VIS, and UV.
- Fe^{2+} complexes; Example: $Fe(H_2O)_6{}^{2+}$ is a typical *high-spin* spectrum.
  - ○ In order to understand the spectra of iron complexes in which the metal ions have d^6 electrons, we must consider an energy-level diagram based upon the energy-level diagram of d^6 configuration (Fig. 5-4).
  - ○ The broad absorption at 1000 nm (in the NIR) is due to $^5T_{2g} \rightarrow {}^5E_g$ (Fig. 5-4):

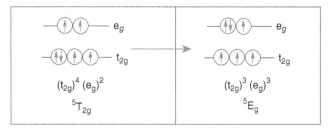

  - ○ This band usually shows some splitting, vibronically induced, caused by Jahn–Teller distortion of ^5E_g (see p. 100).

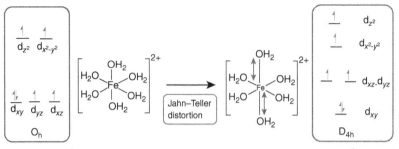

- Fe(CN)$_6^{4-}$, in the *low-spin* state, for the completely filled $(t_{2g})^6$ yields $^1A_1$, the ground state (Figs. 5-1,4).
- Two electronic spin-allowed excitations are expected (Fig. 5-4).

$$^1A_1 \rightarrow {}^1T_1(t_{2g})^5(e_g)^1 \quad \text{at } 320 \text{ nm}$$

$$^1A_1 \rightarrow {}^1T_2(t_{2g})^5(e_g)^1 \quad \text{at } 270 \text{ nm}$$

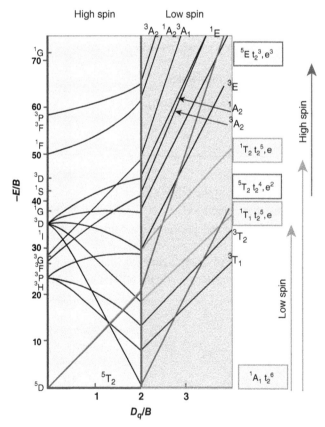

FIGURE 5-4 Energy-level diagram (Tanabo–Sugano) for d^6 ions in an octahedral field. (After Tanabe and Sugano, 1954.)

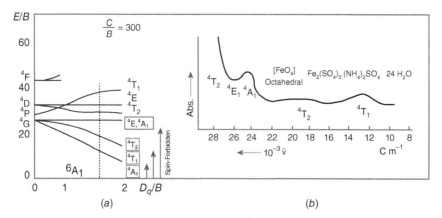

FIGURE 5-5 (*a*) Possible electronic excitations of $^6A_1$ state in order of increasing energy for high-spin d^5 metal in O_h field. (*b*) The consequent bands affirming the lowest excitations in the electronic absorption spectra of $Fe(H_2O)_6^{3+}$ in ferric ammonium sulfate. (Data from Angelici, 1973. Reproduced by permission of Elsevier.)

Excitation to $^1T_1$, $^1T_2$ states gives the configuration $(t_{2g})^5(e_g)^1$ in which the odd spins in t_{2g} and e_g configurations are spin opposed.

- Fe^{3+} complexes
  - High-Spin, the ground state of $3d^5$ is $^6A_{1g}$ $(t_{2g})^3(e_g)^2$, e.g., $Fe(OH_2)_6^{3+}$, (Fig. 5-5). The electronic spectrum of $Fe(H_2O)_6$ in ferric ammonium sulfate exhibits absorption bands at 794 nm ($\varepsilon = 0.05$), 540 nm ($\varepsilon = 0.01$), and 411 and 406 nm ($\varepsilon = 1.3$).
  - Ligand field splitting, Δ_o, is larger for Fe^{3+} than for Fe^{2+}. Therefore, the tendency toward a low-spin ground-state structure is greater.
  - $^4T_{1g}$, $^4T_{2g}$ and the degenerate pair ^4E_g, $^4A_{1g}$ are the lowest electronic excited states (Fig. 5-5).
  - All transitions of $^6A_{1g}$ complexes are *spin forbidden*, and weak absorption bands are expected:

$$^6A_{1g} \rightarrow {}^4T_{1g} \qquad 794\,nm: \qquad \varepsilon = 0.05$$

$$^6A_{1g} \rightarrow {}^4T_{2g} \qquad 540\,nm: \qquad \varepsilon = 0.01$$

$$^6A_{1g} \rightarrow {}^4E_g, {}^4A_{1g} \qquad 411, 406\,nm: \qquad \varepsilon = 1.3$$

The molar extinction coefficients of these bands are very low. Note:

$$d \rightarrow d: \qquad 10^{-2} \le \varepsilon < 5 \times 10^2$$

- It is important to point out the differences in the spectrum of octahedral $Fe^{3+}(H_2O)_6$ and tetrahedral $Fe^{3+}(H_2O)_4$.
  - It is possible to use an energy diagram for T_d as well as O_h. However, splitting of the d orbital is the inverse of that for T_d, and omitting the g subscript means

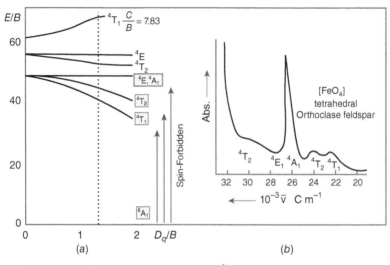

FIGURE 5-6 (*a*) Possible electronic excitations of $^6A_1$ state in order of increasing energy for high-spin d^5 metal in T_d field. (*b*) The consequent bands affirming the lowest excitations in the electronic absorption spectra of [Fe(III)O$_4$] in orthoclase feldspar. (Data from Angelici, 1973. Reproduced by permission of Elsevier.)

it is noncentrosymmetric (Fig. 5-6). The electronic spectrum of Fe(O)$_4$ in orthoclase feldspar shows absorption bands at 444 nm ($\varepsilon = 0.73$), 418 nm ($\varepsilon = 0.76$), and 377 nm ($\varepsilon = 4.1$).

○ From the calculated energy diagram (Fig. 5-6) the spectrum of $Fe^{3+}(H_2O)_4$ consists of three transitions:

$$^6A_1 \rightarrow {}^4T_1 \qquad 444 \text{ nm}: \quad \varepsilon = 0.73$$

$$^6A_1 \rightarrow {}^4T_2, \qquad 418 \text{ nm}: \quad \varepsilon = 0.76$$

$$^6A_1 \rightarrow {}^4E, {}^4A_1 \qquad 377 \text{ nm}: \quad \varepsilon = 4.1$$

○ The molar extinction coefficients of these bands are 10 times greater than the respective bands observed for $Fe^{3+}(H_2O)_6$.

○ With N-donor and CN$^-$ ligands, low-spin ground states are obtained, Fe(CN)$_6^{3-}$ and Fe(en)$_3^{3+}$, which establish the $^2T_{2g}(t_{2g})^5$ ground state (Fig. 5-7).

○ The following two absorptions can often be observed as shoulders:

$$^2T_2 \rightarrow {}^4T_1 \qquad 500 \text{ nm, spin forbidden}$$

$$^2T_2 \rightarrow {}^4T_2 \qquad 348 \text{ nm, spin forbidden}$$

○ Other transitions to the lower set of spin doublets arise, $(t_{2g})^4(e_g)^1$:

$$^2T_2 \rightarrow {}^2A_2, {}^2T_1 \qquad (t_{2g})^4(e_g)^1, 326 \text{ nm, spin allowed}$$

$$^2T_2 \rightarrow {}^2E \qquad 274 \text{ nm, spin allowed}$$

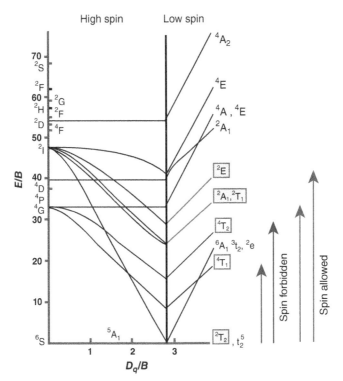

FIGURE 5-7 Energy-level diagram for d^5 ions in an octahedral field. (After Tanabe and Sugano, 1954.)

These can be seen with ligands which are not associated with low-energy charge transfer. For example, the spectrum of $Fe(CN)_6^{3-}$ also shows the low-energy charge transfer band $\pi(CN) \rightarrow Fe^{3+}$.

MÖSSBAUER SPECTROSCOPY OF IRON IONS

Nuclear Resonance Absorption

Discuss the differences between the electronic and nuclear resonance absorption.

- In electronic resonance absorbance, *light* emitted by a given atom during the transition of *electrons* from an excited state to the ground state can be absorbed by atoms of the same kind.
- Example: When white light is transmitted through sodium vapor, the light of the wavelength corresponding to the D-line of sodium will be absorbed.
- Similarly, atomic *nuclei* also have excited states of discrete energy.
- γ-*Radiation* emits in the transition of an atomic nucleus from one definite energy level to another.
- Resonance absorption can also be expected in γ-radiation.

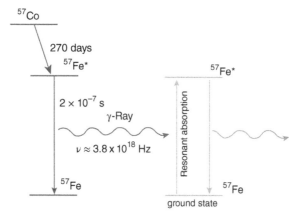

FIGURE 5-8 Transitions involved in Mössbauer effect using ^{57}Fe.

Basic Principles

What are the bases of Mössbauer spectroscopy?

- The basis of Mössbauer spectroscopy is the resonant absorption of a γ-ray photon by a nucleus.
- A typical experiment uses a sample of ^{57}Co as a primary source of an isotope of iron, ^{57}Fe.
- The isotope ^{57}Co decays slowly ($t_{1/2} = 270$ days) and forms the excited ^{57}Fe*.
- ^{57}Fe* decays very rapidly to the ground state ^{57}Fe ($t_{1/2} = 2 \times 10^{-7}$ s) and emits a γ-ray, which is the photon used in Mössbauer spectroscopy (Fig. 5-8).
- This photon strikes a ground-state ^{57}Fe nucleus (the sample) in *another* part of the spectrometer.
- The γ-photon emitted by ^{57}Fe* and absorbed by ^{57}Fe has a frequency of about 3.5×10^{18} Hz.
- The photon is absorbed, if its energy matches the ^{57}Fe $\xrightarrow{E_\gamma}$ ^{57}Fe* energy separation E_γ (this is the resonant absorption step) but will not be absorbed if the energies do not match.
- When ^{57}Fe* decays and emits the photon, the atomic nucleus undergoes *recoil* according to the law of conservation of momentum.
- This recoil gives the deexcited nucleus a velocity of about 10^2 m/s.
- Thus, the energy of the emitted radiation will be reduced by the recoil energy R:

$$R = \frac{E_\gamma^2}{2mc^2}$$

where m is the mass of the nucleus and c is the speed of light.

- Therefore, to *compensate* for the recoil energies, the *Doppler effect* is very suitable.

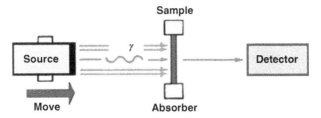

FIGURE 5-9   A Mössbauer spectrometer.

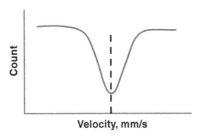

FIGURE 5-10   Characteristic Mössbauer spectrum.

- The energy of γ-quanta increases when the source and the absorber are moving toward each other and decrease in the opposite case (Fig. 5-9). By moving the source or the absorber at velocity: v, the Doppler effect causes a shift:

$$E_d = \pm \frac{v\, E_\gamma}{c}$$

where E_d is the energy change of a photon associated with the source moving relative to the sample and $c = 3.00 \times 10^{11}$ mm/s.

- If the source or the absorber is moved at a suitable velocity, the recoil energy difference can be obtained.

- The absorption of γ-rays is plotted as a function of source velocity in Fig. 5-10. The peak corresponds to the source velocity at which maximum γ-ray absorption by the sample occurs.

- Positive velocities indicate that photon frequency must be increased to allow the absorber species to resonate, and therefore, $^{57}Fe^*$–^{57}Fe separation is greater in the absorber than in the source.

Quadruple Splitting

Why are Mössbauer spectra often split into several lines?

- The degeneracy of nuclear energy levels for nuclei with $I > \frac{1}{2}$ is removed by a nonsymmetric electron or ligand distribution.

- If the spins of all particles are paired, there will be no net spin and the nuclear spin quantum number I will be zero (indicated by A below).

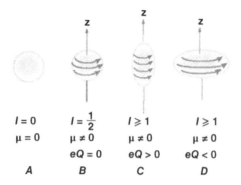

- When $I = \frac{1}{2}$, there is one net unpaired spin and this unpaired spin generates a nuclear magnetic moment. The distribution of positive charge in a nucleus of this type is spherical (B).
- When $I \geq 1$, the nucleus has spin associated with it and the nuclear charge distribution is nonspherical (C).
- In this case, the nucleus has a quadrupole moment eQ, where e is the electrostatic charge and Q is a measure of the deviation of the nuclear charge distribution from spherical symmetry.
  - For a spherical nucleus, $eQ = 0$.
  - $eQ > 0, eQ \to + ve$: The charge is oriented along the direction of the principal axis.
  - $eQ < 0, eQ \to - ve$: The charge accumulates perpendicular to the principal axis.
- The *allowed* orientations of the nuclear magnetic moment vector in a magnetic field are indicated by the nuclear spin angular momentum quantum number m_I.
  - m_I takes values $I, I-1, \ldots, -I$.
  - If $I = \frac{1}{2}$, $m_I = \pm\frac{1}{2}$; if $I = \frac{3}{2}$, $m_I = \pm 1/2, \pm 3/2$.
- The energies of the various quadrupole nuclear states, E_m, is given by

$$E_m = \frac{e^2 Qq \left[3m_I^2 - I(I + 1)\right]}{4I (2I - 1)}$$

where q is the field gradient. So E_m is related to m_I^2.

- In ^{57}Fe, the nuclear spins of the unexcited and excited states are different, $I(^{57}\text{Fe}) = \frac{1}{2}$, but $I(^{57}\text{Fe}^*) = \frac{3}{2}$. For ^{57}Fe*, $I = \frac{3}{2}$, m_I can have values of $\frac{3}{2}, \frac{1}{2}, -\frac{1}{2}, -\frac{3}{2}$.
  - For $m_I = \pm\frac{3}{2} \to E_m = +e^2 Q_q/4$: doubly degenerate set of quadrupole energy states
  - For $m_I = \pm\frac{1}{2} \to E_m = -e^2 Q_q/4$: doubly degenerate set of quadrupole energy states
  - $\Delta E = e^2 Q_q/4 - (-e^2 Q_q/4) = e^2 Q_q/2$
- Therefore, the splitting does *not remove* the + or − degeneracy of the m_I levels, and we obtained a different level for each $\pm m_I$ (Fig. 5-11).

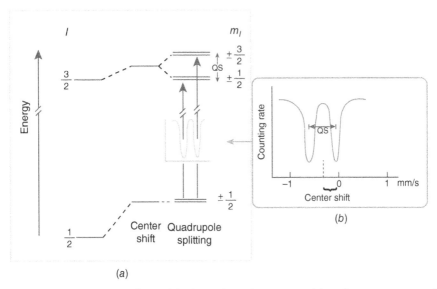

(a)

FIGURE 5-11   Influence of noncubic electronic environment on (a) nuclear energy states of ^{57}Fe and (b) Mössbauer spectrum.

- The ground state, $I(^{57}\text{Fe}) = \frac{1}{2}$, does not split but the excited state is split, leading to two peaks in the spectra.
- The splitting of the excited state will not occur in the spherically symmetric or cubic field.
- However, the splitting will occur only when there is a field gradient at the nucleus caused by *asymmetric p- or d-electron distribution* in the compound.
- If the t_{2g} set and the e_g set of orbitals in octahedral transition metal ion complexes have equal populations in the constituent orbitals, the quadrupole splitting will be zero.
  - Low-spin Fe(II) complexes (t_{2g}^{6}) and high-spin Fe(III) complexes ($t_{2g}^{3}e_g^{2}$) will not give rise to a quadrupole splitting (Fig. 5-12), e.g., $K_4Fe(CN)_6, 3H_2O$, and $FeCl_3$ [see spectra in Kerler (1962) and Kerler and Neuwirth (1962)].

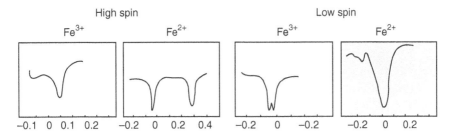

FIGURE 5-12   Typical quadrupole splitting in Mössbauer spectra of Fe(II) and Fe(III) complexes.

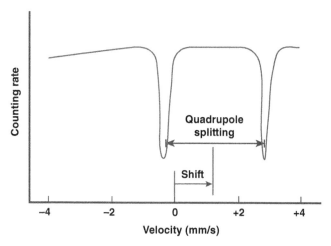

FIGURE 5-13 Isomer shift and quadrupole splitting.

○ However, quadrupole splitting is expected for Fe(III), e.g., $K_3Fe(CN)_6$, and Fe(II) in the weak field, e.g., $FeSO_4 \cdot 7H_2O$.

○ The splitting forms a picture of the symmetry of the electron distribution near the nucleus.

Isomer Shift

Why does the resonance shift relative to the source frequency in Mössbauer spectroscopy?

- The change in position of the resonance is called the *isomer shift* (Fig. 5-13).
- The excited- and ground-state nuclei differ in radius to a significant extent.
- Therefore, the electrostatic interaction with the surrounding electrons changes on excitation.
- Changes in the electron cloud of an atom disturb the electrostatic interactions and produce small alteration in the energy levels of the nucleus.
- The isomer shift measures the difference in the chemical environment of the two nuclei.
- In ^{57}Fe, the nuclear spins of the unexcited and excited states are different, $I(^{57}Fe), = \frac{1}{2}$ but $I(^{57}Fe^*), = \frac{3}{2}$.
- ^{57}Fe has a *spherical* distribution of the charge, but $^{57}Fe^*$ is concentrated at the poles.
- The effect is significant only for the electrons in s orbitals, which can approach the nucleus.
- Only *s electrons* have a finite probability of overlapping the nuclear charge density.

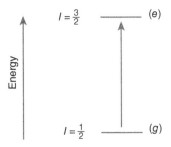

- It should be remembered that:
  - p, d, and other electron densities can influence *s-electron* density by screening the s density from the nuclear charge.
  - A decrease in the number of d electrons causes a marked increase in the total s-electron density at the iron nucleus.
- So the energy of the nucleus changes (δE) by an amount proportional to the change in the nuclear radius (δR) and the s-electron density at the nucleus [$\psi_s^2(0)$],
  - $\delta E \propto \psi_s^2(0)\delta R$
  - Assume δR is the same in the source and absorber and the s-electron density might be different for the two cases because the ^{57}Fe nuclei are in a different chemical environment, and

$$\delta E \propto \{[\psi_s^2(0)]_{absorber} - [\psi_s^2(0)]_{source}\}\delta R$$

where $[\psi_s^2(0)]_{source}$ and δR are constant for a given nucleus.
- By measuring the isomer shift it is possible to assess the *extent* to which s orbitals are involved in the bonding.
- For ^{57}Fe, an increase in 4s density decreases the isomer shift, while an increase in 3d density increases the isomer shift.

ESR SPECTRA OF IRON (III)

ESR and Magnetic Susceptibility

What are the advantages of using ESR technique to probe the magnetic properties of the iron–proteins?

1. Experimental consideration:
   (a) For $Fe(NH_4)(SO_4) \cdot 12H_2O$, we need 50-µmol sample, i.e., ~24 mg:

$$\chi_{net} = \chi_{para} + \chi_{Dia}$$

where χ is the magnetic susceptibility:
$$\chi_{Fe^{3+}+HS} = 14{,}700 \times 10^{-6}$$
$$\chi_{Dia} = -350 \times 10^{-6}(2.4\% \text{ of } \chi_{Fe^{3+}+HS})$$
$$\chi_{para} \sim 98 - 100\% \text{ of } \chi_{net}$$

where the χ_{Dia} correction must be estimated (about $\pm 20\%$ error). The amount of error in χ_{Dia} ($\pm 20\%$ of $350 \times 10^{-6} = \pm 70 \times 10^{-6}$) causes insignificant error in the χ_{net} (14350×10^{-6}).

(b) For Fe^{+3} protein, molecular weight $\sim$40,000 Da, we need 50-μmol, i.e., $\sim$2000 mg ($\sim$2 g):

$$\chi_{\text{para}} = +14{,}700 \qquad \chi_{\text{Dia}} = -10{,}000 \pm 20\%$$

$$\chi_{\text{net}} \sim 5000 \qquad (\text{cut } 3\times)$$

A $\pm 20\%$ error in the estimate of χ_{Dia} (20% of $\chi_{\text{Dia}} = \pm 2000$) causes about 13.6% error in χ_{para} and 8–10% in μ—a significant error.

Therefore, we need to use a more sensitive probe of the spin state. For ESR only 0.1 mL of 10–4 m $\sim$ 0.5 mg are required to probe the magnetic properties of the iron proteins.

2. ESR detects only unpaired electrons, so protein and solvent are *invisible*.

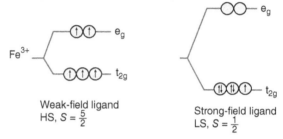

g-Values of Ferric Ion

What are the factors that affect the g-values, and how will they influence the ESR spectra of Fe^{3+}?

- In ESR, different energy states arise from the alignment of electron spin moments ($m_s = \pm\frac{1}{2}$ for a free electron) relative to the applied field, Zeeman effect (see p. 105).
- A fixed energy difference between the parallel and antiparallel spin states is commonly used in ESR.
- The ESR experiment is generally carried out at fixed frequencies (e.g., 9 GHz = X-band, 12.5 GHz = Q-band).
- Scan H_{app} until splitting fits $E = h\nu$:

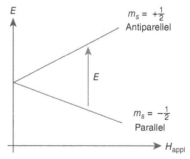

Alignment of electron spin relative to applied field $H : M_s = \Sigma\, m_s$, $|\Delta M_s| = 1$

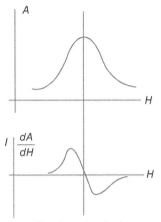

Spectrometer display,
(A is the absorbance, $I = dA/dH$)

$$\iint I \, dH \propto \text{ number of spins}$$

and

$$E_{ms} = g\beta H_{field} \, m_s,$$
$$E = E_{1/2} - E_{-1/2}$$

then:

$$E = h\nu = g\beta H_{field}$$

where β is the Bohr magneton, is approximately equal to the intrinsic electron magnetic dipole moment.

$$\beta = \frac{eh}{4\pi m_e c}$$

Assume H is the field experienced by e^-

$$(H_{felt}) = H_{app} \pm H_{local}$$

and g is the Landé g factor (proportionality constant). Then

$$H = \frac{714.44}{g} \nu$$

where, for free e^-, $g = 2.0023$.

g Value and H_{local}

- Due to the chemical environment, the local field, other e^-, and so on, $H_{applied}$ shifted, and $H_{applied} \neq H_{felt}$. That is,

$$H_{felt} = H_{app} + [H_{local}]$$

The induced shift due to H_{local} can be calibrated with DPPH (diphenylpicryl-hydrazyl free radical), Acac (pentane-2,4-dione anion), or VO(Acac), which have $g = 2.0023$. If $\nu = 9.500$ GHz ($0.3\,cm^{-1}$), $H_{app} = 3389.5$ GHz, and $H_{felt} = 3389.5$ GHz, the calculated $[H_{local}] = 0$, or

$$H_{felt} = H_{app} \pm ([H_{local}] = 0)$$
$$H_{felt} = H_{app}$$

When $H_{app} \neq H_{felt}$, the observed $g \neq 2.0023$, and the induced shift can be assigned.

Other Factors That Influence g Value

- In a sample of high-spin Fe^{3+}, the ESR spectrum indicates a signal at $H_{app} = 1542$ GHz and $\nu = 9.26$ GHz,

$$g = \frac{714.44}{H} \nu$$

where $g = 4.290$, while for free e^-, $g = 2.0023$, the variation in g value may be due to:

1. e^- Orbital angular momentum (*spin–orbital coupling*)
2. Other unpaired electron (*zero field splitting*)
3. Orientation of e^- orbital *vs.* applied field
4. Contact with nuclei of nonzero spin (hyperfine splitting, A)

- For Fe^{3+} ($3d^5$) the important factor is multiple unpaired e^-, which is possible in the high-spin case.

$$S_z = +\tfrac{5}{2} \qquad S_z = +\tfrac{3}{2} \qquad S_z = +\tfrac{1}{2}$$

- Electron energies are modulated by crystal field splitting ($10\,cm^{-1}$) but in the ESR regime are due to the spin contributions.
- The induced energy difference in the ESR regime is due to five unpaired electrons of Fe^{3+} and is given by

$$E = g\beta H S_z + D[S_z^2 - \tfrac{1}{3}S(S + 1)]$$

The constant D decreases the sensitivity of a given Fe^{3+} compounds energy level to the possible spin microstates (S_z values).

○ For example, for $S_z = \pm\frac{1}{2}$ in O_h, $S = S_{\text{High Spin}} = \frac{5}{2}$:

$$S_z^2 - \frac{1}{3}S(S+1) = \frac{1}{4} - \frac{1}{3}\left(\frac{5}{2}\right)\left(\frac{5}{2}+1\right)$$

$$= \frac{1}{4} - \frac{35}{12} = -\frac{32}{12} = -\frac{8}{3}$$

$$E = -\frac{1}{2}g\beta H - \frac{8}{3}D$$

○ For $S_z = \pm\frac{3}{2}$ in O_h, $S_{\text{High Spin}} = \frac{5}{2}$:

$$S_z^2 - \frac{1}{3}S(S+1) = \frac{9}{4} - \frac{1}{3}\left(\frac{5}{2}\right)\left(\frac{5}{2}+1\right)$$

$$= \frac{9}{4} - \frac{35}{12} = -\frac{8}{12} = -\frac{2}{3}$$

$$E = -\frac{3}{2}g\beta H - \frac{2}{3}D$$

○ For $S_z = \pm\frac{5}{2}$ in O_h, $S_{\text{High Spin}} = \frac{5}{2}$:

$$S_z^2 - \frac{1}{3}S(S+1) = \frac{25}{4} - \frac{1}{3}\left(\frac{5}{2}\right)\left(\frac{5}{2}+1\right)$$

$$= \frac{25}{4} - \frac{35}{12} = \frac{40}{12} = +\frac{10}{3}$$

$$E = -\frac{5}{2}g\beta H + \frac{10}{3}D$$

○ Figure 5-14 indicates that the high-spin Fe^{3+} ion is expected to show resonance with $g \neq 2$ and often more than one band. *Note*: For high spin in the O_h symmetry above, lower symmetry coordination environments will also affect g values, e.g., by splitting resonances.

• For low-spin Fe^{3+}, the possible spin microstates, S_z, is equal to $S = \frac{1}{2}$:

$$E = g\beta H S_z + D[S_z^2 - \frac{1}{3}S(S+1)]$$

$$= \pm\frac{1}{2}g\beta H + 0$$

Then we might expect a single resonance (only transition between $+\frac{1}{2}$ and $-\frac{1}{2}$ will be observed) $g \sim 2$. For example, for $K_3Fe(CN)_6$

$$g \simeq 1.79 \quad \text{(low spin)}$$

But for Fe^{3+} in soda glass

$$g \simeq 4.5 \quad \text{(high spin)}$$

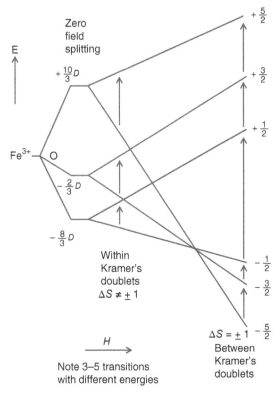

FIGURE 5-14   Expected resonances due to multiple unpaired electrons of high-spin Fe^{3+} ion.

IRON BIOAVAILABILITY

Explain the uptake of iron and subsequent distribution during metabolism.

The uptake of iron and subsequent controlled distribution and metabolism are summarized in Scheme 5-1.

Iron: Most Abundant Transition Metal in Biological Systems

Discuss the biological role, the dominated oxidation state, and the difficulties of the bioavailability of iron. Why is iron used as the ethylenediaminetetraacetic acid (EDTA) complex rather than a simple salt for supplying iron for plants in basic soil?

- Iron is the most abundant transition metal in mammals. The human body contains 4.2–6.1 g. An increase in Fe leads to an increase in the probability of infection.
- Injury: Fe release assists bacteria.
- Newborn human: scavenge and secrete Fe to resist infection.

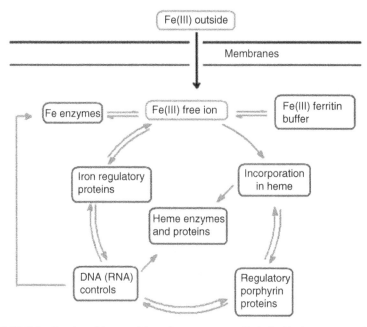

SCHEME 5-1 Uptake of iron and its subsequent controlled distribution and metabolism.

- Daily exchange: ~ 1 mg.
- $T_{\frac{1}{2}}$: five years.
- Normal oxidation states: Fe^{2+} and Fe^{3+}.
- In air, $Fe^{2+} \rightarrow Fe^{3+}$, except in acidic solution.
- Fe^{3+} is a strong Lewis acid, while Fe^{2+} falls between hard and soft acids.
- Ligand atom preferred: N, O, S.
- Figure 5-15 shows the abundance of iron and other metals, and the abundance ratios in seawater and Earth's curst.
- In general, abundances are greater for elements with an even atomic number than for neighbors with odd atomic numbers, as seen in Fig. 1-15.
- The elements having water-soluble compounds occur as cations or simple anions and have high concentrations in seawater (Table 5-2).

Why is Fe much less abundant in seawater than in Earth's crust (Fig. 5-15)?

- There are various levels of consideration for actual bioavailability.
- Fe occurs as:
    (i) Mineral oxides
    (ii) Sulfides

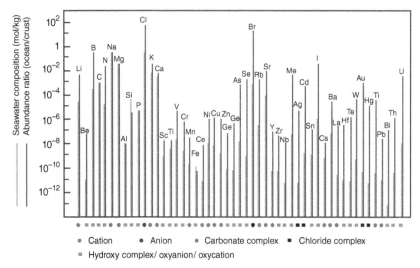

FIGURE 5-15 Concentration of elements in seawater and abundance ratio (ocean/crust), showing major species present. (Data from Cox, 1989.)

- In an aqueous environment (hydrosphers):

$$O_2 + 4H^+ + 4e^- \rightarrow 2H_2O \qquad E^0 = +1.229\text{ V}$$

$$Fe^{2+} \rightarrow Fe^{3+} + e^- \qquad E^0 = -0.771\text{ V}$$

$$4Fe^{2+} + O_2 + 4H^+ \rightarrow 4Fe^{3+} + 2H_2O \qquad E^0 = +0.458\text{ V}$$

$$E^{0'} = \begin{cases} E^0 - 0.059\,\text{pH} & \text{at } 25°\text{C} \\ E^0 - 0.061\,\text{pH} & \text{at } 37°\text{C} \end{cases}$$

TABLE 5-2 Inorganic Speciation of Some Trace Metals in Marine Systems at 25°C, 1 atm Pressure, 3.5% salinity, and pH 8

Metal → Main species
$Al^{3+} \rightarrow Al(OH)_3$ 100%
$Cr^{3+} \rightarrow Cr(OH)_3$ 100%
$Fe^{3+} \rightarrow Fe(OH)_3$ 100%
$Mo^{4+} \rightarrow MoO_3{}^{2-}$ 100%
$Cd^{2+} \rightarrow Cd^{2+}$ 3%, $CdCl^+$ 97%
$Cu^{2+} \rightarrow Cu^{2+}$ 9%, $CuCO_3$ 79%, $Cu(OH)^+$ 3%, $CuSO_4$ 1%
$Mn^{2+} \rightarrow Mn^{2+}$ 58%, $MnCl^+$ 37%, $MnSO_4$ 3%, $MnCO_2$ 1%
$Fe^{2+} \rightarrow Fe^{2+}$ 69%, $FeCl^+$ 20%, $FeCO_3$ 5%, $FeSO_4$ 4%, $Fe(OH)^+$ 1%
$Ca^{2+} \rightarrow Ca^{2+}$ 58%, $CaCl^+$ 30%, $CaCO_3$ 5%, $CaSO_4$ 5%, $Ca(OH)^+$ 1%
$Ni^{2+} \rightarrow Ni^{2+}$ 47%, $NiCl^+$ 34%, $NiCO_3$ 5%, $NiSO_4$ 4%, $Ni(OH)^+$ 1%
$Pb^{2+} \rightarrow Pb^{2+}$ 3%, $PbCl^+$ 47%, $PbCO_3$ 41%, $Pb(OH)^+$ 9%, $PbSO_4$ 1%
$Zn^{2+} \rightarrow Zn^{2+}$ 48%, $ZnCl^+$ 35%, $Zn(OH)^+$ 12%, $ZnSO_4$ 4%, $ZnCO_3$ 3%

Source: Data from Turner et al., 1981.

○ Therefore at pH 7 and 25°C, $E^{0'} = 0.045$ V,

$$\Delta G° = -nFE^{0'} = -RT \ln K$$
$$= -4 \times 96500 \times 0.045 = -8.3 \times 298 \times \ln K$$
$$K = 1.1 \times 10^3$$

and Fe^{+3} is the *dominant* oxidation state.

○ $Fe^{3+} + OH^- \rightarrow Fe(OH)_3 \downarrow$ (Table 5-2). But $Fe(OH)_3$ has $k_{sp} = 10^{-38}$. So, for pH 7,

$$K_{sp} = [Fe^{3+}][OH^-]^3$$
$$K_{sp} = [Fe^{3+}][10^{-7}]^3 \qquad [Fe^{+3}] = 10^{-17} \text{ m/L}$$

- Therefore, there is *very low bioavailability* despite the applicable abundance.
- The hydrolysis of Fe^{+3} yields polymer formation and finally precipitates; Fe^{3+} is insoluble at neutral pH in distal water. Only by suitable ligands is it possible to sustain soluble Fe^{3+} in neutral and basic solutions. Suitable ligands are EDTA, $N(CH_2COOH)_3$ and citrate.
- Almost all microorganisms need iron for growth. In bacteria, iron is usually dissolved and transported by non-protein-chelating ligands, called siderophores.
- Plants can synthesize chelating agents to form soluble Fe through cell membranes.
- Iron may present in vivo as Fe^{2+} or Fe^{3+} or may be redox between the two.
- On the other hand, Fe^{2+} is soluble to 0.1 M at pH 7, but Fe^{2+} is spontaneously oxidized to Fe^{3+} in air.
- Exercise: Given the above data and $K_{sp} = 10^{-14}$ for Fe^{+2}, work out $\Delta G^{o'}$ for $Fe^{+2} \rightarrow Fe^{+3}$ at pH 7, $pO_2 = 0.2$ atm at 25°C.
- Note the similar nature of Al^{+3}, which is toxic, so the ability to select Fe^{+3} over Al^{+3} is important. The pO_2 and pH, for example, are not always applied.
- The ancient atmosphere (prebiotic) had little or no O_2 but had much CO_2, NH_3, and H_2S, which favor a low oxidation state (O_2 favors a high oxidation state).
- M^{n+} will more likely precipitate as sulfide than hydroxide, so bioavailability depended on the k_{sp} of sulfide, rather than that of hydroxide (Fig. 5-16).
  ○ Cu^{2+}/Cu^+: great disadvantage in bioavailability
  ○ Fe, Ni: average disadvantage
  ○ Mn, Mg: no disadvantage, used in photosynthesis
- The k_{sp} of sulfide may be the basis of a good deal of original selection of elements.

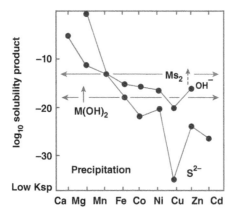

FIGURE 5-16 Solubility products of hydroxide and sulfide. Below the horizontal lines are indicated the solubility products, which give precipitate at pH 7 when $[M] = 0.1$ mM; the arrows indicate the direction of the increase in solubility. (Data from Fraústo da Silva and Williams, 2004.)

- S^{-2} (even H_2S in acid)

$$H_2S \leftarrow S^{2-} + 2H^+$$

$$\uparrow \text{at low pH}$$

  results in low availability for the heavy/soft/class B metal ions:
  - Mo, Cu^+, Ag^+, Cd^{+2}, Sn^{+2}, Pb^{+2}, As^{+3}, Sb^{+3}, Bi^{+3}, plus rare transition metals
  - Then in the great photosynthetic revolution, e.g.,

$$Cu^+ + (O_2 \text{ or } S^{-2}) \rightarrow Cu^{+2} \text{ (aq)}$$

  becomes available as does Fe^{+3}
- Some metal ions form solute thioanions:

$$Mo^{+4} + 2S^{-2} \rightarrow MoS_2 \downarrow$$

$$MoS_2 + S^{-2} \rightarrow MoS_4{}^{-4}$$

$$(MoS_4{}^{-4}, WS_4{}^{-n}, VS_4{}^{-n})$$

- However, Zn^{+2}, Mn^{+2}, Co^{+2} (Ni^{+2}) sulfides may be much less bioavailable (Fig. 5-16).
- All of these elements are appreciably abundant in seawater and are abundant in the human body (Figs. 5-15 and 5-17).
- The elemental composition of humans replicates that of the surroundings. This shows that living organisms are open systems that closely connect with their environments.
- The physiology of advanced organisms greatly counts on the essential elements Fe, Cu, P, Zn, etc. (Fig. 5-17).

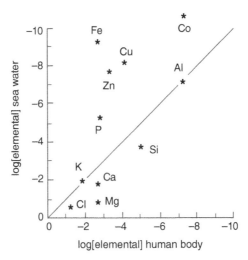

FIGURE 5-17 Elemental composition of humans and surroundings. (Data from Fraústo da Silva and Williams, 2004.)

SIDEROPHORES

Role and Chemical Features

Give a definition for siderophores, and explain their biochemical roles.

- Almost all microorganisms need iron for growth. In bacteria, iron is usually dissolved and transported by non-protein-chelating ligands, which are called siderophores.
- *Siderophores* are iron-carrying ligands (phytosiderophores in plants, or plant iron carriers).
- Siderophores are low-molecular-weight compounds synthesized by microbes, which are involved in cellular iron transport.
- Due to low $[Fe^{3+}]$ availability (10^{-17} m) and the hydrophobic nature of cell wall, Fe^{3+} cannot diffuse through the cell wall membrane (Fig. 5-18). Bacteria and fungi have Fe ionophores for chelating and transport of Fe^{3+} and probably partial modes for absorption of Fe by ingestion (mammals).
- Siderophores, or iron-carrying ligands:
  - Stabilize and capture Fe in the environment
  - Transport Fe through the cell wall
  - Release Fe for metabolic use
- Siderophores must be strong chelating agent competes with insolubility of $Fe(OH)_3$ and with the environment inside the organism.

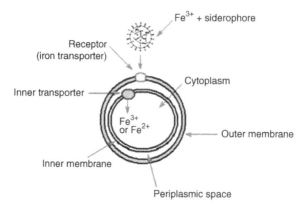

FIGURE 5-18   Iron uptake into cell.

Classes and Examples

What are the main classes of siderophores? Give examples for each class.

1. Hydroxamates class
   (a) Ferrichrome family
   (b) Fluopsin/thioformin family
   (c) Ferrioxamine family
   (d) Rhodotorulic acid family
   (e) Aerobactin family
   (f) Mycobactin family
   (g) Fusarinine family
2. Phenolate (α-dihydroxyaromatic) class
   (a) 2,3-Dihydroxybenzoylglycine (itoic acid)
   (b) 2,3-Dihydroxy-N-benzoyl-L-serine
   (c) 2-N,6-N-di(2,3-dihydroxybenzoyl)-L-lysine
   (d) Enterobactin
3. α-Hydroxycarboxylate class
   (a) Rhizoferrin
   (b) Mugineic acid
   (c) Pseudobactin (contains three types of binding)

1. *Hydroxamates*
   (a) Ferrichome family
      Classify, and discuss the main features and the coordination sites of ferrichromes. What is the proposed mechanism for Fe^{3+}-transfer into the cell?

TABLE 5-3   Members of Ferrichrome family

Ferrichrome	$R = CH_3$, $R_1 = R_2 = R_3 = H$
Ferrichrysin	$R = CH_3$, $R_1 = R_2 = CH_2OH$, $R_3 = H$
Ferricrocin	$R = CH_3$, $R_1 = H$, $R_2 = CH_2OH$, $R_3 = H$
Ferrichrome C	$R = CH_3$, $R_1 = H$, $R_2 = CH_3$, $R_3 = H$
Ferrichrome A	$R = -CH{=}C(CH_3)-CH_2-COOH$ (trans), $R_1 = R_2 = CH_2OH$, $R_3 = H$
Ferrirhodin	$R = -CH{=}C(CH_3)-CH_2-CH_2-COOH$ (cis), $R_1 = R_2 = CH_2OH$, $R_3 = H$
Ferrirubin	$R = -CH{=}C(CH_3)-CH_2-CH_2-COOH$ (trans), $R_1 = R_2 = CH_2OH$, $R_3 = H$
Albomyci δ_1	$R = CH_3$, 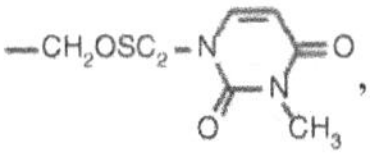 , $R_2 = CH_2-OH$, $R_3 = CH_2OH$

Source: Data from Raymond (1977).

- Members of the ferrichrome family (Table 5-3) are commonly produced by:
  - Ascomycetes (fungi)
  - Basidiomycetes (fungi)
  - Fungi *Imperfecti* (*Penicillium*)
  - Yeast (*Crypotococcus melibiosus*)
  - *Ustilago sphacrogena* (smut fungus)

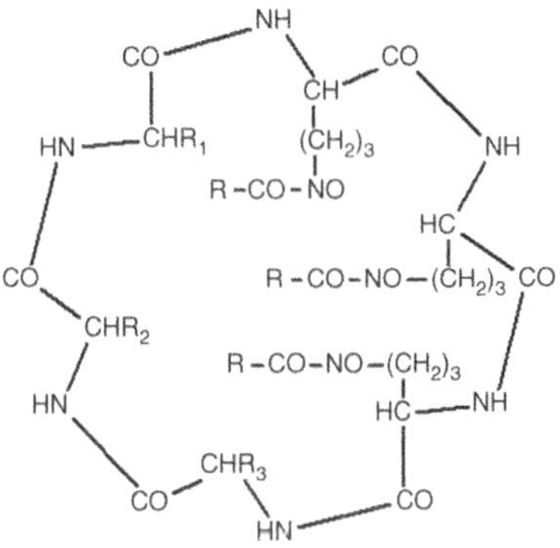

Free ferrichrome

- The ferrichrome family belongs to the hydroxamate class:

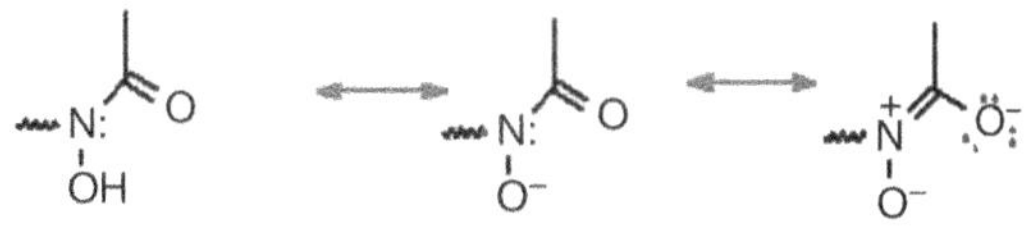

Hydroxamic acid

- Ferrichrome is a macrocyclic molecule with dangling ligands (Table 5-3).
- Fe^{3+} is hard, class A and so uses anionic O donors, and forms high-spin Fe^{III}-tris-hydroxamate, $[Fe(Ferr)]^0$, $K_f^{III} = 10^{29}$:

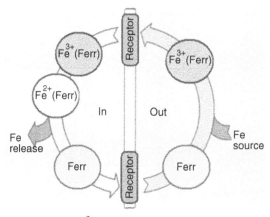

- The siderophores are too bulky to pass through the water-filled channels of the outer membrane so they are recognized there by receptor proteins and are moved across the inner membrane to the cytoplasm by an ATP-driven reaction.
- *Escherichia coli* picks ferrichrome, Fe^{3+}(Ferr), through its outer membrane by an active iron transporter protein called Fhua A. Ferrichrome is then transported into cytoplasm by Fhua B.
- Inside the cell, Fe^{3+}(Ferr) is reduced enzymatically to Fe^{2+}(Ferr), which is known to dissociate [to Fe^{2+} + free(Ferr)] much more easily than does the oxidized form:

$$Fe^{3+}(Ferr) + e^- \leftrightarrows Fe^{2+}(Ferr)^- \qquad K_f^{II} = 10^8 \qquad E^0 = -0.45 \text{ V}$$

So Fe is released intracellularly:

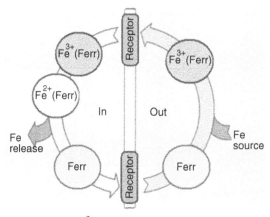

- The properties of Fe^{3+}(Ferr) are compatible with an oxido reduction mechanism for metal transport.
- When ligand Ferr is generated with C^{14}, Fe^{3+}(Ferr-C^{14}) is taken into cell, suggesting the above mechanism.
- $Cr(Ferr)^0$, which cannot release $Cr^{2+/3+}$, can be used to show that C^{14} is not just of free Ferr.

(b) Fluopsin/thioformin family

Fluopsin is thiohydroxmic acid, which is producd by *Pseudomonas Fluorescens*, suggest the binding sites and the possible structure.

- *Pseudomonas fluorescens* has fluopsin/thioformin.
- *Pseudomonas*: rod-shaped aerobic bacteria commonly found in soil, water, and decaying matter, including some plants and animals.
- A thiohydroxamic acid forms tris(N-methyl formothiohydroxamato)iron (III), which is purple, high-spin Fe^{3+}, with $E_{1/2} \sim -0.31$ V. It is an anionic shift vs. the $[FeO_6]^0$ system (-0.45 V) because the S donor favors Fe^{3+} less than Fe^{2+}.

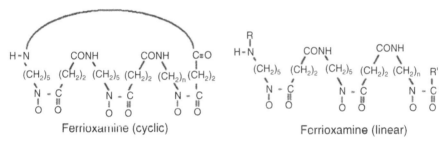

(c) Ferrioxamine family

Cyclic and acyclic free ferrioxamines, which are produced by *Arthobacter* and *Bacillus megaterium*, bind Fe^{3+} ion to permeate the cell membrane. What are the expected coordination features of the formed Fe^{3+}-ferrioxamine?

Ferrioxamine (cyclic)   Ferrioxamine (linear)

- A hydroxamate siderophores, ferrioxamines are cyclic and acyclic:

Ferrioxamine (cyclic)   Ferrioxamine (linear)

(d) Rhodotorulic acid family

Rhodotorulic and reto-rhodotorulic are siderophores produced by *Arthrobacter* **and** *Salmonella typhimurium*:

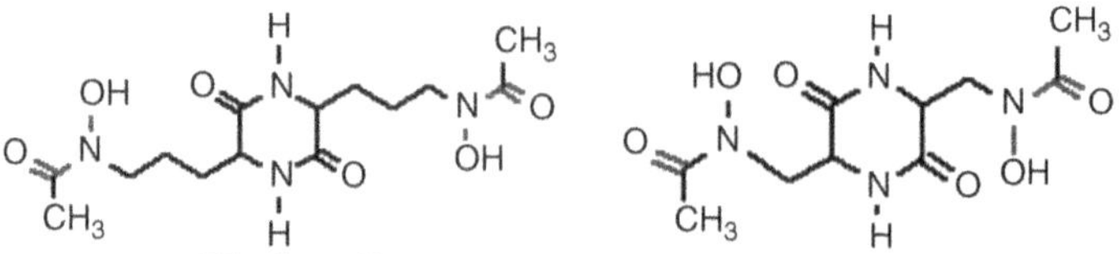

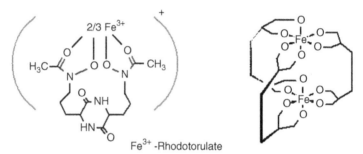

Rhodotorulic   Reto-rhodotorulic acids

Suggest the coordination sites and draw the mechanism used to permit Fe^{3+} **into the cell.**

- Rhodotorulic acid is a tetradentate-acyclic compound, RtH_2.
- To satisfy the six coordinate sites, $1\frac{1}{2}\,Rt^{2-}$ (charge also) is needed to form $Fe_2\,Rt_3$:

Fe^{3+}-Rhodotorulate

Diketopiperazine acts as link between hydroxamate arms.

- Fe^{3+} and Ga^{3+} bind to RtH_2 and go into the cell, but Cr^{3+} cannot. Ga^{3+} and Fe^{3+}: $t_{1/2} \sim 1$ s, Cr^{3+}: $t_{1/2} \sim 10^4$ s, typical of Cr^{3+} slow kinetics—kinetically inert. This excludes the possibility of the catalytic Fe^{2+}/Fe^{3+} mechanism.
- *Rhodotorula pilimanae* binds Fe^{3+} ion and form $Fe_2\,Rt_3$. $Fe_2\,Rt_3$ delivers Fe^{3+} into the cell through its channel "taxi mechanism":

$$M_2Rt^3 + H^+ \rightarrow 2M^{3+} + RtH_2$$

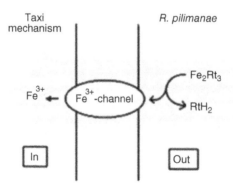

Coprogen

Discuss the source and formation of coprogen and draw the expected chemical structure.

- Dimerum + N-acetyl fusarinione + Fe^{3+} → coprogen

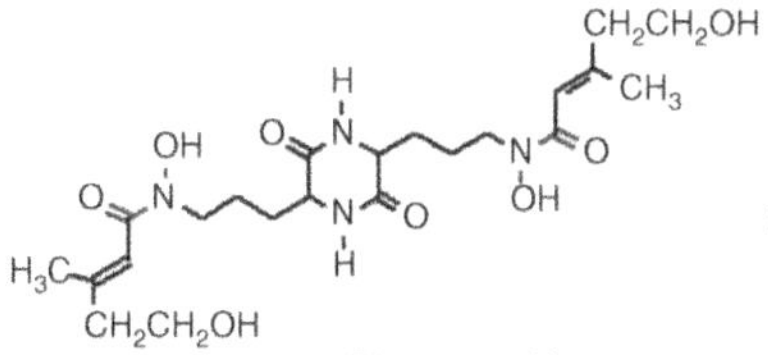

Dimerum acid

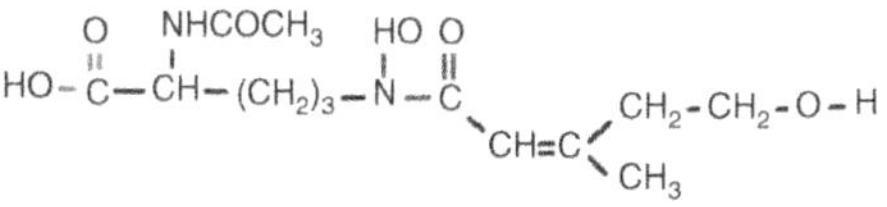

N-Acetyl Fusarinione

Dimerum acid is a diketopiperazine of the trans isomer of fusarinione.

- Source of coprogen: *Arthrobacter terregens, Arthrobacter, Microbacterium,* and *Pilobolus kleinii*
- Coprogen: Acyclic, Fe^{3+} complex of a molecule of dimerum acid esterifies on one side of the primary hydroxyl group by a residue of N-acetyl fusarinine.

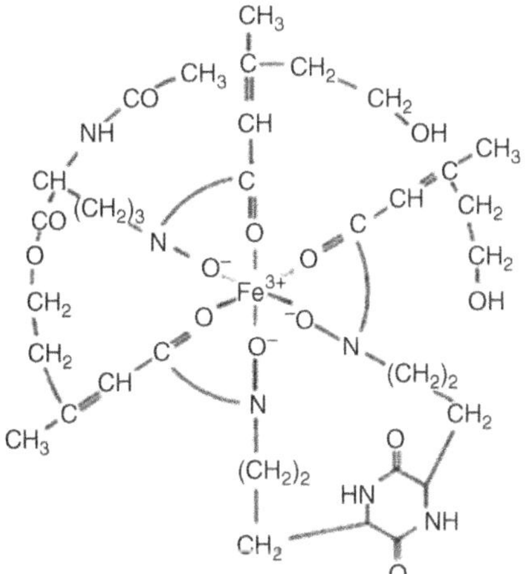

Coprogen

(e) Aerobactin family

The source of aerobactin or deferrischizokinen is *Salmonella typhimium*:

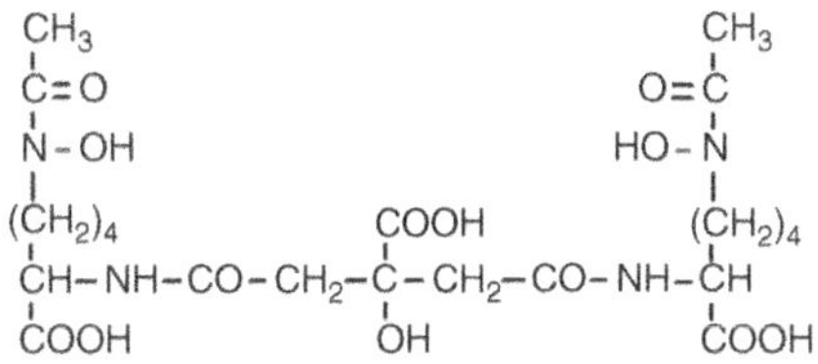

Aerobactin

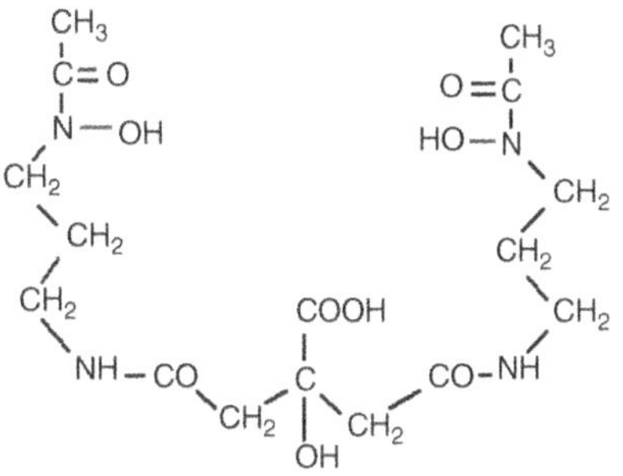

Deferrischizokinen

Suggest the coordination sites in Fe^{3+}-aerobactin or Fe^{3+}-deferrischizokinen:

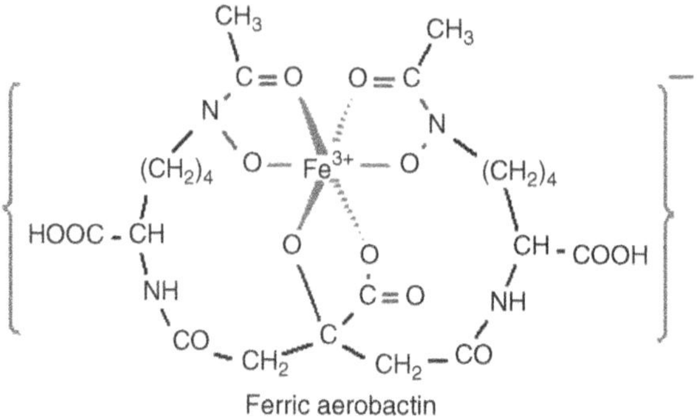

Ferric aerobactin

(f) Mycobactin family

$$Mycobactin + Fe^{3+} \rightarrow Fe^{3+}\text{-mycobactin}$$

Mycobactin is produced by *Arthrobacter terregens, Arthrobacter, Mycobacterium johnei, and B. megaterium.*

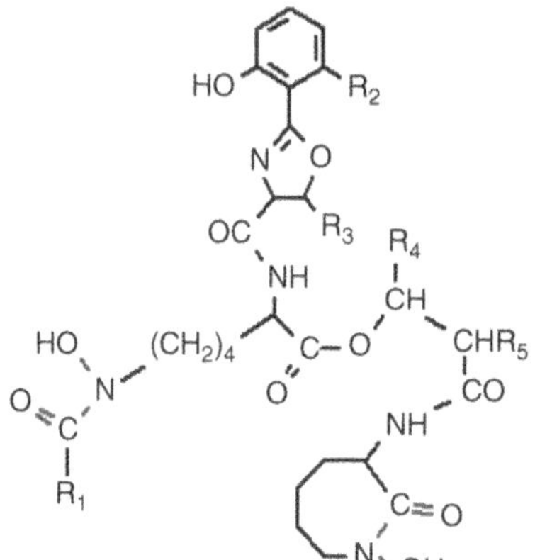

Mycobactin

Propose the expected coordination sites in Fe^{3+}-mycobactin

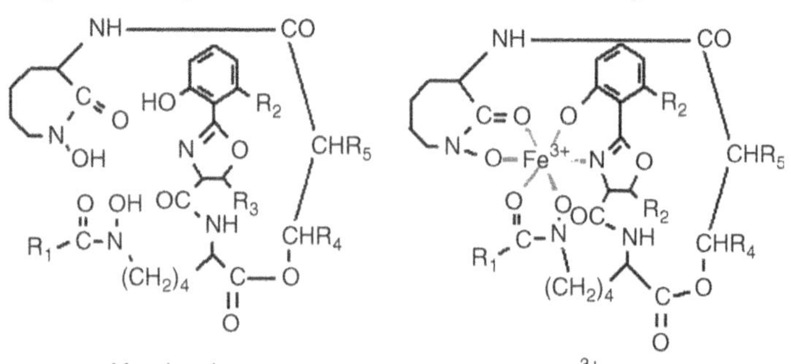

Mycobactin Fe^{3+}-mycobactin

(g) Fusarinine family

$$3 \text{ Fusarinine} \rightarrow \text{fusarinine C (cyclo)}$$

Fusarinine $C + Fe^{3+} \rightarrow$ ferrifusigen, which is produced by *Arthrobacter:*

Suggest the coordination sites in ferrifusigen:

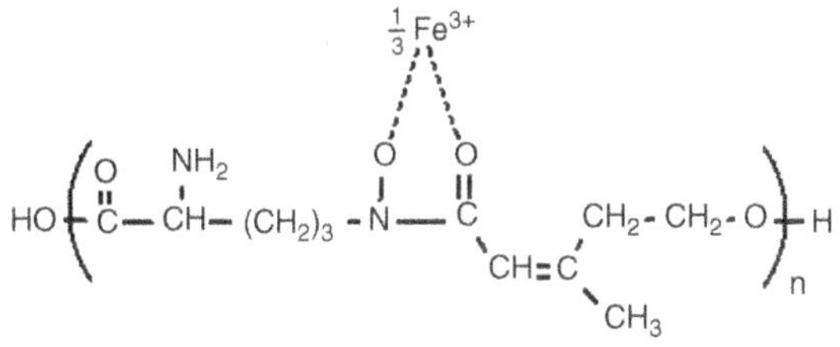

Fusarinines
Fusarinine: n = 1
Fusarinine B: n = 2
Fusarinine C (Deferrifusigen): n = 3, cyclo

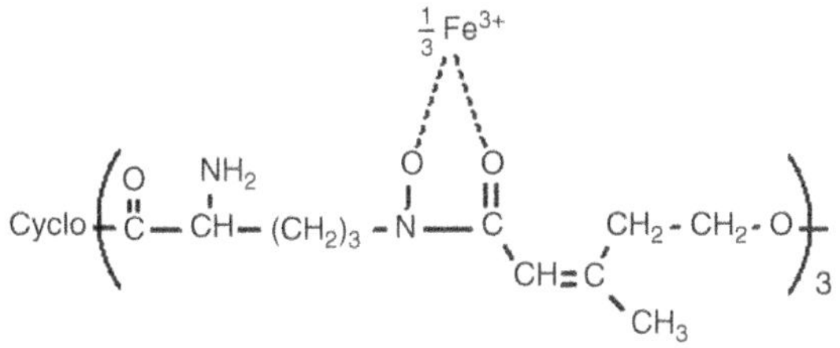

2. *Phenolates*

Name some naturally occurring phenolate siderophores.

(a) 2,3-Dihydroxybenzoylglycine (itoic acid) facilitates iron transport in *Bacillus subtilis*:

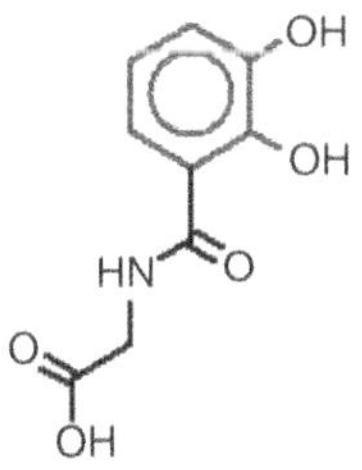

(b) 2,3-Dihydroxy-*N*-benzoyl-L-serine has been obtained from *Aerobacter aerogenes* and *S. typhimurium*:

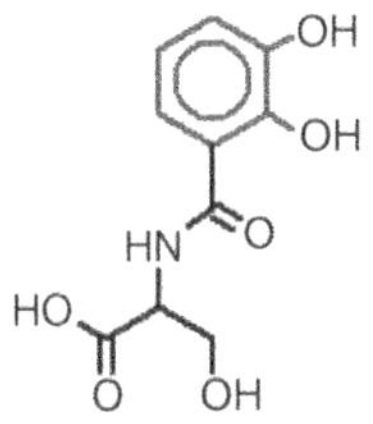

(c) 2-*N*,6-*N*-Di(2,3-dihydroxybenzoyl)-L-lysine is involved in iron transport in *Azotobacter vinelandii*:

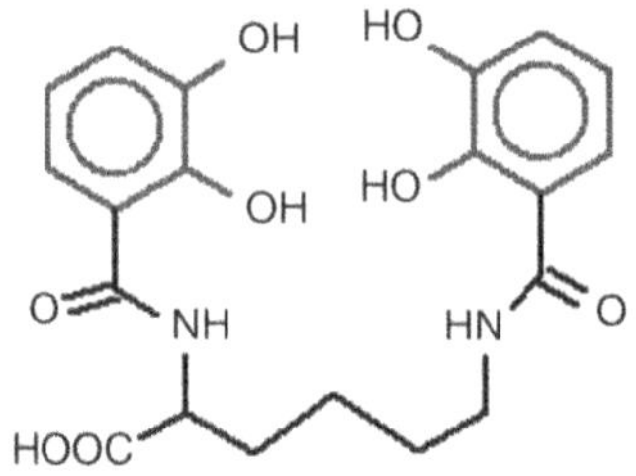

(d) Enterobactin

Three molecules of dihydroxybenzoserine can give the cyclic triester enterobactin (siderophores), which is produced by *S. typhimurium*:

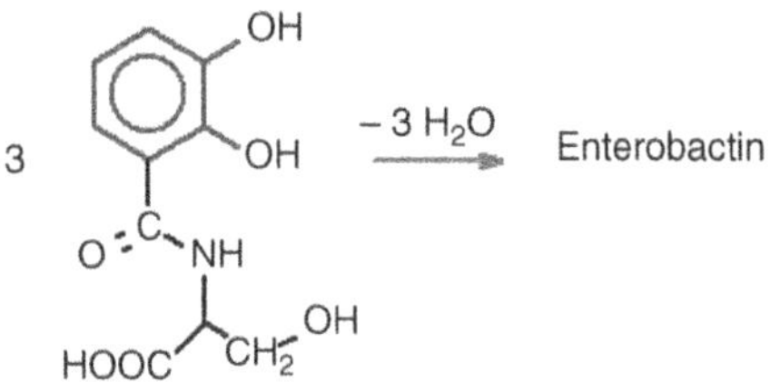

DHBS

Draw the chemical structure of enterobactin and Fe^{3+}-enterobactin. What is the proposed mechanism of the iron release and suggest some of the related siderophores.

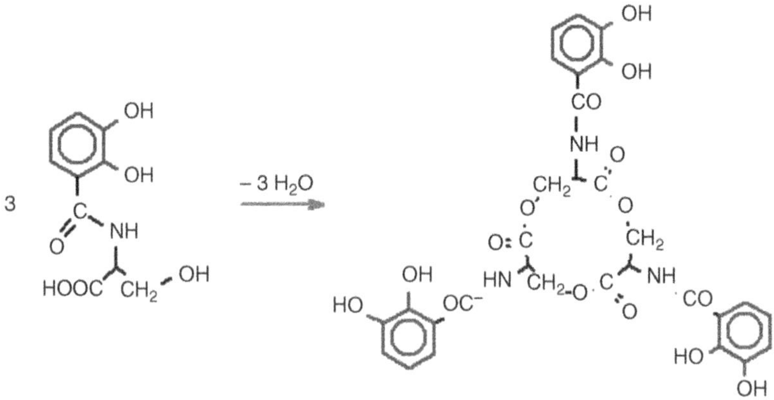

Enterobactin

From enteric bacteria, enterobactin is a cyclic triester.

- The thermodynamic stability of $Fe^{III}(Enb)]^{3-}$ is notable, $\log K = 52$ (Table 5-4).

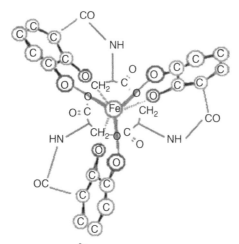

Fe^{3+}–Enterobactin

- Iron release: There are two possible routes:
  - (i) Redox route:

$$Fe^{III}(Enb)^{3-} + H^+ \rightarrow Fe^{III}(HEnb)^{2-} \quad E^0 = 750\,mV \quad \text{at pH 7}$$
$$Fe^{II}(Enb)^{4-} + H^+ \rightarrow Fe^{II}(HEnb)^{3-} \quad E^0 = 300\,mV \quad K = 10^{10}$$

The reduction potential for Fe^{3+}/Fe^{2+}–enterobactin complex is pH dependent, and release of Fe^{2+} is much easier when compared with Fe^{3+}, so release of Fe^{3+} by redox route is now possible.
  - (ii) Hydrolysis route: Hydrolysis of Enb in the metal complex will minimize $K_f^{III,II}$:

$$Fe^{3+}(Enb)^{3-} + H_2O \rightarrow Fe(DHBS)_3$$
$$Fe(DHBS)_3 \rightarrow Fe^{3+} + 3DHBS$$

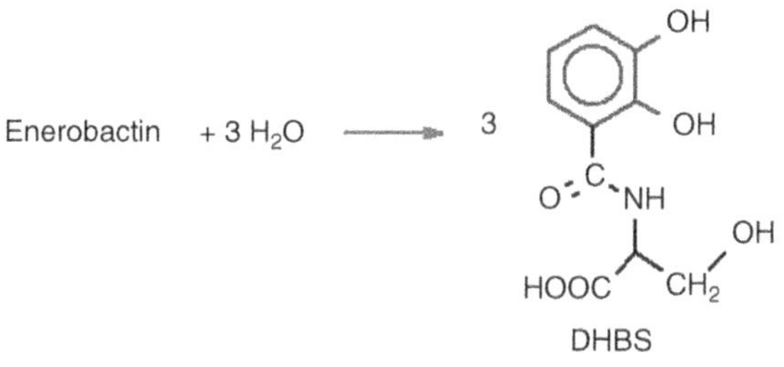

Enerobactin $+\ 3\,H_2O \longrightarrow$ 3

DHBS

- $Fe^{III}(Enb)^{3-}$ probably not 3-anion in vivo, $[Fe^{2+}(Enb)Ca_2]^0$ may be formed

- Model compound: $Fe(Cat)_3^{3-}$ with one of the 3 K^+:

[$Fe^{3+}(Cat)_3Ca]^-$ is a structural model to dihydroxybenzoglycine: from *A. vinelandii*.

Exercise:

The following table summarizes the stability constants of natural occurring siderophores. Arrange the following siderophores so that the top extracts the iron ion from the subsequent one.

Siderophores	log K	Siderophores	log K
Ferrioxamine	32.5	Coprogen	30.2
Ferrioxamine B	30.5	Ferricrocin	30.4
Ferrichrysin	30.3	Ferrichrome A	32.0
Ferrichrome	29.1	Actobactin	22.5
Rhodotorulic	31.2	Enterobactin	52

- Free enterbactin + Fe^{3+}-ferrioxamine → Fe^{3+}-enterbactin + free ferrioxamine

TABLE 5-4   Stability Constants of Naturally Occurring Siderophore Complexes

Siderophore	log K
Enterobactin	52
Ferrioxamine E	32.5
Ferrichrome A	32.0
Rhodotorulic	31.2
Ferrioxamine B	30.5
Ferricrocin	30.4
Ferrichrysin	30.3
Coprogen	30.2
Ferrichrome	29.1
Actobactin	22.5

3. *α-Hydroxycarboxylates*
    (a) Rhizoferrin (bacterial and fungal)
    (b) Mugineic acid (phytosiderophore)
    (c) Pseudobactin (from rhizobacteria)

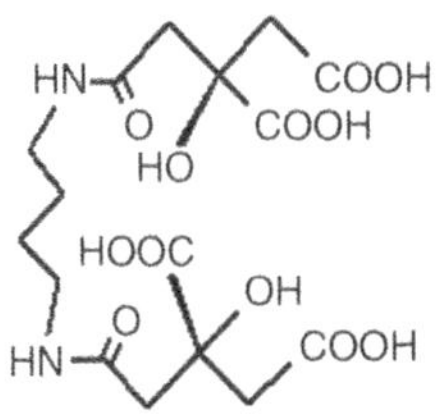

HN—

COOH
COOH
HO

HOOC

OH

HN— COOH

Bacterial rhizoferrin

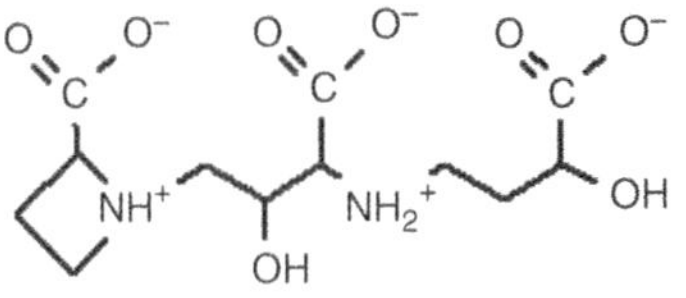

Mugineic acid

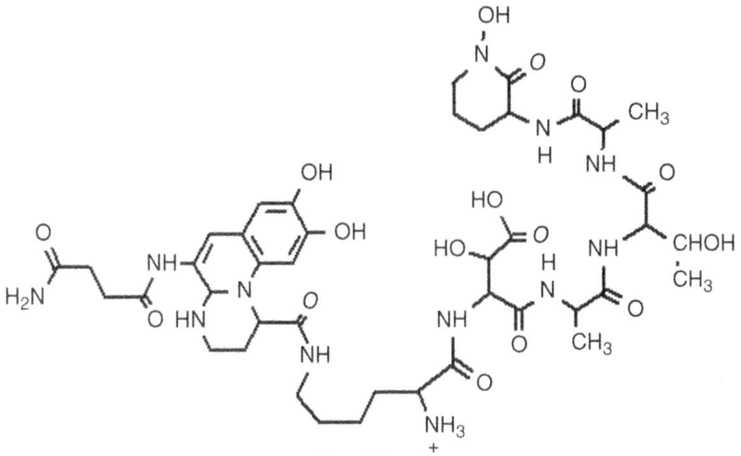

Pseudobactin

Suggest the coordination sites in rhizoferrin, mugineic acid, and pseudobactin.

- ○ Rhizoferrin has two OHs and four carboxylates that presumably coordinate to Fe^{3+}. Mugineic acid is produced by plants for iron metabolism.

- ○ Pseudobactin contains hydroxymate, o-dihydroxy aromatic, and α-hydroxylcarboxylic—all three types of chelating centers:

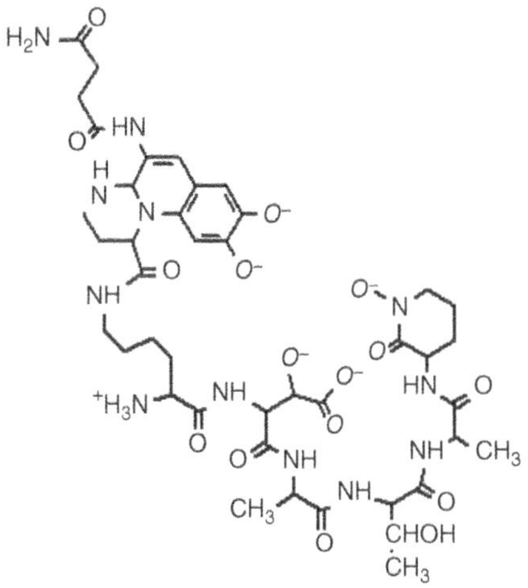

o This siderophore is a linear hexapeptide with chelating side groups likely to form an O_6 octahedron around Fe(III).

o Pyochelin, mycobactin, enterobactin, pseudobactin, catecholates, etc., are O donors, and form high-spin Fe^{3+} complexes.

Synthetic Siderophores

Classify the following synthetic siderophores and find the coordination sites

- Synthetic MECAM, CYCAM uptake through membrane:

MECAM   3,3,4-CYCAM

- MECAM and 3,3,4-CYCAM are dihydroxy aromatic, the six OHs presumably coordinating to Fe^{3+}.

IRON STORAGE AND TRANSFER PROTEINS

Iron Storage Proteins

How is iron stored in mammals? Identify and characterize a common form of iron–storage protein.

- Fe^{3+} increase leads to an increase in the probability of infection.
- Fe overload from transfusion is treated with desferrioxamine-B (Desferal).
- The synthetic siderophores MECAM and CYCAM are used for Fe removal.
- For Fe^{3+}, $q/r = 4.6\,e/Å$, similar to Pu^{4+}, $q/r = 4.2\,e/Å$. The bioavailability and mobility of Pu^{4+} and the formation of soluble Pu^{4+}(sidophore) complex cannot be excluded. Therefore removal is needed; otherwise Pu^{4+} is incorporated into the liver. Under reducing conditions, stable Pu^{3+}(siderophore) complexes are unlikely to persist.
- In mammals iron is stored in the liver in two forms: *ferritin* and *hemosidrin*
- Hemosidrin is not well understood and is probably the product of the breakdown of ferritin.

FERRITIN

- Ferritin (Ft) is an iron storage protein.
- It is found in most organisms.

- In mammals the primary residence is the spleen.
- Ferritin is a water-soluble protein.
- The 1° structure of Ft from horse, human liver, human spleen, rat liver, and chicken blood has a similar sequence (80–90%).
- Apoferrtin has 174 peptides, molecular weight ~20,000 Da, and is considerably helicity elongated and globular.
- Ferritin is made up of 24 subunits of H and L types.
- The 24 subunits are arranged in a hollow spherical shell of external diameter about 124 Å and inner cavity of around 70–80 Å in diameter (Fig. 5-19).
- The protein shell accommodates the iron core. This inner cavity is filled with an inorganic micellar core (aggregate of molecules colliding) made up of an oxohydroxophosphato-iron (III) complex of composition $(FeO \cdot OH)_8(FeO.OPO_3H_2)$ (Fig. 5-20). The iron is between two close-packed layers of oxygen (strip or sheet form).

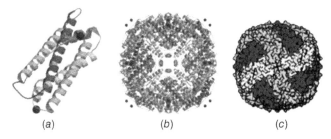

(a)   (b)   (c)

FIGURE 5-19 Crystal structure of human ferritin, PDB 1FHA (Lawson et al., 1991); (a) polypeptide subunit; (b, c) 24 subunits arranged in a hollow spherical shell.

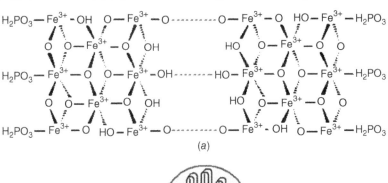

(a)

(b)

FIGURE 5-20 (a) Iron between two close-packed layers of oxygen (strip form), 9 Fe/P; (b) Folding of strip into micelle. (Modified from Heald et al., 1979.)

- The sheet is terminated by binding the phosphate groups, giving a width of 60 Å.
- The length of the strip depends on the amount of iron in the micelle. The micelle may contain up to 4500 iron atoms.
- The octahedral coordination is established by Mössbauer spectroscopy and extended X-ray fine structure (EXAFS).
- X-ray crystallography indicates mononuclear, dinuclear, and trinuclear species of Fe bound to the polypeptide. The degree of polymerization enhances as iron ions move to the center.

Binding and Release of Iron Ions

In iron–storage proteins, what is the oxidation state of iron, and explain the binding and the release of iron?

- The protein shell has six channels 10 Å in diameter (Fig. 5-21) that occur at the trimer interfaces.
- These channels are used to utilize the uptake and release of iron:

$$Fe^{3+} + Ft \rightarrow FtFe^{3+} \qquad K_f = 10^4$$

where K_f is misleadingly low because simple Fe^{3+} is not true factor bound.
- The iron in the ferritin is either Fe^{3+} or Fe^{2+}, but it is stored as Fe^{3+}–oxide/hydroxide. Therefore Fe^{2+} needs to be oxidized:

$$2\,Fe^{2+} + O_2 \xrightarrow{\text{H-ferroxidase}} Fe^{3+}-O-O-Fe^{3+}$$

$$Fe^{3+}-O-O-Fe^{3+} + 2\,Fe^{2+} \rightarrow 4\,Fe^{3+} + 2\,H_2O$$

- Fe^{3+} is stored in Ft and is released when needed, maybe by reductive chelation (flavoprotein):

$$FeO \cdot OH + 2H^+ + e^- \rightarrow Fe^{2+} + OH^- + H_2O$$

Reduction may subsequently occur as Fe is now transported as transferrin-Fe.

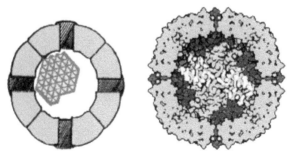

FIGURE 5-21 Protein channels connecting inner core to outer environment, PDB 1FHA (Lawson et al., 1991.)

TRANSFERRIN

Iron Transport Proteins

How can iron be transported in mammals? Identify and characterize a common form of iron–transport protein.

- Transferrin: Fe transport and storage
- A class of iron binding molecules that includes:
  - Transferrin (T_f) in serum
  - Lactoferrin (L_f) in milk
  - Ovotransferrin (conalbumin) in egg white
- Serum transferrin is the transport protein that carries iron from the breakdown sites of hemoglobin in the cell of the spleen and liver back to the site in bone marrow for the synthesis of hemoglobin.
- Serum transferrin accounts for ~4 mg of body iron in humans and delivers ~40 mg of body iron in humans.
- Humans with a genetic disability to synthesize transferrin suffer from both iron deficiency anemia and toxic effects of iron overload.
- Transferrin is glycoprotein, mucoprotein, and protein–polysaccharide compound, molecular weight ~79,000 Da.
- Examples of glycoprotein linkages:

Linkage of *N*-acetylglucosamine   Linkage of *N*-acetylgalactosamine
      to asparagine   to threonine

- Consists of a single-polypeptide chain of 168 amino acids, which is folded into two compact regions (two lobes). Each lobe consists of two similar domains N_1, N_2 and C_1, C_2. In lactoferrin (L_f, Fig. 5-22):

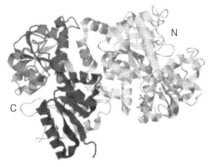

FIGURE 5-22   Structure of human diferric lactoferrin refined at 2.2 Å resolution, PDB 1LFG (Haridas et al., 1995).

- ○ N-terminal half is at the top.
- ○ C-terminal half is at the bottom.
- ○ Each lobe binds one Fe^{3+}.
- ○ Number of disulfide bridge: 6 in the N-lobe and 10 in the C-lobe, promoting pH-induced structural alternations.
- ○ There are two glycosylation sites, one in C-lobe and one in N-lobe.

Conditional Binding and Structural Analysis

What are the conditions of iron–binding? Identify the metal–binding sites of transferrin.

- The coordination environments were first revealed by various spectroscopics and confirmed by crystal structure analysis. The structure about Fe^{3+} is a 5-coordinate tetragonal pyramid (Fig. 5-23).
- The environment involves two phenolate oxygens from tyrosine, an oxygen from the *synergistic*, carbonate anion, nitrogen from histidine, and oxygen from a carboxylate group of aspartate.
- Binding synergism
  - ○ K-values depend not only on pH but also on $[Cl^-]$ and [carboxylate].
  - ○ Iron binding is only possible when an anion is also bound. In the absence of an appropriate anion, iron is not bound at all.
  - ○ The naturally occurring anion carbonate $[CO_3^{2-}/HCO_3]$ as well as other anions such as oxalate, salicylate, malonate, citrate, glyconate, pyrovate, nitrilotriacetate, and EDTA (N-lobe only) will also activate the metal binding site.

$$2 \, Fe^{3+} + 2 \, CO_3^{2-} + L_F \leftrightarrows L_F(FeCO_3)_2$$

  where K_N, $K_C \sim 10^{21}$ and CO_3^{2-} enhances the binding of Fe^{3+} by $>10^3$.

  - ○ Binding *synergism* between CO_3^{2-} and Fe^{3+} is due to the simultaneous (in space) binding to each other as well as to the protein.
- Metal binding sites
  - ○ Tyrosyl residue

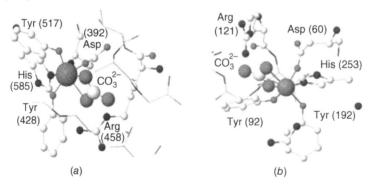

(a)   (b)

FIGURE 5-23 (*a*) C-terminal of human serum diferric transferrin, PDB 3QYT (Yang et al., 2012). (*b*) N-terminal of human lactoferrin, PDB 1LFG (Haridas et al., 1995).

(a) Adding Fe^{3+} to apoprotein over a range of pH values indicated that the pK_a of the metal binding group was near 11.2, suggestive of tyrosyl residues [phenolic OH glycoprotein (gp)] (Fig. 5-23).

(b) Chemical modification of tyrosyl leads to loss of metal binding properties.

(c) Comparison of NMR spectra of the iron and gallium complexes of transferrin shows broadening of the tyrosyl proton resonances in the proximity of paramagnetic Fe^{3+}.

(d) Binding Tb^{3+} to transferrin is associated with an increase in the intensity of fluorescence of Tb^{3+} by a factor of 10^5. This was attributed to energy transfer from a tyrosyl ligand, and the increase suggests two tyrosyl residues.

○ Histidyl residue

(a) The involvement of histityl residues in metal binding is supported by resonance Raman spectra.

(b) The ligand hyperfine structure in ESR spectra of Cu^{2+}-transferrin is consistent with a nitrogen donor group.

(c) Chemical modification of histidine residues confirms their role in the active site (Fig. 5-23).

Mechanism for Iron Delivery

Characterize the delivery of iron via transferrin. What is the suggested oxidation state of iron during transfer through cells?

• Each domain is able to bind one Fe(III) ion (Fig. 5-24).

• The crystal structure of iron-free human serum transferrin provides insight into interlobe communication and receptor binding:

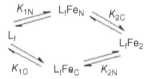

○ Serum transferrin reversibly binds Fe^{3+} ion in each of the two lobes.

(a) The constants K_{1N}, K_{1C}, K_{2N}, K_{2C} are dependent on temperature, pH, ionic strength, and concentration of anions.

(b) K_1, $K_2 \sim 10^{18}$, K_N/K_C: $0.01 < \text{ratio} < 2$.

(c) K_N/K_C increases as pH changes from 6 to 9, mostly $K_N > K_C$.

○ The two Fe^{3+} of transferrin-complex are delivered to cells by a receptor-mediated process that is pH dependent.

○ The binding and release of iron are associated with a large conformational variation in which two subdomains of each lobe close or open with a rigid twisting movement around a hinge (Fig. 5-24).

• Delivery of iron to cells requires binding of two iron-containing human transferrin molecules to the specific *transferrin receptor* on the cell surface, forming a vesicle inside the cell (endosome) (Fig. 5-25).

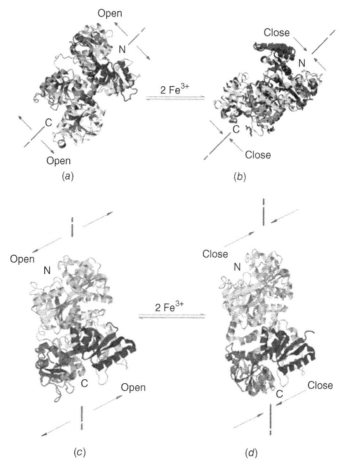

FIGURE 5-24 (*a*) Apo-human serum transferrin, PDB 2HAV (Wally et al., 2006). (*b*) Diferric bound human serum, PDB 3QYT (Yang et al., 2012). (*c*) Human apolactoferrin, PDB 1CB6 (Jameson et al., 1998). (*d*) Human diferric lactoferrin, PDB 1LFG (Haridas et al., 1995). (See the color version of this figure in Color Plates section.)

- Binding of human transferrin induces global conformational changes within the transferrin receptor.
- The transferrin receptor undergoes pH-induced movements that influence the stability of Fe^{3+}-transferrin interaction, and iron is released from human transferrin within the endosome.

Release Mechanism

Release and uptake of iron commonly involve Fe^{3+}/Fe^{2+} redox. How can the mode of ligation influence the Fe^{3+}/Fe^{2+} redox cycle? Use a model compound.

Does the protein play a special role in controlling $E^{0\prime}(Fe^{3+}/Fe^{2+})$? Or how can the binding of the phenolate (Tyr) affect $E^{0\prime}$?

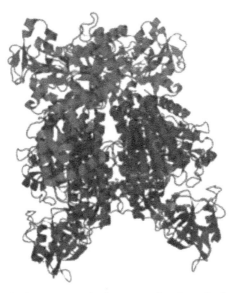

FIGURE 5-25 Complex between transferrin receptor 1 and transferrin with iron in N-Lobe at room temperature, PDB 3S9N (Eckenroth et al., 2011). (See the color version of this figure in Color Plates section.)

- Release and uptake may involve Fe^{3+}/Fe^{2+} redox.
- Consider the redox cycle:

$$L_f\text{-}Fe^{3+} + e^- \xrightarrow{\Delta G_1^\circ} L_f\text{-}Fe^{2+} \qquad E^{\circ\prime} < -400 \text{ mV}$$

$$\Big\downarrow \Delta G_2^\circ \qquad\qquad \Big\uparrow \Delta G_4^\circ \qquad\qquad \updownarrow$$

$$L_f + Fe^{3+} + e^- \xrightarrow{\Delta G_3^\circ} L_f + Fe^{2+} \qquad E^{\circ\prime} < +770 \text{ mV}$$

Let ΔG_1° and ΔG_3° be given by:

$$\Delta G^\circ = -nFE^{0\prime}, \text{ while}$$

$$\Delta G_2^\circ = +RT \ln K_{Fe}^{3+} \qquad K_{Fe}^{3+} \sim 10^{21}$$

$$\Delta G_4^\circ = -RT \ln K_{Fe}^{2+} \qquad K_{Fe}^{2+} < 10^2$$

For

$$Fe^{3+} \xrightarrow[K_1^{III}]{L} Fe^{3+}L \xrightarrow[K_2^{III}]{L} Fe^{3+}L_2 \xrightarrow[K_3^{III}]{L} Fe^{3+}L_3 \xrightarrow{etc.} Fe^{3+}L_i$$

and

$$Fe^{2+} \xrightarrow[K_1^{II}]{L} Fe^{2+}L \xrightarrow[K_2^{II}]{L} Fe^{2+}L_2 \xrightarrow[K_3^{II}]{L} Fe^{2+}L_3 \xrightarrow{etc.} Fe^{2+}L_i$$

where K values are the stepwise formation constants, while β is the overall formation constant:

$$\beta_1 = K_1 \qquad \beta_2 = K_1 K_2 \qquad \beta_3 = K_1 K_2 K_3 \qquad \beta_4 = K_1 K_2 K_3 K_4$$

$$E_{Obs} = E^\circ - \frac{RT}{nF} \ln \left(\frac{1 + \sum \beta_i^{III}[L]^i}{1 + \sum \beta_i^{II}[L]^j} \right)$$

For lactoferrin K_1's only are important, and at standard condition of $[L]=1M$:

$$E_f = E^o - \frac{RT}{nF} \ln \left(\frac{1 + K_1^{III}}{1 + K_1^{II}} \right)$$

- When a ligand binds to a ferric ion stronger than it binds to a ferrous ion ($K_1^{III} > K_1^{II}$), the value of E_f drops ($\downarrow$) versus that of E°.
- This applies for L_f $Fe^{3+/2+}$. T_f and L_f are designed to complex Fe^{3+} strongly. Tyr-phenolate has a K advantage for Fe^{3+} (hard acid carrying a negative charge). That is, consider that phenolate lowers the E_f of $Fe^{3+/2+}$.

Spectral and Structural Analysis

The ESR spectrum of transferrin shows a strong line near 1500 G and a weaker line near 700 G at 77K with microwave frequency of 9.15 GHz, while the electronic spectrum exhibits a band at 465 nm ($\epsilon \sim 5000$) (Figs. 5–26 and 5–27). Show the implications of the mode of ligation on the ESR and the electronic spectra of transferrins. Use a model to support your answer.

- The structure about the Fe^{3+} of transferrin is a 5-coordinate tetragonal pyramid. The ferric ion is coordinated to CO_3^{2-}, two Tyr, one His, and one Asp (NO_5 coordination sphere $\rightarrow$ high-spin O_h Fe^{3+}).
- The ESR spectrum of human diferric transferrin (Fig. 5-26) shows a strong line near 1500 G ($g = 4.3$) that arises from the middle Kramer's doublet. The weaker line near 700 G ($g = 9.7$) represents a composite of transitions from the highest and lowest doublets.

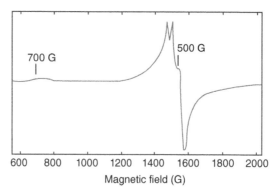

FIGURE 5-26 ESR spectrum of human diferric transferrin. (Brock, 1985. Reproduced by permission of Wiley.)

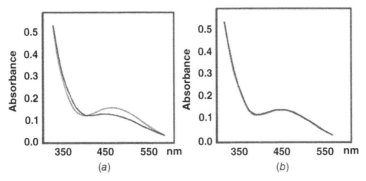

FIGURE 5-27 (*a*) Visible spectra of N-terminal (black) and C-terminal (gray) monoferric bovine transferrin. (*b*) Visible spectra of bovine diferric transferrin and an equimolar mixture of the N- and C-terminals. (Brock, 1985. Reproduced by permission of John Wiley & Sons.)

- In electronic spectra of bovine transferrin (Fig. 5-27):
  - Band at 465 nm, $\varepsilon \sim 5000$ is too strong to be d → d transition. This is a charge transfer (CT) band. d → d: $10^{-2} \le \varepsilon \le 5 \times 10^2 \, M^{-1} \, cm^{-1}$, but the nature of the 245-nm band is not clear.
  - To understand the implication of the ligand to metal charge transfer band, LMCT, consider MX_4^{2-} (T_d, $M = Co^{2+}$, Ni^{2+}, $X = Cl, Br, I$). Note lowest ν (longer λ) bands:

$$\begin{array}{llll}
 & Cl^- & Br^- & I^-: \quad \varepsilon = 1500 - 4500 \, M^{-1} \, cm^{-1} \\
Co^{2+} & 240 & 290 & 400 \quad nm \\
Ni^{2+} & 285 & 370 & 520 \quad nm
\end{array}$$

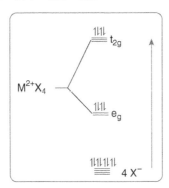

  - Consider Ni^{2+} vs. Br^-:
    - (a) Atomic orbitals of Br^- are at lower energy than those of Ni^{2+} due to contraction across the period.
    - (b) Transition involves e^- excitation from an orbital with mostly ligand character to an orbital with mostly metal character: LMCT.
    - (c) Ground state $= Ni^{2+} - Br^- \xrightarrow{\ h\nu\ }$ Excited state $= Ni^+ - Br$

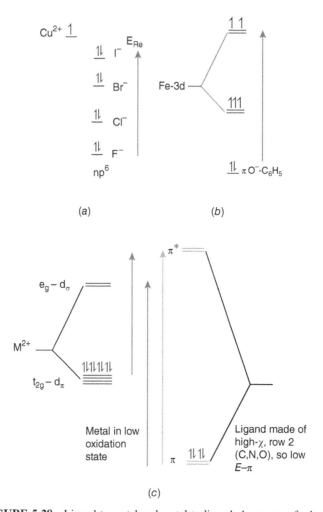

FIGURE 5-28 Ligand-to-metal and metal-to-ligand charge transfer band.

o This transition occurs at lower energy when:

(a) It is easier to remove e^- from the ligand, i.e., higher E ligand orbital.

(b) It is easier to oxidize the ligand (reducing agent).

(c) The ligand is of lower electronegativity. Therefore, λ-order $Cl^- < Br^- < I^-$. This leads to the *optical electronegativity scale* $\chi^L_{opt.}$:

L	F^-	OH_2	Cl^-	Br^-, CN^-, RS^-
χ^L_{opt}	3.9	3.5	3.0	2.8

(d) It is easier to place e^- on the metal.

(e) The lower-E metal orbital is empty.

So the λ-order is $Co^{2+} < Ni^{2+}$, which leads to the optical electronegativities, χ_{opt}^{m}:

M^{2+}	Zn^{2+}	Fe^{2+}, Co^{2+}	Ni^{2+}	Co^{3+}	$Cu^{2+}, HS\ Fe^{3+}$
χ_{opt}^{M}	1.2	1.8	2.1	2.3	2.4

o $\nu_{CT} = 30{,}000\ (\chi_{opt}^{L} - \chi_{opt}^{m})\ cm^{-1}$

(a) Biological ligands of low χ_{opt}^{L}: thiolate Cys, thioether RS^{-}, and phenolate Tyr

(b) Biometallic of high χ_{opt}^{m}: Cu^{2+}, Fe^{3+}, and Mn^{4+}

(c) λ_{CT}: $F^{-}-Cu^{2+} < Cl^{-}-Cu^{2+} < Br^{-}-Cu^{2+} < I^{-}-Cu^{2+}$ (Fig. 5-28a).

o $L_f\text{-}Fe_2$: Phenolate $\rightarrow$ Fe^{3+}, LMCT (Fig. 5-28b). CT is L $\rightarrow$ M when the L orbital has e^{-} and there is a vacancy (hole) in the M orbital at higher E. Also it is possible for the inverse of this, e.g., M (α-Diimine)$_3^{2+}$.

o Fe $(BiPy)_3^{2+}$, Ru $(Phen)_3^{2+}$ (Fig. 5-28c):

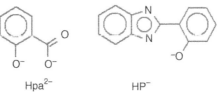

Ligand:	$\pi \rightarrow \pi$	(UV)	$\sim$250 nm
LMCT:	$\pi \rightarrow d$	near UV	$\sim$350 nm
MLCT:	$d \rightarrow \pi^{*}$	visible	$\sim$500 nm (overwhelms $\varepsilon_{d\text{-}d}$)

o The similarity between the spectrum of diferric transferrin with that of a model compound $[Fe^{3+}\ (Hpa)(Bp) \cdot 2H_2O]$ confirms $Fe^{3+}-NO_5$ moiety.

$$Fe^{3+}(Hpa)(Bp) \cdot 2H_2O \qquad E = -800\ mV \quad \text{Deep red}$$
$$\lambda = 500\ nm \qquad \varepsilon = 1300 \qquad Tyr^{-} \rightarrow Fe^{3+}$$

Hpa^{2-} HP^{-}

DIOXYGENASE IRON PROTEINS

Reactions of Oxygen Molecules

Summarize the reactions of oxygen molecules that are catalyzed by metal ion.

Reactions of oxygen may be classified in three general ways (Scheme 5-2).

- Insertion reaction
  - o One oxygen atom enters the substrate or
  - o two oxygen atoms enter the substrate.

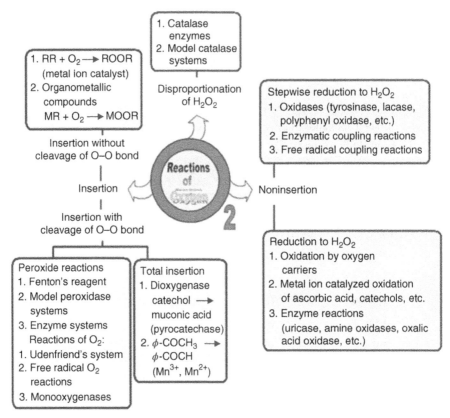

SCHEME 5-2   Three general behaviors of oxygen reactions. (Data from Martell and Taqui Khan, 1973. Reproduced by permission of Elsevier.)

- Oxidizing agent
  - Oxygen undergoes a four-electron reduction to water or
  - oxygen undergoes a two-electron reduction to hydrogen peroxide.

Organic Compounds and Oxygen Molecule

Why are direct reactions between organic compounds and oxygen molecules being unfavorable? How can these reactions be proceeded in biological systems?

- From the valance shell electron pair repulsion (VSEPR) theory, the direct reactions between the organic compounds and oxygen molecules are kinetically unfavorable because oxygen exists as a triplet ($\uparrow\uparrow$, $^3O_2$, the arrows represent electron spin) in the ground state and the carbon compounds exist as singlets ($\uparrow\downarrow$, $^1C_{carbon}$).
- Biological substrates and products resulting from the substrate oxidations usually have singlet ground states, containing no unpaired electrons, $^1C_{biological\ substrates}$:

$$\tfrac{1}{2}\,{}^3O_2 + {}^1C_{\text{biological substrate}} \longrightarrow {}^1CO_{\text{biological product}}$$

- It is impossible for this reaction to take place in one fast concerted step:

$$\tfrac{1}{2}\,{}^3O_2 + {}^1C_{\text{biological substrate}} \longrightarrow {}^3CO_{\text{biological product}}$$

$$^3CO_{\text{biological product}} \xrightarrow[\text{40 – 70 kcal/mol}]{\text{high activation barrier}} {}^1CO_{\text{biological product}}$$

or

$$^3O_2 + 22.5\ \text{kca/mol} \longrightarrow {}^1O_2$$

$$^1O_2 + {}^1C_{\text{biological substrate}} \longrightarrow {}^1CO_{\text{biological product}}$$

- This makes the reaction difficult without some kind of activation. In a biological system this difficulty is avoided through the enzymatic pathway by reacting dioxygen either with the unpaired d electrons of a transition metal ion complex or with stable organic free radicals.

Oxygen Fixation

What are the roles of dioxygenases in the biological systems?

- *Oxygenase*, such as catechol-dioxygenase, or pyrocatechase, catalyzes the insertion of oxygen in organic substrate (oxygen fixation). *Note:* both atoms of the oxygen molecule are incorporated directly into the reaction product (Scheme 5-3).

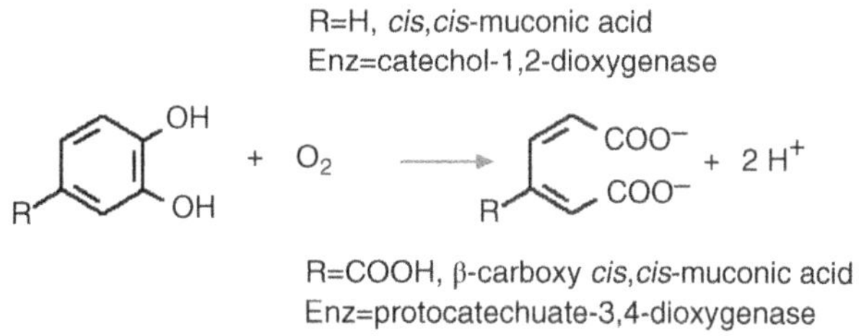

R=H, *cis,cis*-muconic acid
Enz=catechol-1,2-dioxygenase

R=COOH, β-carboxy *cis,cis*-muconic acid
Enz=protocatechuate-3,4-dioxygenase

SCHEME 5-3 O_2 insertion by "intradiol" mechanism.

- *Pyrocatechase* is an iron phenolate enzyme.
- Pyrocatechases catalyze the cleavage of catechol through insertion of dioxygen.
- There are two kinds of *catechol dioxygenases:*
  - Fe^{3+}-Intradiol O_2-ase
    - (a) Catechol-1, 2-dioxygenase (CTD)
    - (b) Protocatechuate-3,4-dioxygenase (metapyrocatechase) (PCD)
  - Fe^{2+}-Extradiol O_2-ase: Catechol-2,3-dioxygenase

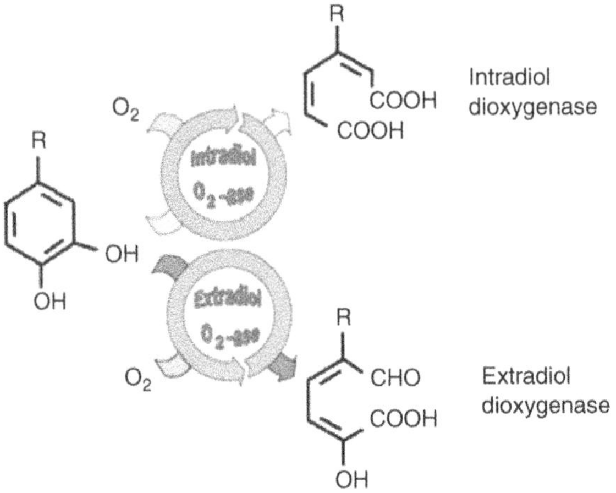

- Pyrocatechases play a crucial role in the degradation of natural products.
- Aerobic microorganisms have the ability to degrade stable organic compounds such as lignin, alkaloids, flavanoids, and terpenes.
- These enzymes are also used to accommodate the oxidation of industrial chemicals and pollutants.

Products of Enzymatic Oxygen Insertion

Write the possible products for the following enzymatic oxygen insertion:

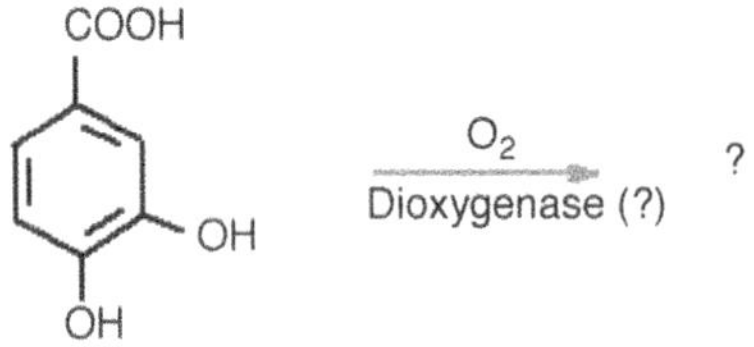

Specify the used enzyme and the source of each of these enzymes.

- The active sites of such enzymes are designed to yield specific products.
- Three enzymes are presented (Scheme 5-4).
  1. PCD 3,4-oxygenase in:
     Pseudomonas fluorescence

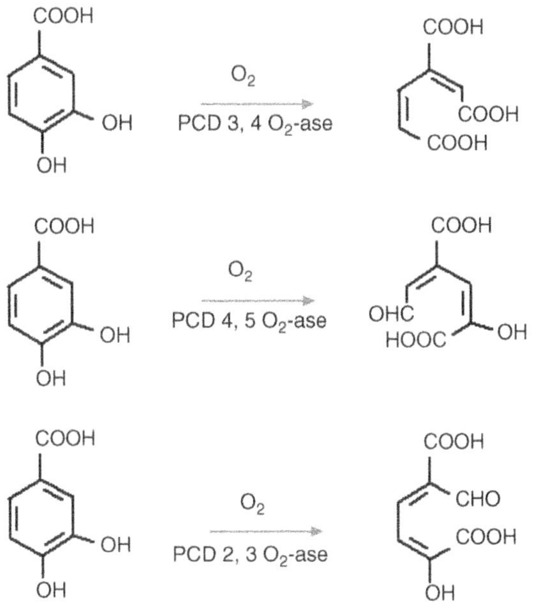

SCHEME 5-4 Dioxygenase reactions catalyzed by PCD 3,4-, 4,5-, and 2,3-oxygenases. (From Wood, 1980. Reproduced by permission of John Wiley & Sons.)

> *Pseudomonas aeruginosa*
>
> *Pseudomonas putida*
>
> *Pseudomonas arvilla*

- PCD 3,4-oxygenase extracted from the cell of *Pseudomonas* sp. (bacteria) when grown on (fed) *p*–hydroxybenzoate as the sole carbon and energy source.
- PCD 3,4-oxygenase is composed of nonidentical subunits α and β.
- The native enzyme dissociated into an $\alpha_2\beta_2$ structure and each $\alpha_2\beta_2$ contains only one iron active center.
- Units of $\alpha_2\beta_2$ are associated to form the crystalline Protein, $(\alpha\beta Fe^{3+})_{12}$.
- The structure of protocatechuate 3,4-dioxygenase from *P. aeruginosa* (now reclassified as *P. putida*) has been refined to an *R* factor of 0.172 to 2.15 Å resolution (Fig. 5-29). The structure is a highly symmetric $(\alpha\beta Fe^{3+})_{12}$ aggregate. Molecular weights: $\alpha = 23$ kDa, $\beta = 26$ kDa, 12 $\alpha\beta = 590$ kDa.

2. PCD 4,5-oxygenase is extradiol and is isolated from *Pseudomonas testosterone*. This enzyme catalyzes:

- The cleavage of the aromatic to give α-hydroxy-γ-carboxy-*cis*, *cis*-muconic semialdahyde, consuming one O_2 molecule.
- The cyclization of α-hydroxy-γ-carboxy-*cis*, *cis*-muconic semialdehyde with NH_4^+ ion to give 2,4-lutidinic acid (2,4-dicarboxlic pyridine), which helped establish that 4,5-fission of the aromatic ring of protocatechuic acid, PCA, had occurred.

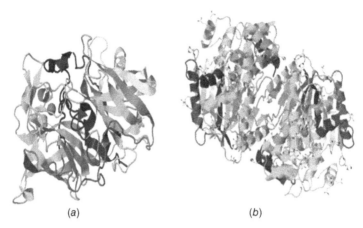

(a)   (b)

FIGURE 5-29 (*a*) Structure of protocatechuate 3,4-dioxygenase from *P. aeruginosa* at 2.15 Å resolution, PDB 2PCD (Ohlendorf et al., 1994). (*b*) Structure protocatechuate 4,5-dioxygenase complexed with 3,4-dihydroxybezoate, PDB 1B4U (Sugimoto et al., 1999).

3. PCD 2,3-oxygenase is extradiol and is isolated from *Bacillus macerans*.

Spectral Analysis

(a) *Pseudomonas arvilla* 3,4-CTD-ase is a pink enzyme and shows a strong band at ~ 450 nm ($\varepsilon = 3000$–$4000 \, M^{-1} \, cm^{-1}$), similar to the optical spectra of transferrins. Addition of substrate in the absence of oxygen shows an increase in the intensity at 480 nm. On admission of oxygen to the mixture of the substrate and the enzyme, a new band appears at 520 nm.

(b) Resonance Raman spectrum indicates signals at ~ 1170, 1270, 1500, and $1600 \, cm^{-1}$.

(c) The ESR spectra shows resonances at $g = 9.67$, 4.26. $^{17}OH_2$ causes broadening of ESR lines.

(d) The CN^- probe titration method shows Fe^{3+} goes to low-spin ESR and 2 CN^- binds to Fe^{3+}.

Use the above information to discuss the chemistry of protocatechuate-3,4-dioxygenase and write your conclusion.

• The similarity in the optical spectrum of PCD 3,4-oxygenase to that of various Fe^{3+}-tyrosinate proteins (e.g., transferrins) indicated that tyrosine may be coordinated to the active iron center; therefore, LMCT is explained as tyrosine $\rightarrow Fe^{3+}$.

• *Pseudomonas arvilla* CTD-ase is a pink enzyme, due to a strong band at ~ 450 nm ($\varepsilon = 3000$–$4000 \, M^{-1} \, cm^{-1}$, d $\rightarrow$ d: $10^{-2} < \varepsilon < 10^2 \, M^{-1} cm^{-1}$), which assigned to $PhO^- \rightarrow Fe^{3+}$, LMCT.

• Substrate binding also changes UV-Vis; substrate and oxygen bind to the iron center.

• This assignment is confirmed by examining the resonance Raman spectrum, which shows the characteristic tyrosine vibrational modes (~ 1170, 1270, 1500, and $1600\,cm^{-1}$). These bands have been assigned as a C–H bending vibration and a C–O and two C–C stretching vibrations of the phenolate ligand.

- The ESR spectra:
  - ○ Suggest there is only one type of Fe^{3+} center in the enzyme.
  - ○ Show resonances at $g = 9.67$, 4.26, characteristic of high-spin Fe^{3+}.
  - ○ Show a typical Fe^{3+}-phenolate protein.
  - ○ Suggest that $^{17}OH_2$ causes broadening of ESR lines due to the nonzero nuclear spin moment of ^{17}O. Conclude that there are coordination site(s) on Fe^{3+} that are accessible to H_2O.
- ESR titration method is used for counting "free" coordination sites; *free* means that the site is not occupied by amino acid side chains.
- The CN^- probe titration method shows Fe^{3+} goes to low-spin ESR. It no longer broadens when two CN^- bind to Fe^{3+}, implicating the replacement of $^{17}OH_2$.
- Therefore, there are four amino acid ligands and two H_2O ligands in the Fe^{3+} enzyme.

Structural and Chemical Characterization of Active Site

What do you conclude from Fig. 5-30, which represents the active site structure of protocatechuate 3,4-dioxygenase before and after the addition of oxygen and dihydroxybenzoate?

A view of the active site of the PCD from *P. aeruginosa* is given in Fig. 5-30.

- Tyr^{447} (Tyr^{147} beta) and His^{462} (His^{162} beta) are axial ligands of the trigonal bipyramid, whereas Tyr^{408} (Tyr^{108} beta) and His^{460} (His^{160} beta) form part of the equatorial plane.
- An equatorial coordination site is thus available for binding of solvent.

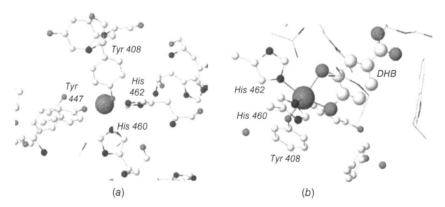

(a)   (b)

FIGURE 5-30   (*a*) Active site of protocatechuate 3,4-dioxygenase from *P. aeruginosa* at 2.15 Å resolution, PDB 2PCD (Ohlendorf et al., 1994). (*b*) Active site of protocatechuate 3,4-dioxygenase complexed with 3,4-dihydroxybezoate, PDB 3PCA (Orville et al., 1997).

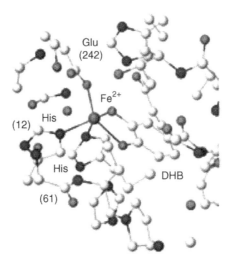

FIGURE 5-31   Structure of protocatechuate 4,5-dioxygenase complexed with 3,4-dihydroxy-benzoate, PDB 1B4U (Sugimoto et al., 1999).

- Substrate complex formation is associated with a dissociation of the endogenous axial tyrosinate Fe^{3+} ligand, Tyr^{447} (Tyr^{147} beta) (Fig. 5-31), and an equatorial coordination site is available for binding of solvent.

- This implies that an O_2 analogue can occupy the cavity and that electrophilic O_2 attack on PCD is begun from this site.

- The substrate complex seems to play a crucial role in the activation of substrate for O_2 attack.

Reaction Mechanism

What are the proposed mechanisms that explain the action of dioxygenases?

- ESR studies of the crystal structure of catechol 1,2-dioxygenase (1,2-CTD) and protocatechuate 3,4-dioxygenase (3,4-PCD) indicate:
  - These two enzymes have different molecular weights and subunit compositions but hold very similar active site structures and function by very similar mechanisms.
  - There is only one type of high-spin Fe^{3+} center.
  - Four amino acid ligands and two coordination sites are accessible to H_2O.
  - Fe^{3+}-intradiol enzymes have two His's and two Tyr's.
  - Reactions with substrate analogues induce spectral shifts of the iron chromophore, implying that the substrate binds directly to the iron center in the course of the enzymatic reaction.
  - Substrate binds to Fe before O_2 reacts in any way.
  - Substrate complex formation is associated with a dissociation of the endogenous axial tyrosinate Fe^{3+} ligand.

○ The equatorial coordination site is available for binding of solvent and suggests electrophilic O_2 attack on PCD.

- Similarity in the positions of the methyl resonances in ^1H NMR spectra of the model complexes with those of the substrate 4-methylcatechol bound to the enzymes CTD and PCD indicates that substrate is bound to CTD in a monodentate fashion and to PCD in a bidentate fashion.

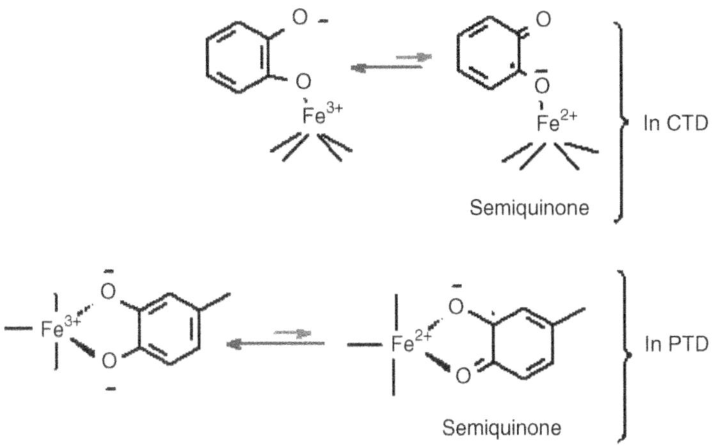

Semiquinone

In CTD

Semiquinone

In PTD

- Studies of the oxidation of the ferric complex of 3,5-di-t-butyl-catechol with different ligands L have been useful in exploring the mechanistic possibilities for these enzymes.

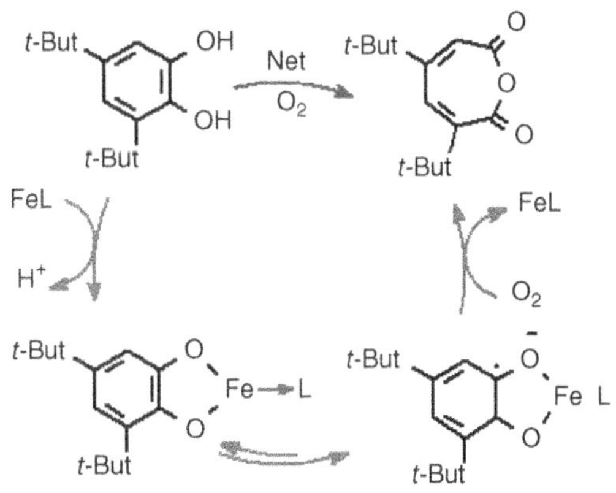

The studies show that decreasing the electron donation of L increases the reactivity of these complexes toward O_2. Thus the reactivity of these complexes is enhanced by an increase in the contribution of the minor *semiquinone*.

- The general scheme can be proposed:

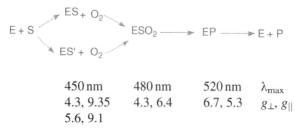

$$450\,nm \quad\quad 480\,nm \quad\quad 520\,nm \quad \lambda_{max}$$
$$4.3, 9.35 \quad\quad 4.3, 6.4 \quad\quad 6.7, 5.3 \quad g_\perp, g_\parallel$$
$$5.6, 9.1$$

- These studies have led to the mechanisms summarized in Schemes 5-5 and 5-6.
- In these proposed mechanisms, the catechol substrate coordinates to the Fe^{3+} center in either a monodentate or a bidentate fashion, replacing a coordinated solvent.
- Both OH's of catechol are deprotonated, the catechol reduces Fe^{3+} to Fe^{2+}, and forms ortho semiquinone.
- O_2 binds across the Fe^{2+} and the free radical on the catechol ring to give a peroxy derivative of the Fe^{3+}-catechol.

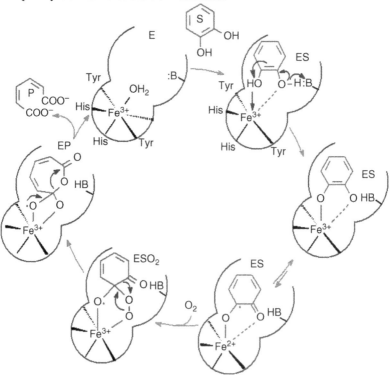

SCHEME 5-5 Proposed reaction mechanism for CTD 1,2-oxygenase. (Modified from Valentine, 1994, and Wood, 1980. Reproduced by permission of John Wiley & Sons.)

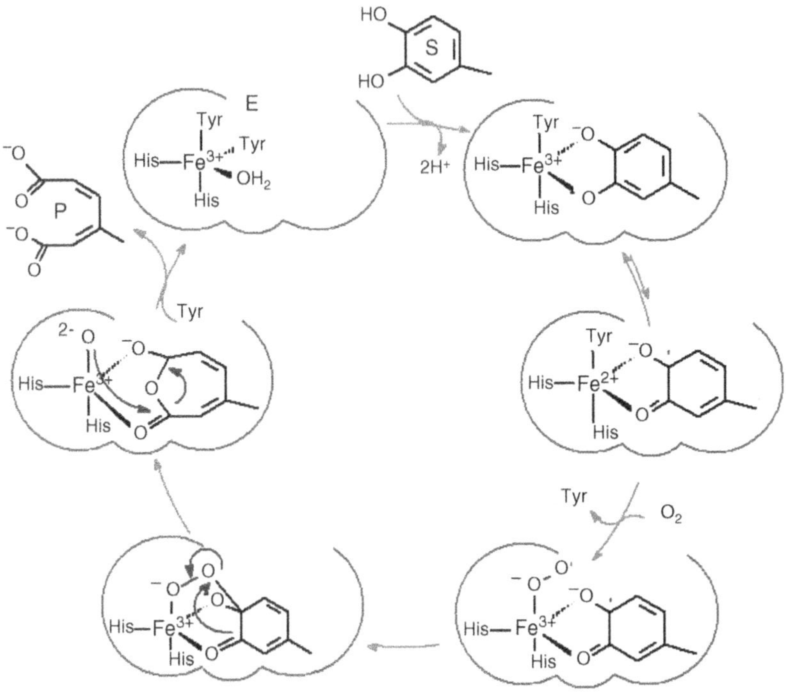

SCHEME 5-6 Proposed mechanism of intradiol dioxygenase (3,4-PCD). (Modified from Ochiai, 2008. Reproduced by permission of Elsevier.)

o This peroxy species is rearranged to give anhydride intermediate, which is analogous to well-characterized reactions that occur when catechols react with alkaline hydrogen peroxide.

- The *extradiol* dioxygenase is an Fe^{2+}-dependent enzyme and is catalyzed by protocatechuate 4,5-dioxygenase, or protocatechuate 2,3-dioxygenase (Scheme 5-7).
  o Fe^{2+}-extradiol enzymes have two His's and one Glu (Fig. 5-31).
  o The substrate binds to Fe^{2+} through the two OH's, but the binding is not the same; the one with the proton removed binds more strongly than the other, unlike the intradiol case, where both OH's of catechol are deprotonated.
  o Substrates that have unadjacent OH groups are dioxygenated in an extradiol manner. Therefore, concurrent coordination of the two OH's on adjacent positions may not be required in the extradiol mechanism.
  o O_2 is assumed to bind to Cat-Fe^{2+}, yielding Cat-Fe^{3+}–$O_2^{\bullet-}$. Then the catechol reduces Fe^{3+} back to Fe^{2+}, and semiquinone is formed (Scheme 5-7).

Intradiol Versus Extradiol Mechanism

Why do catechol 1,2-dioxygenase and protocatechuate 3,4-dioxygenase follow the intradiol mechanism, while the protocatechuate 4,5-dioxygenase or protocatechuate 2,3-dioxygenase follow the extradiol mechanism?

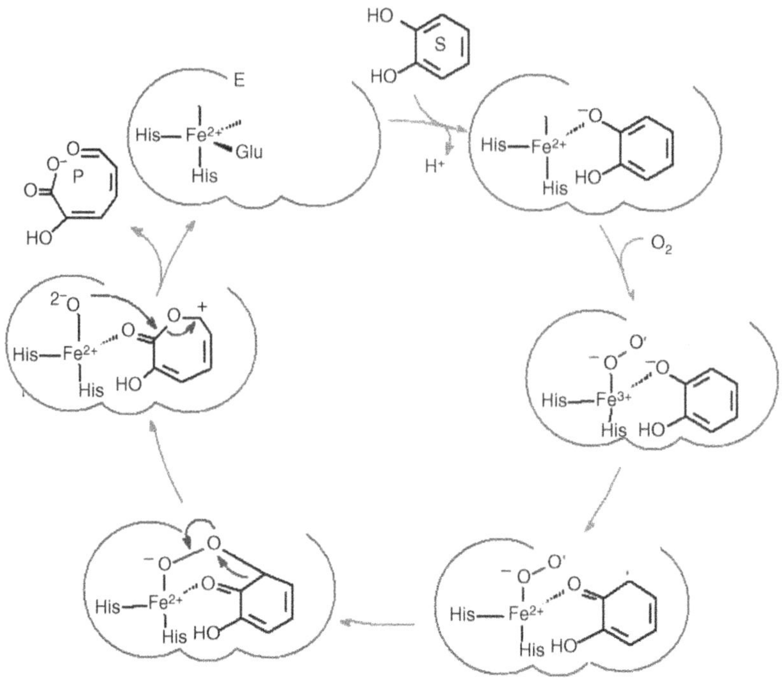

SCHEME 5-7 Proposed mechanism of extradiol dioxygenase. (Modified from Ochiai, 2008. Reproduced by permission of Elsevier.)

It depends on:

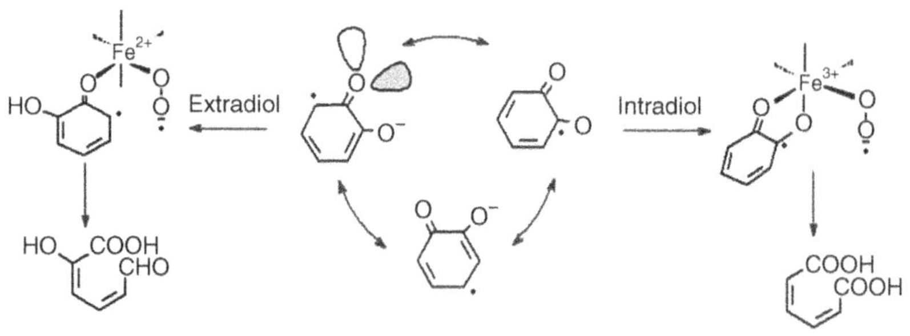

Extradiol and intradiol mechanisms (modified from Ochiai, 2008)

- The initial oxidation state: Fe^{2+} or Fe^{3+}
- The spin localization of the *ortho*-semiquinone and its relationship to the free-radical end of $O_2^{\cdot-}$.
- The semiquinone has three resonance structures and a carbonyl O in sp^2. Fe^{3+} chelates *both* O's while Fe^{2+} binds to *one* of the O's and the odd electron is in the nearest position to the free-radical end of $O_2^{\cdot-}$.

Oxygenase Mimics

Design a model system to mimic the action of oxygenase.

Model system to mimic action of oxygenase, see Scheme 5–8.

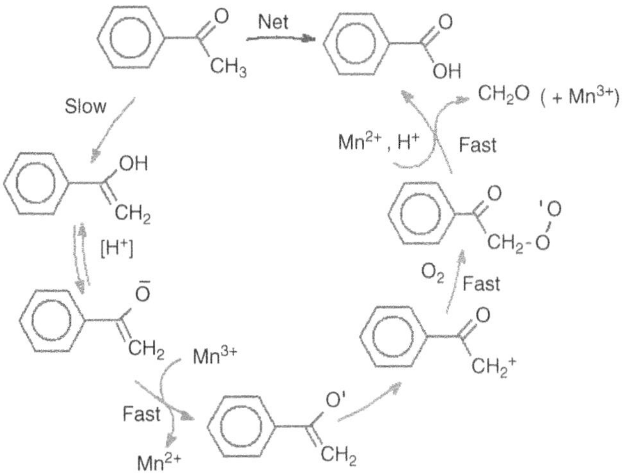

SCHEME 5-8   Oxidation of acetophenone to benzoic acid catalized by Mn^{2+}/Mn^{3+}. (Based on Hayaishi et al., 1955.)

IRON–SULFUR PROTEINS

Classification

What are the main classes of iron–sulfur proteins? Give examples.

- Roles:
  - Electron transfer in photosynthesis process
  - Nitrite reduction in spinach nitrite reductase
  - Steroid hydrolysis in adrenal steroid hydroxylases (adrenodoxin)
  - Oxidative phosphorylation in mitochondaria
- Iron–sulfur proteins can be classified as simple and complex (Schemes 5-9 and 5-10).
- Complex iron–sulfur clusters can perform other enzymatic functions besides electron transfer.

RUBREDOXIN

Roles, Sources, and Characterization

What are the biological roles, sources, and the main characteristics of rubredoxin?

- Rubredoxin is a bacterial e^- transfer protein.
- Rubredoxins are the simplest class of iron–sulfur proteins

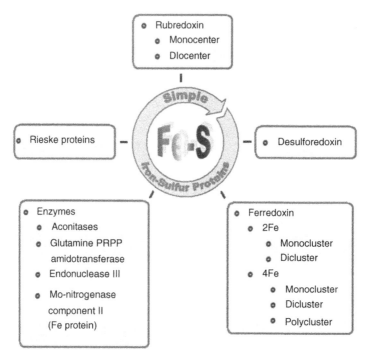

SCHEME 5-9 Iron–sulfur proteins by type and number of prosthetic centers.

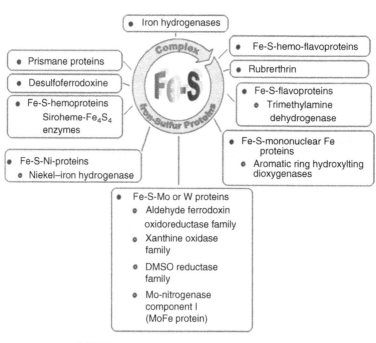

SCHEME 5-10 Complex iron-sulfur proteins.

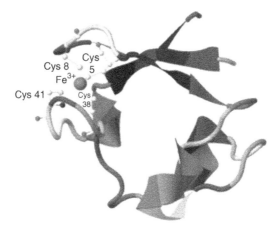

FIGURE 5-32 Neutron crystallographic analysis of rubredoxin mutant at 1.6 Å resolution, PDB 1IU5 (Chatake et al., 2004). (See the color version of this figure in Color Plates section.)

- Source is from anaerobic (proceed in absence of O_2):
  - *Clostridium pasteurianum*, $E° = -570$ mV
  - *Pyrococcus furiosus*
  - *Clostridium thermosaccharolyticum*
  - *Thermodesulfobacteruim commune*
- All examples:
  - Have Fe–S ratio 1:4, one Fe and four cysteines per mole of protein
  - Have molecular weight ~ 6000
  - Have 54 amino acid residues
  - Lack arginine and histidine residues and have a large amount of carboxylic and aromatic residues
  - Have iron bounded by four cysteine sulfur atoms in approximately tetrahedral stereochemistry (Fig. 5-32).
  - Have *red* oxidized protein and colorless reduced protein.
  - Have iron recycling between Fe^{3+} and Fe^{2+}

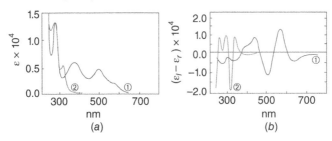

FIGURE 5-33 Spectral properties of *Peptostreptococcus elsdenii* rubredoxin, (1) refers to Fe^{3+} and (2) to Fe^{2+}-rubredoxin: (*a*) absorption spectra; (*b*) circular dichroism spectra. (From Orme-Johnson, 1973. Reproduced by permission of Elsevier.)

Structural and Ligation Features of Rubredoxin

The absorption and circular dichroism spectra of the oxidized and reduced rubredoxin have been recorded (Fig. 5-33), and the circular dichroism of Fe^{2+}-rubredoxin also shows one band at $6250 \, cm^{-1}$ with $\Delta\varepsilon/\varepsilon = 0.05$.

The protein in the oxidized form yields ESR signals at 4.3 and near 9.4, while there was no signal seen from the reduced protein. The effective magnetic moment was found to be 5.85 BM for Fe^{3+} and 5.05 BM (Bohr magneton) for Fe^{2+}.

From the X-ray and structural determination of the oxidized form, show the following:

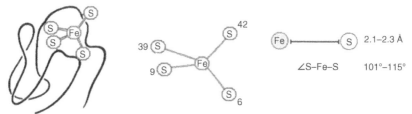

Structure of *C. pasteurianum* rubredoxin, Fe = red, S-Cys = yellow.

Use of absorption, circular dichroism, ESR spectra, magnetism, and X-Rays to determine structural and ligation features of rubredoxin.

- Bleaching on reduction suggests oxidized color due to LMCT.
- The low Kuhn anisotropy factors ($\Delta\varepsilon/\varepsilon$) for the observed bands make it unlikely that magnetically allowed transitions occur ($d \rightarrow d$).
- Therefore, two $S \rightarrow Fe$ bands in the region 600–300 nm may assign in Fe^{3+}-rubredoxin.
- The very long λ band at $6250 \, cm^{-1}$ (1630 nm, $\Delta\varepsilon/\varepsilon = 0.05$) in Fe^{2+}-rubredoxin suggests magnetically allowed $d \rightarrow d$ transitions in the weak field, Δ_t.
- The *small* Δ_t proposes a high-spin Fe^{2+} in the *tetrahedral field*, and this transition may assign as $^5E \rightarrow {}^5F$. This confirms the maintenance of this stereochemistry on reduction.
- The protein in the oxidized form yields ESR signals at 4.3 and near 9.4, while there was no signal seen from the reduced protein.
- This type of signal is characteristic of high-spin Fe^{3+} ($S = \frac{5}{2}$) in the rhombic environment.
- The effective magnetic moment was found, 5.85 BM for Fe^{3+} and 5.05 BM for Fe^{2+}, in agreement with Fe^{3+} ($S = \frac{5}{2}$) and Fe^{2+} ($S = \frac{4}{2}$), where

$$\mu_{eff} = 2\sqrt{S(S+1)}$$

  The calculated μ_{eff} is 5.9 BM for Fe^{3+} and 4.89 BM for Fe^{2+}.
- X-ray and structural determination of the oxidized protein shows that the protein has an irregular folded structure, with no α-helix but with some antiparallel pleated sheet conformation.
- The four-cysteine sulfur atoms in approximately tetrahedral stereochemistry bind the iron (Fig. 5-32).

Metal Removal

How can the central metal ion be removed from rubredoxin?

- Removal of Fe is possible (apoenzyme): The iron atom can be released from the protein by:
  1. Acidification, but not H_2S is evolved and no labile sulfur is present.
  2. Being "exposed" to:
     (a) *Ellman's reagent* [5,5′-dithiobis-(2-nitrobenzoate)], which often reacts to convert the SH groups of the proteins into a mixed-disulfide polypeptide:

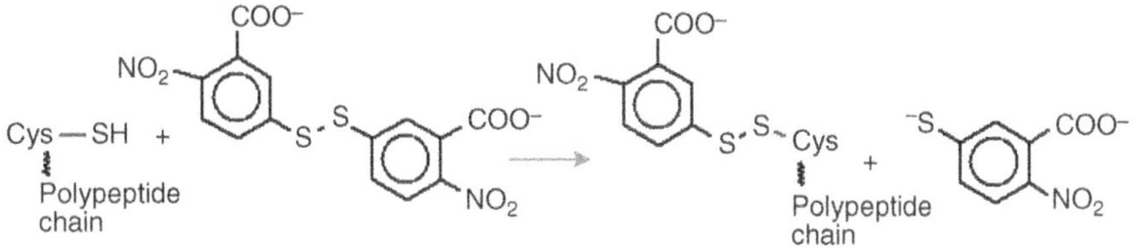

   The intensive absorbance at 412 nm makes Ellman's reagent very useful for quantitative determination of SH groups.

     (b) *Organic mercury compounds*, which are highly specific reagents for protein SH groups, e.g., chloromercuridinitrophenol is used as a reagent for SH groups.

- Reincorporation of the iron atom into aporubredoxins will only occur in the presence of thiols, e.g., mercaptoethanol:

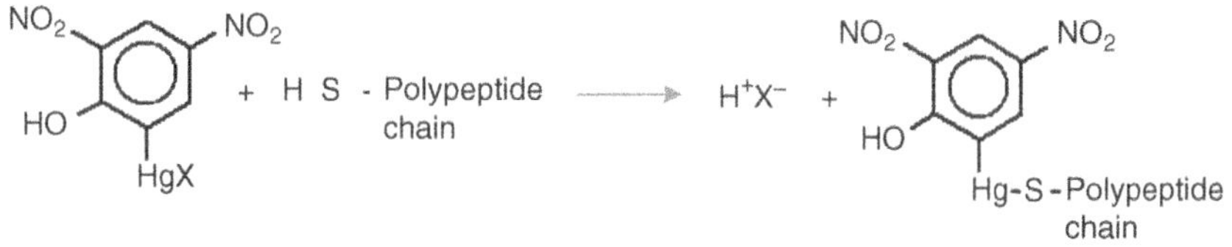

Model compounds

Design a model compound for rubredoxin.

- S_2-*o*-xyl reacts with Fe^{3+} and gives [Fe $(S_2$-*o*-xyl$)_2]^-$.
- It is a high-spin FeS_4 site.

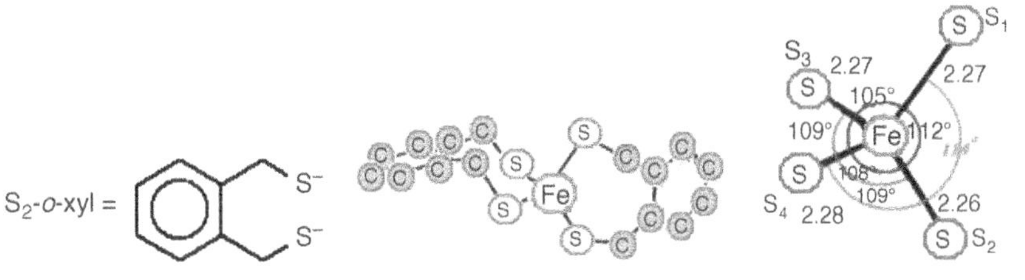

Structure of [Fe(S_2-*o*-xyl)$_2$]⁻ anion

- It Illustrates a chair conformation of the chelate rings.

- There is a resemblance in the bond lengths and electronic spectra between rubredoxin and [Fe $(S_2$-o-xyl)$_2$]$^-$ (see Fig. 1 and Fig. 2, p.2870, Lane, et al. 1975). The results substantially designate [Fe(S_2-O-syl)$_2$] as a synthetic analog of the Fe^{3+}(S-Cys)$_4$ center in oxidized rubredoxin proteins.

FERREDOXINS

Roles and Types

What are the biological roles and the main types of ferredoxins?

- Biological roles of ferredoxin (Fd)
  - Electron transfer and electron storage protein
  - Catalyzes oxidation–reduction between +350 and –600 mV (hydrogen electrode = –420 mV)
- Ferredoxins contain labile S^{2-}, giving H_2S on treatment with mineral acids.
- It is widely distributed (plants, animals, bacterial):
  - *C.pasteuriaium, A. viralandii*
  - *P. putida, Spirulina maxima*
  - Green plants: chloroplasts
- Simultaneous binding of Fe ion, S^{2-} may reflect (prebiotic) strategy for Fe uptake.
- Several types exist:

$$2Fe–2S \qquad 4Fe–4S \qquad 7Fe–8S \qquad 8Fe–8S$$

2Fe–2S FERREDOXINS

Sources and Chemical Characteristics

What are the simplest form, sources, structural, and chemical characteristics of ferredoxins?

- The simplest form is 2Fe–2S.
- This protein is isolated from *plants* (spinach, alfalfa, and parsley). It is found in chloroplasts and their role in photosynthesis is well-established.
- A 2Fe–2S protein from a *nonplant* source is putidaredoxin, molecular weight 12,500, from *P. putida*. It acts as a one-electron transfer agent in the enzyme complex that catalyzes the hydroxylation of camphor.
- An example of this enzyme from an *animal* source is adrenodoxin from the adrenal cortex. It is part of a multienzyme complex that catalyzes the hydroxylation of steroids.
- Sources of 2Fe–2S
  - Spinach: molecular weight 10,600, $E^0 = -420$ mV
  - *Microcystis*: molecular weight 10,300
  - *Azotobacter*: molecular weight 21,000, $E^0 = -350$ mV

- The two iron atoms and the two atoms of sulfur are released upon acidification or treatment with mercurials (yield apoprotein).
- Addition of Fe^{3+} or Fe^{2+} salts, sulfide, and a mercaptan to solutions of apoprotein at neutral pH results in formation of holoprotein.
- Substitution of selenide for sulfide leads to a holoprotein containing labile selenium.
- It has complicated UV-Vis spectra:
  - The absorption spectra show bands near 320, 420, and 460 nm, with a shoulder near 550 nm.
  - The absorption in the 420–460-nm region declines ~50% on reduction.

$$Fd_{ox} + 1e^- \leftrightarrows \qquad Fd \qquad E^0 = -0.40 \text{ V}$$

$$\text{Red} \qquad\qquad \text{orange red}$$

  - Generally, a broad maximum near 550 nm is found to be the only pronounced feature of spectra of the reduced form.
- The 2Fe–2S proteins exhibit ESR signals ($g_z \approx 2.02$, $g_y \approx 1.95$, $g_x \approx 1.91$) in the reduced form.
- No ESR signals and no magnetic susceptibility are observed in the oxidized samples at low temperature.
- The ESR and optical absorption parameters of these proteins underlie similarity of their chromophores and emphasize the essential symmetries of the 2Fe–2S class.

Iron Ions and Active Center

How many iron atoms are in the active–center of Ferredoxins (2Fe–2S)?

- Looking for the electron hyperfine structure in signals from proteins enriched in ^{57}Fe ($I = \frac{1}{2}$) has been used to determine whether one or both iron atoms are present in the electron–accepting center.
- No resolved nuclear hyperfine structure was observed, and the spectrum only exhibits broadened ESR signals.
- A trace computed on the assumption that both iron atoms were present in the center and that both contributed 14 G isotropic, the hyperfine structure is *superimposed* on the experimental spectrum.
- No calculated trace assuming one iron atom to be present in the center would match the observed spectra.
- Therefore, the 2Fe of the protein participates in a binuclear iron complex and can accept a single electron.
- X-ray of *Aphanothece sacrum* Fd (Fig. 5-34) is comparable with that of $[(S\text{-}o\text{-xyl})_2(Fe^{3+})_2S_2]^{2-}$.
- Then, the structure can be regarded as a bitetrahedral one (Fig. 5-34). The X-ray of $[(S\text{-}o\text{-xyl})_2(Fe^{3+})_2S_2]^2$ is comparable with that of *A. sacrum* Fd:

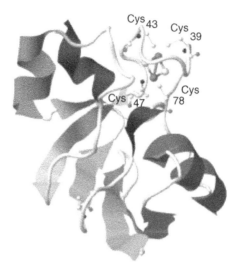

FIGURE 5-34 Refined structure of plant-type [2Fe–2S] ferredoxin I from *A. sacrum* at 1.46 Å resolution, PDB 3AV8 (Kameda et al., 2011).

$[(S\text{-}o\text{-}xyl)_2(Fe^{3+})_2S_2]^{2-}$ (Mayerle et al., 1975.)

Sulfur Ligation

Prove that cysteines and labile sulfur atoms are ligated to iron in the active–center of 2Fe–2S ferredoxins.

- Samples of 2Fe–2S proteins enriched by ^{33}S ($I = \frac{3}{2}$) are prepared by combinations of chemical exchange and growth techniques.
- The ESR spectra of these samples are compared.
- Broadening occurs when either the labile sulfur or the protein-bound (Cyst or Met) sulfur atom is ^{33}S. Therefore both types are present in the center.

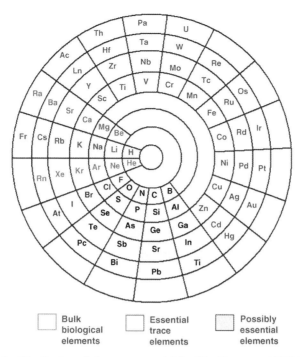

FIGURE 1-1 Distribution of elements essential for life (Cotton and Wilkinson, 1980).

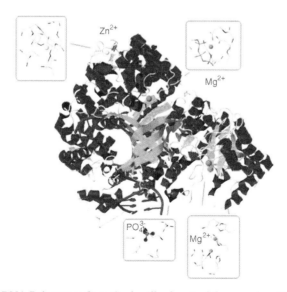

FIGURE 2-8 DNA Polymerase from *Geobacillus kaustophilus* complex at 2.4 Å resolution, PDB 3F2B: Mg^{2+} = violet, Zn^{2+} = yellow (Evans et al., 2008).

Chemistry of Metalloproteins: Problems and Solutions in Bioinorganic Chemistry, First Edition.
Joseph J. Stephanos and Anthony W. Addison.
© 2014 John Wiley & Sons, Inc. Published 2014 by John Wiley & Sons, Inc.

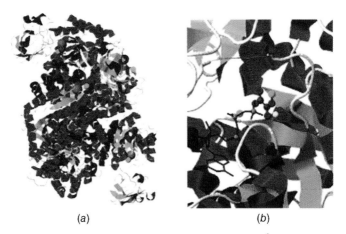

FIGURE 2-9 (a) Pyruvate kinase from rabbit muscle at 2.7 Å resolution, PDB 1AQF (Larsen et al., 1997). (*See text for full caption.*)

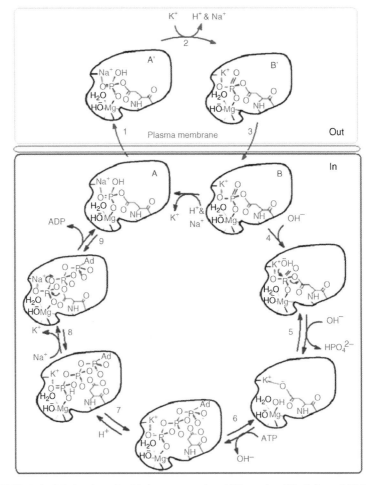

SCHEME 2-1 Mechanism for Na⁺ transport by ATPase (modified from Mildvan and Grisham, 1974).

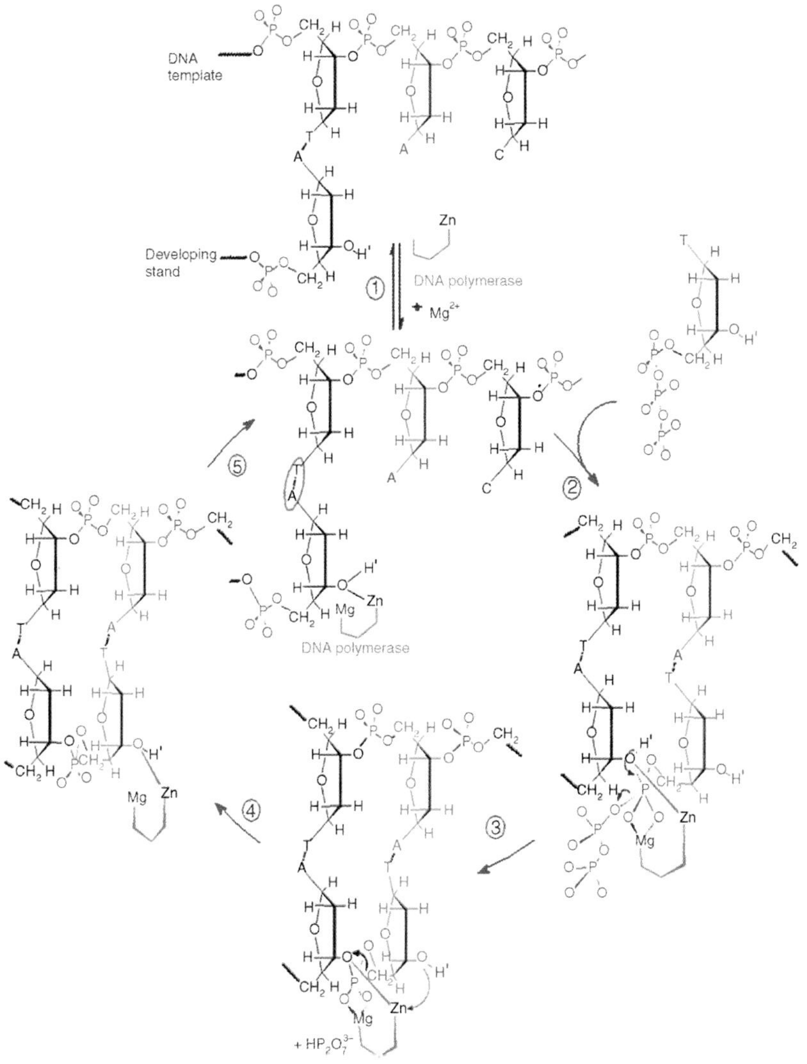

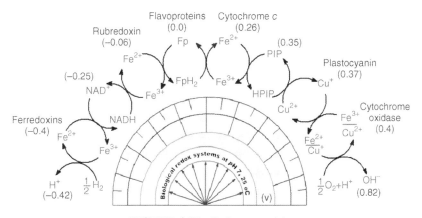

SCHEME 2-3 DNA polymerase (modified from Mildvan and Grisham, 1974).

FIGURE 4-17 Redox potentials.

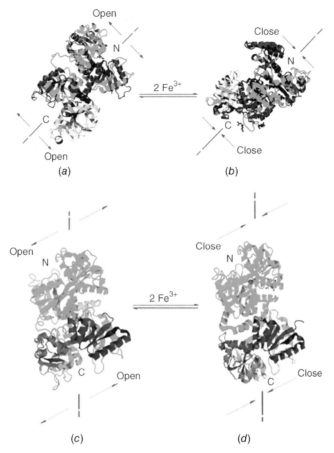

FIGURE 5-24 (*a*) Apo-human serum transferrin, PDB 2HAV (Wally et al., 2006). (*See text for full caption.*)

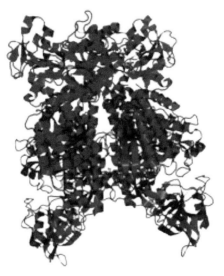

FIGURE 5-25 Complex between transferrin receptor 1 and transferrin with iron in N-Lobe at room temperature, PDB 3S9N (Eckenroth et al., 2011).

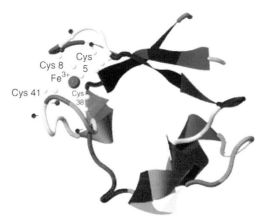

FIGURE 5-32 Neutron crystallographic analysis of rubredoxin mutant at 1.6 Å resolution, PDB 1IU5 (Chatake et al., 2004).

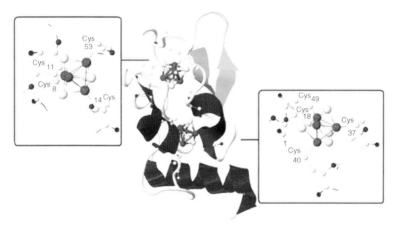

FIGURE 5-37 Structure of 2[4Fe–4S] ferredoxin from *P. aeruginosa*, PDB 2FGO (Giastas et al., 2006).

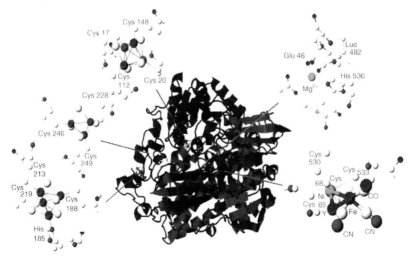

FIGURE 5-40 Crystal structure of oxidized form of Ni–Fe hydrogenase, PDB 2FRV (Volbeda et al., 1996).

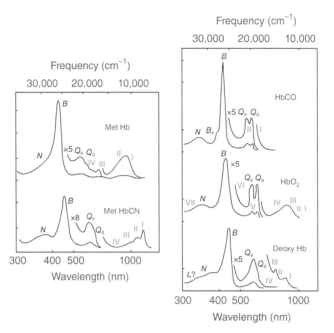

FIGURE 5-69 Solution absorption spectra of hemoglobin derivatives. (Data from Eaton and Hofichter, 1981. Reproduced by permission of Elsevier.)

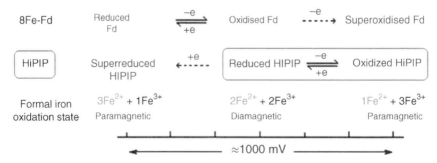

SCHEME 5-11 Fd cycling between upper two and HiPIP between bottom two. (Based on Cotton and Wilkinson, 1980, reproduced by permission of John Wiley & Sons.)

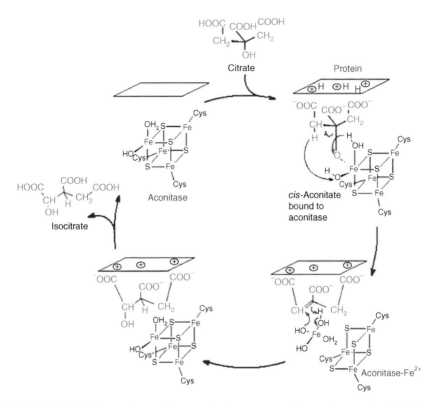

SCHEME 5-14 Three point attachments of substrate to aconitase and one to the labile iron, aconitase–Fe^{2+}– substrate. (Based on Mayes, 1977.)

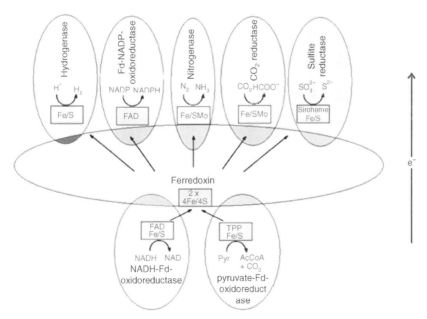

SCHEME 5-15 The soluble ferredoxin acts as electron carrier between reactants (bottom) and terminal enzymes (top). (Modified from Thauer and Schönheit, 1982.)

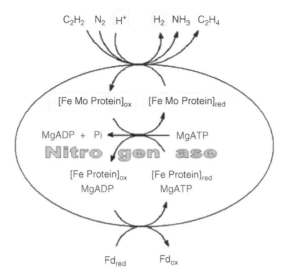

SCHEME 5-18   Nitrogenase reactions.

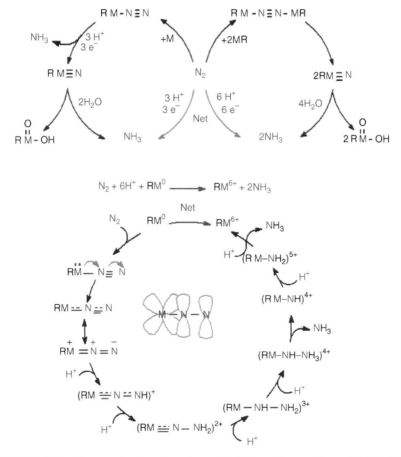

SCHEME 5-20   Nitrogen fixation through metal nitride as intermediate. (Modified from Owsley and Helmkamp, 1967, and Allen and Senoff, 1965.)

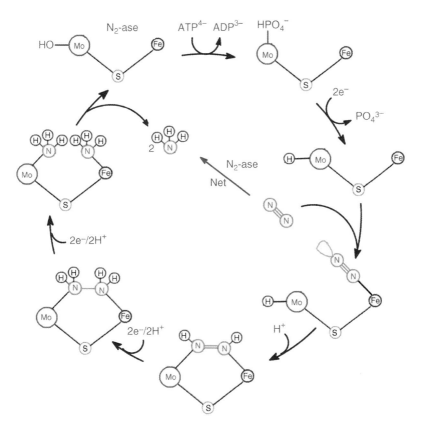

SCHEME 5-24 Nitrogen fixation through substrate complexation. (Modified from Hardy et al., 1971.)

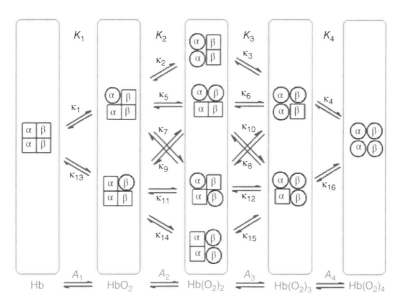

SCHEME 5-30 Oxygen binding by hemoglobin (red are oxygenated and blue squares are deoxygenated).

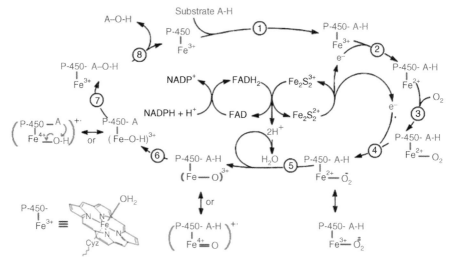

Catalytic cycle of cytochrome P-450 hydroxylase, modified from Sono et al., 1996.

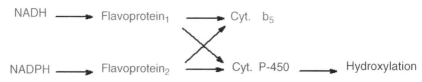

SCHEME 5-42 The electron transport chain in liver P-450 does not require the iron-sulfur protein Fe_2-S_2.

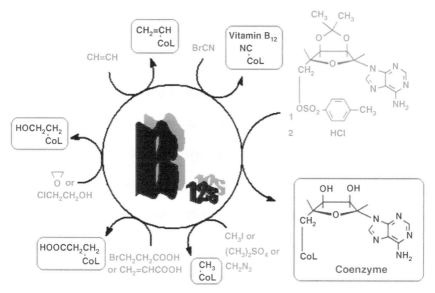

SCHEME 6-7 Nucleophilic reactions of B_{12s}. (Data from Smith et al., 1962; Johnson et al., 1963; and Müller and Müller, 1962, 1963.)

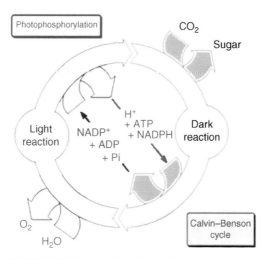

FIGURE 7-5 Overview of steps in photosynthesis.

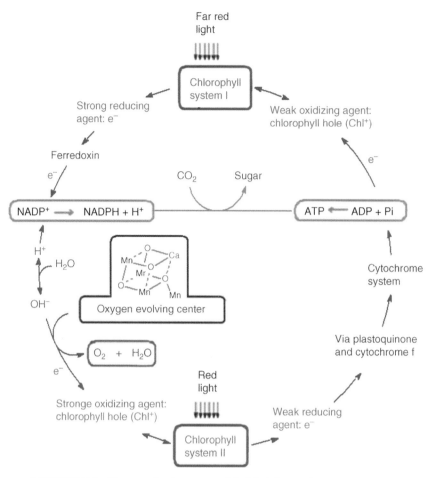

SCHEME 7-2 Photosynthesis process. (Modified from Johnson et al., 1969.)

- The chemical exchange of selenide $[^{80}Se(I = 0), {}^{77}Se(I = \frac{1}{2})]$ for labile sulfur of the proteins is a useful tool in ESR investigation.
- A comparison of signals from proteins incorporating these isotopes indicates that both ^{77}Se atoms contributed to the observed nuclear hyperfine splitting at $g = 2.03$.
- This suggests that both labile sulfur atoms are ligands in the 2Fe–2S cluster in the native protein.

Valence-Trapped Dimers

Why are 2Fe–2S ferredoxins described as valence-trapped dimers? Explain the consequent unusual magnetism.

- Note two Fe but one e^- only:

$Fe^{3+} Fe^{3+}$	$\rightarrow$	$Fe^{3+} Fe^{2+}$	$\rightarrow$	$Fe^{2+} Fe^{2+}$
Ox	$\rightarrow$	Red		
		Ox	$\rightarrow$	Red
d^5, d^5		d^5, d^6		d^6, d^6
Fd_{ox}		Fd_{red}		

No ESR signal	Show ESR spectrum with $g = 1.93$ (low spin, Fe^{3+})
400–500 nm $S \rightarrow Fe$, LMCT	Rather similar, but lower in intensity (Fe^{3+})
Paramagnetic at 300 K but diamagnetic at low temp. (<50 K)	Paramagnetic at 4–300 K with one e^- unpaired at 4 K

- Therefore:
  OX: $\quad d^5 d^5 \xrightarrow{e}$ Red: $\quad d^5 d^6$
  $$d^5 + d^6 \rightarrow \text{has 1 } e^- \text{ unpaired}$$

- Mössbauer spectroscopy of the ^{57}Fe-enriched proteins suggests that the *oxidized* samples at low temperature exhibit a *single* quadrupole split pair of lines (QS ≈ 0.6) and isomer shift (IS ≈ -0.08).
- Therefore, both ferric ions are *equivalent* structurally.
- Since the oxidized proteins are diamagnetic at these temperatures, Mössbauer spectra are compatible with low-spin iron—Fe^{2+} (unlikely) or antifferromagnetically coupled Fe^{3+} atoms.

- Mössbauer spectra of *reduced* ferredoxins show a more complex behavior, that is, *two* quadrupole splitting pairs of lines are observed.
- The two iron atoms became nonequivalent ($QS \approx 0.6$ and $IS \approx -0.1$ for Fe^{3+} and $QS \approx -3$ and $IS \approx 0.2$ for Fe^{2+}).
- The added electron appears to reside entirely in one of the pair of metal atoms and is therefore described as a *valence-trapped dimer*.

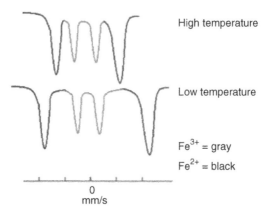

High temperature

Low temperature

Fe^{3+} = gray
Fe^{2+} = black

0
mm/s

- Reduction places the e on one (end) of the FeSSFe moiety:

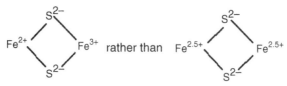

- The more widely split pair of lines have a quadrupole splitting characteristic of high-spin Fe^{2+}, while the narrow pair could be due to high- or low-spin Fe^{3+}.
- In addition, the reduced protein has an effective magnetic moment equivalent to the $S = \frac{1}{2}$ case.
- Based on these data, the paramagnetic center is a high-spin Fe^{2+} ($S = 4/2$) and a high-spin Fe^{3+} ($S = 5/2$) atom, which are antiferromagnetically coupled to give a net electron spin of 1/2 in the ground state and is the one which gives rise to the ESR signal.

Exchange Coupling Constant, J, in Antiferromagnetic and Ferromagnetic couplings

(a) **How can S^{2-} bridges affect the magnetic spin coupling in 2Fe–2S ferredoxins? If the splitting energy between the singlet and the triplet states, E, is equal to $9800\ cm^{-1}$, and the electron pairing π is equal to $10,000\ cm^{-1}$, calculate the exchange coupling J.**

(b) **What is the difference between positive and negative exchange coupling and how do J and temperature influence μ_{eff}?**

(c) **If $J = -150\ cm^{-1}$, and $KT = 200\ cm^{-1}$ at 200 K, for $Cu_2(CH_3COO)_4(H_2O)_2$ calculate μ_{eff}.**

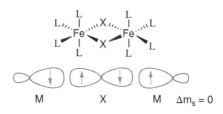

FIGURE 5-35 Bridge-mediated, spin polarization model.

(d) **If $-2J = 182\,\text{cm}^{-1}$ for Fe^{3+}–Fe^{3+} and $-2J = 98\,\text{cm}^{-1}$ for Fe^{3+}–Fe^{2+} in the oxidized and reduced 2Fe–2S ferredoxins, what do you conclude from these values?**

- For two $S^0 = \frac{1}{2}$ ions, spin pair is predicted by the ZFS model:

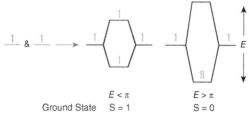

where E is the splitting energy and π is the electron pairing usually $E \approx \pi$.
  When $E > \pi$, $S = 0$ is the ground state and $S = 1$ is now an excited state, higher by $E - \pi$.

- Electronic properties, via magnetic/spin coupling mediated by the S^{2-} bridges can be described as in Fig. 5-35. Note the small splitting between levels in the bridge–mediated model.

- The splitting is usually described by the magnetic exchange integral, J,

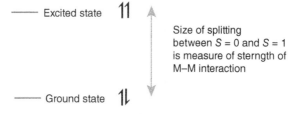

where J in energy units is called the exchange coupling constant or exchange parameter. (Note: It is to be distinguished from the quantum number J.)

- The interaction between the metal orbitals on the metal centers can be described for a pair i and j by the Hamiltonian

$$\hat{H} = -2J\,\hat{S}_i\hat{S}_j$$

If

$$\hat{S} = \hat{S}_1 + \hat{S}_2 \qquad \hat{S}^2 = \hat{S}_1^2 + \hat{S}_2^2 + 2\hat{S}_1\hat{S}_2$$

then

$$2\hat{S}_1\hat{S}_2 = \hat{S}^2 - \hat{S}_1^2 - \hat{S}_2^2$$

If

$$\hat{S}^2\psi = S(S+1)\psi$$

then

$$-2J\,\hat{S}_1\hat{S}_2\psi = -J[S(S+1) - S_1(S_1+1) - S_2(S_2+1)]\psi$$

where

$$S_1 = S_2 = \tfrac{1}{2} \quad S = 0, 1$$

- For $S=0$ (*ground* state) we obtain $\tfrac{1}{2}$

$$-2J\,\hat{S}_1\hat{S}_2 = -J\left[0 - \frac{3}{4} - \frac{3}{4}\right] = +\frac{3}{2}J$$

while for $S=1$ (*excited state*)

$$-2J\,\hat{S}_1\hat{S}_2 = -J\left[2 - \frac{3}{4} - \frac{3}{4}\right] = -\frac{1}{2}J$$

Then the difference between excited and ground states:

$$-2J = E - \pi \quad \text{for} \quad S_1 = S_2 = \tfrac{1}{2}$$

- If $\pi = 9800\,\text{cm}^{-1}$ and $E = 10,000\,\text{cm}^{-1}$, then

$$-2J = E - \pi = 200\,\text{cm}^{-1}$$
$$-2J = 200\,\text{cm}^{-1} \qquad J = -100\,\text{cm}^{-1}$$

- J (or $-2J$) is a measure of the *coupling strength*.
- When $-2J$ is positive:

$$J < 0$$

Singlet state ($S=0$) below triplet state ($S=1$)

antiferromagnetic coupling/exchange

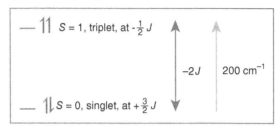

- When $-2J$ is negative:

  $$J > 0$$

Triplet state is the ground state

Ferromagnetic coupling/exchange

- Experimental: Cu^{2+}-acetate (d^9, $S = \frac{1}{2}$), $-2J$ of $300\,cm^{-1}$, $J < 0$

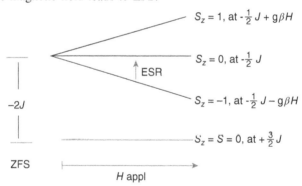

- In magnetic field, and vary temperature
- The magnetic field leads to ZFS.

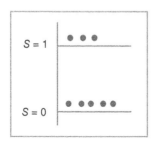

- If $KT = 200\,cm^{-1}$ at 200 K, molecules have thermally accessible energy levels.
- The Boltzmann distribution of molecules between the states is as follows:

$$\frac{N_1}{N_0} = e^{-\left(\frac{2J}{KT}\right)}$$

where, N_1 is the number of electrons in the exciting state (the unpaired electrons), and N_0 is the number of the electrons in the ground state (paired electrons).

$$N_1 = N_0 e^{2J/KT} \quad (J \propto N_1)$$

For example, for $Cu_2(ACO)_4(OH_2)_2$, $J = -150\,cm^{-1}$.

- When $T \to 0\,K$, $N_1 \to 0$, 100% in $S = 0$ state. Diamagnetic because spin coupling, and no ESR spectra are observed.
- At $T \to \infty$, $N_1 = N_0 \to 50\%$ in $S = 0$ and 50% in $S = 1$. For $S^0 = \frac{1}{2}$:

$$\mu_{eff} \propto \sqrt{\frac{N_0}{N_0 + N_1}}$$

$$\mu_{eff} = \sqrt{\frac{1}{2}} \cdot \mu$$

$$\mu_{eff} = \frac{1}{\sqrt{2}} \cdot \mu = \frac{1}{\sqrt{2}} \cdot 2\sqrt{s(s+1)} = \sqrt{2} \cdot \sqrt{s(s+1)}$$

$$= 1.414 \times 0.866 = 1.2\,BM/Cu^{2+}$$

- Now back to ferrodoxin:

$$Fd_{ox} : S^0 = \frac{5}{2}, \frac{5}{2} \qquad\qquad \text{ground state, no ESR}$$
$$Fd_{red} : S^0 = \frac{5}{2}, \frac{4}{2}\ (Fe^{3+}, F^{2+}) \quad \text{ground state is } S = \frac{1}{2}$$

so ESR spectra are observed.

  ○ *Spirulina maxima has*

$$Fe^{3+}, Fe^{3+} : \quad -2J = 182\,cm^{-1}$$

The exchange coupling constant shows that there is delocalization of the spin through the sulfur atoms and the iron:

$$Fe^{3+}, Fe^{2+} : \quad -2J = 98\,cm^{-1}\ (\text{less delocalized})$$

Reduction places the e on one (end) of the FeSSFe moiety:

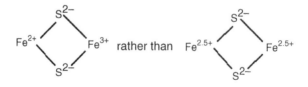

  ○ Moreover, the structure can be regarded as a bitetrahedral one.

4Fe–4S FERREDOXINS

Active Site and Chemical Features

What is the structure of the active site in 4Fe–4S proteins, and the main chemical features?

- These proteins contain an iron–sulfur prosthetic group in which iron and labile sulfur are arranged in a distorted *cubic* structure, with each iron atom linked to the protein through a cysteine residue, $[Fe_4S_4(S\text{-}Cys)_4]^{-n}$, $n = 1, \ldots, 3$ (Fig. 5-36).

- In contrast to the behavior of other iron–sulfur proteins, the iron remains bound to the protein even after acidification has displaced the labile sulfur and destroyed the chromophore.

- There are two classes of proteins:

  (a) High-potential iron–sulfur protein (HiPIP): Sources are *Chromatium* iron protein, *Chromatium* HiPIP iron–sulfur cluster; photosynthesis purple bacteria, *Chromatium vinosum* (HiPIP), *Rhodospharoides gelatinosa*.

- HiPIP contains only a *single cubane cluster*.

- HiPIP is *paramagnetic*, giving ESR signals in the oxidized state ($g_\perp = 2.04$, $g_\parallel = 2.12$ for *C. vinosum* at 28 K).

- On one electron reduction an ESR silent state is obtained, which is diamagnetic at cryogenic temperature.

- On passing from the oxidized to the reduced state, about 20% of the absorbency near 380 nm is lost, and an overlapping triple shouldered structure gives way to a single absorption at 388 nm:

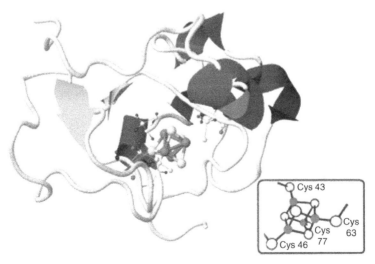

FIGURE 5-36   Three-dimensional solution structure of reduced HiPIP from *C. vinosum*, PDB 1HRR (Banci et al., 1995).

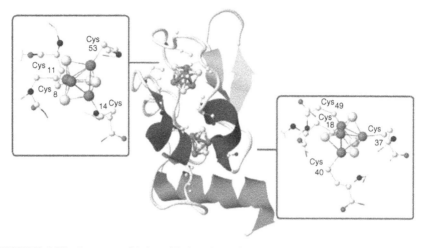

FIGURE 5-37 Structure of 2[4Fe–4S] ferredoxin from *P. aeruginosa*, PDB 2FGO (Giastas et al., 2006). (See the color version of this figure in Color Plates section.)

$$Fd_{ox} + 1e^- \quad \leftrightarrows \quad Fd_{red} \qquad E^{0'} = +0.35 \text{ V}$$
$$S = \tfrac{1}{2} \qquad\qquad\quad S = 0 \qquad\qquad HiPIP$$
$$(g_{//} = 2.12, g_\perp = 2.04) \quad (\text{no ESR ?})$$

(b) Bacterial ferredoxins (Fd): *Clostridium* or *Peptococcus*:

$$Fd_{ox} + e^- \quad \leftrightarrows \quad Fd_{red} \qquad E^{0'} = -0.45 \text{ V}$$
$$S = 0 \qquad\qquad S = 1 \text{ with}$$
$$\qquad\qquad\qquad g < 2$$

- Molecular weight 6000
- Contains two Fe_4S_4 cubes separated by 12 Å (Fig. 5-37).
- During titration of Fd with dithionite two sets of ESR signals are seen, corresponding to two paramagnetic centers.
- The integration of the ESR signals of fully reduced protein showed that two spins were present per molecule.

Redox Potential

Why is there a wild difference in the redox potential between HiPIP and 8Fe–Fd?

- Despite the similarity in the Fe–S chromophore in these two proteins, there is a considerable difference in the redox potential.
- Why do E^0 differ by almost 1 V?
- The bond length and the angles in the HiPIP and Fd_{ox} Fe_4S_4 clusters are the same:

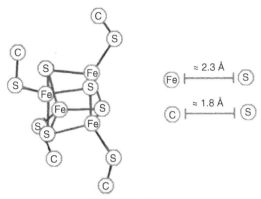

Fe$_4$S$_4$(SCH$_2$Ph)$_4$ core

- It is difficult to consider this difference being due to differences in the protein environments of the two clusters.
- The large difference in the redox potential between HiPIP and 8Fe–Fd may be based on the assumption that the Fe$_4$S$_4$ cluster has three oxidation states (Scheme 5-11).

[Fe$_4$S$_4$(S-Cys)$_4$]$^{-1}$	[Fe$_4$S$_4$(S-Cys)$_4$]$^{-2}$	[Fe$_4$S$_4$(S-Cys)$_4$]$^{-3}$
Fe^{2+}Fe$_3$$^{3+}$	Fe$_2$$^{2+}Fe_2$$^{3+}$	Fe$_3$$^{2+}$Fe^{3+}
21 d e$^-$	22 d e$^-$	23 d e$^-$
	$S=0$	

HiPIP$_{ox}$	⇌	HiPIP$_{red}$	⇌	Superreduced
Superoxide	⇌	Fd$_{ox}$	⇌	Fd$_{red}$ ($E^{0'} = -300$ mV)
+0.12 V	←	[Fe$_4$S$_4$(S-^{t}Bu)$_4$]$^{-2}$	→	−1.18 V

- When HiPIP has the configuration [Fe$_4$S$_4$(S-Cys)$_4$]$^{-2}$, it cannot be reduced further to superreduced HiPIP, [Fe$_4$S$_4$(S-Cys)$_4$]$^{-3}$.
- When Fd has the configuration [Fe$_4$S$_4$(S-Cys)$_4$]$^{-2}$, it cannot be oxidized further to super oxidized Fd, [Fe$_4$S$_4$(S-Cys)$_4$]$^{-1}$.
- In each case, the protein prevents the further reduction or oxidation of each cluster.

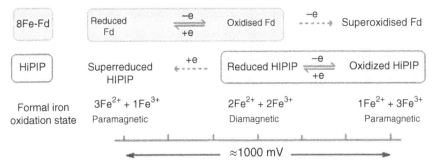

SCHEME 5-11 Fd cycling between upper two and HiPIP between bottom two. (Based on Cotton and Wilkinson, 1980, reproduced by permission of John Wiley & Sons.) (See the color version of this scheme in Color Plates section.)

- The important role of the protein had been confirmed by:
  - (a) The superreduced HiPIP can be formed by using 80% dimethyl sulfoxide (DMSO) to distort the protein environment followed by reduction with sodium dithionite ($Na_2S_2O_4$). The environment of HiPIP is more hydrophobic and less polar than that of Fd and, consequently, the highly charged state, $[Fe_4S_4(S-Cys)_4]^{-3}$, becomes less stable. The superreduced HiPIP has ESR signal at $g = 1.94$ and $g = 2.05$—similar to that of Fd_{red}.
  - (b) Fd can be oxidized to the superoxidized form by ferricyanide. The ESR signals suggest that this oxidized state is equivalent to that in the oxidized HiPIP.
  - (c) The form $[Fe_4S_4(S-Cys)_4]^{-2}$ has no ESR signal, μ $Fe_4 = 1.1$ BM at 500 K, $\mu Fe_4 = 0.2$ BM at 100 K, making it strongly antiferromagnetic.

Core Extrusion Reactions

Complete the following core extrusion reactions and write your conclusions:

Rubredoxin + 2-o-xyl(SH)$_2$ $\longrightarrow$ – – – – – – – – – +apo-rubredoxin

Fd-Haloprotein + RSH $\longrightarrow$ – – – – – – – – – +apoprotein

- Answers:

Rubredoxin + 2-o-xyl(SH)$_2$ $\longrightarrow$ $[Fe(S_2 - o - xyl)_2]^-$ + apo-rubredoxin

$$\text{Haloprotein + RSH} \longrightarrow \begin{cases} [Fe_2S_2(SR)_4]^{2-} \\ \text{and/or} \\ [Fe_4S_4(SR)_4]^{2-} \end{cases} + \text{apoprotein}$$

The labile tetrahedral iron centers illustrate:
- The occurrence of rapid thiolate substitution reactions
- The possibility of protein reconstitution from apoprotein
- The possibility of determining the shape of iron–sulfur cores
- The removal of intact iron–sulfur cores from protein

Active Site Analogues

How can you synthesis active site analogues of non-heme Fe-S proteins?

- There are three types of different active sites in non-heme iron-sulfur proteins; $Fe(S-Cys)_4$ in rubredoxins, $Fe_2S_2(S-Cys)_4$ in 2Fe-2S Fds, and $Fe_4S_4(S-Cys)_4$ in HPIP and 4Fe-4S Fds.
- The synthetic analogs of these active centers such $[Fe(S-R)_4]^-$ and $[Fe(S-R)_4]^{2-}$, $[Fe_2S_2(S-R)_4]^{2-}$, $[Fe_4S_4(S-R)_4]^{2-}$ and $[Fe_4S_4(S-R)_4]^{3-}$ have been prepared and the structures are well established.

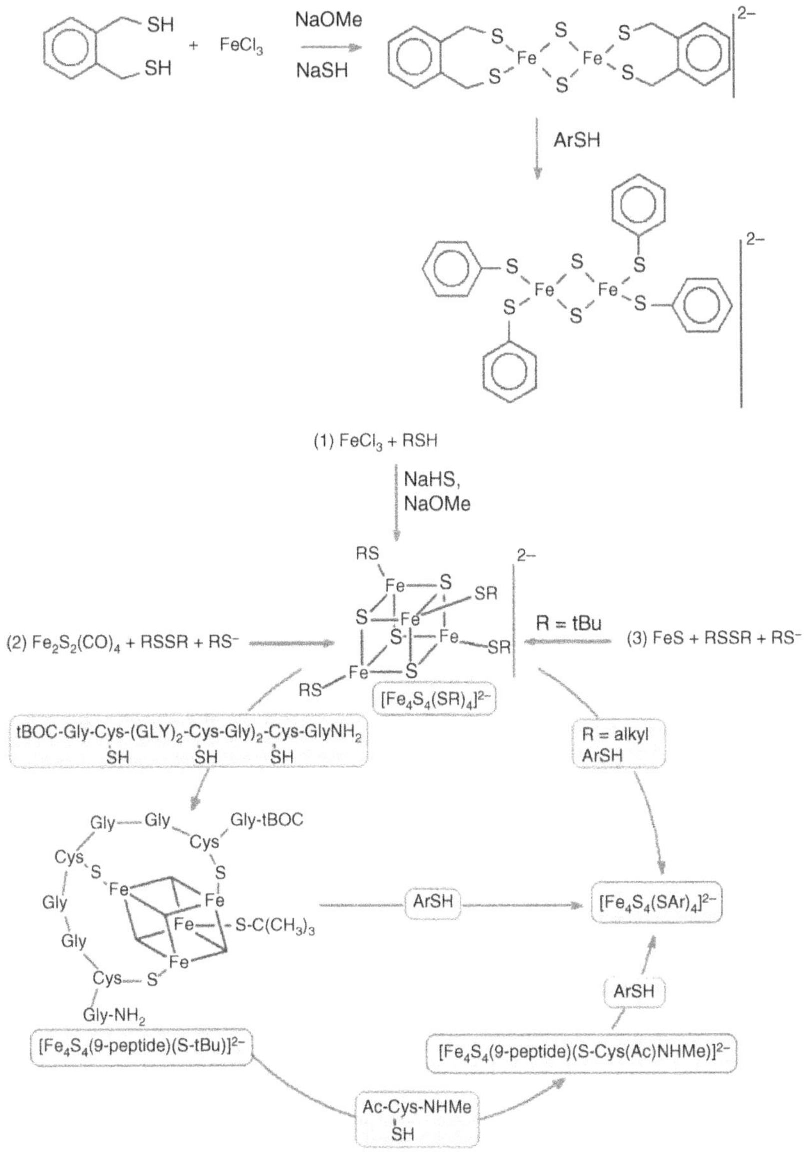

SCHEME 5-12 Analogues for HiPIP and ferredoxins. (Based on Mayerle et al., 1973, 1975; Venkateswara Rao and Holm, 2004; and Que et al., 1974.)

- Scheme 5-12 summarizes some synthetic routs of cysteinate binding active site mimic.
- A number of tetranuclear complexes have been prepared in which glycylcysteinyl oligopeptides are the terminal ligands.
- In addition, the water-soluble complex $[Fe_4S_4(SCH_2CH_2CO_2)_4]^{6-}$, which is isoelectronic with tetranuclear dianions derived from mononegative thiolates, has been synthesized.

ACONITASE

Forms, Roles, and Structures

What is the biological role of aconitase, and the difference between the active and inactive forms?

- Aconitase
  Citrate dehydratase
  Present in mitochondria
  Molecular weight (from beef heart) $\sim 80 \times 10^3$
  7 Fe-ferredoxins, Fe_4S_4 and Fe_3S_4

$$Fe_4S_4 \leftrightarrows Fe_3S_4 \quad + Fe^{2+} \quad \text{(Scheme 5-13)}$$

  $\quad\quad$ Active $\quad$ Inactive

  Fe_3S_4: one Fe corner (and aconitase Cys-S$^-$) removed from Fe_4S_4 cubane structure.

- Aconitase (aconitate hydratase) catalyzes the stereospecific isomerization of citrate to isocitrate via *cis*-aconitate in a nonredox active process.
- The 3D structure of mitochondrial aconitase has been determined and has shown to contain four αβ domains (Fig. 5-38).
- Mössbauer studies of the oxidized form (the inactive type) indicate that the cluster contains three iron atoms, Fe_3S_4.

$$Fd_{ox} + e \leftrightarrows Fd_{red} \quad\quad E^0 = -130 \text{ to} - 420 \, mV$$

$$Fd_{ox} = [Fe_3S_4(Cys)_3]^{2-} \rightarrow Fe_3^{3+}$$

  All display antiferromagnetic coupling (AFC) to give $S = \frac{1}{2}$, $g = 2.01$.

- On the activation of the enzyme with Fe^{2+} ion, the paramagnetic Fe_3S_4 cluster was transformed into a diamagnetic, $S = 0$, form, $2Fe^{2+}2Fe^{3+}$ in $[Fe_4S_4]^{2+}$.
- $[Fe_4S_4]^{2+}$ is prepared by a rapid purification procedure with up to 90% activity.
- $[Fe_4S_4]^{2+}$ transforms into Fe_3S_4 by oxidation.
- The Fe_3S_4 form can also be activated to $[Fe_4S_4]^+$ by the addition of Fe^{2+} and the reducing agent dithionite.

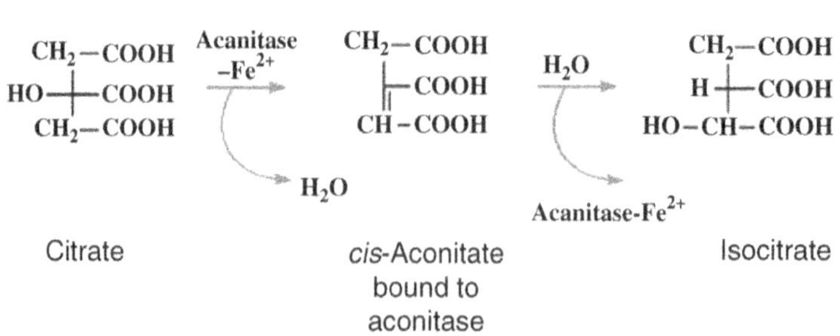

SCHEME 5-13 Biological role of aconitase.

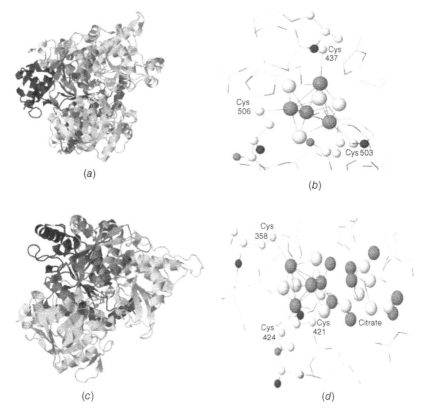

FIGURE 5-38 (*a*) Crystal structure of human aconitase. PDB 2B3Y (Dupuy et al., 2006). (*b*) Active site in human aconitase, PDB 2B3Y (Dupuy et al., 2006). (*c*) Structure of aconitase–citrate complex, PDB 1C96 (Lloyd et al., 1999). (*d*) Fe$_4$S$_4$–citrate complex, PDB 1C96 (Lloyd et al., 1999).

- [Fe$_4$S$_4$]$^+$ has an ESR signal at 2.06, 1.93, and 1.86 typical of those of Fd. *Note*: Fe^{2+}Fe^{3+}, iron high spin → Fe^{2+} ($S_0 = \frac{4}{2}$), Fe^{3+} ($S_0 = \frac{5}{2}$). Therefore, in ferromagnetic coupling (FC), $S = \frac{9}{2}$ and in AFC $S = \frac{1}{2}$:

$$\text{Fe}^{3+}\text{Fe}^{3+} \rightarrow \text{AFC } S = 0 \qquad \text{Fe}^{2+}\text{Fe}^{2+} \rightarrow \text{AFC } S = 0$$

- It is not known which state of [Fe$_4$S$_4$] is present during enzyme turnover?
- The forms of aconitase may convert as follows:

- In all the conversions, it is never necessary to add S^{2-}.

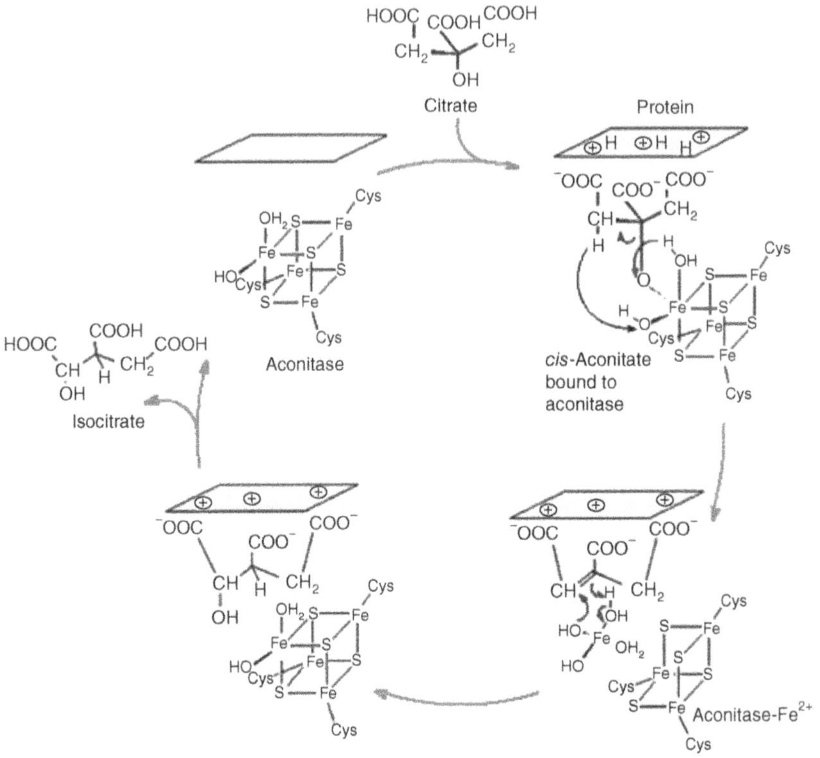

SCHEME 5-14 Three point attachments of substrate to aconitase and one to the labile iron, aconitase–Fe^{2+}– substrate. (Based on Mayes, 1977.) (See the color version of this scheme in Color Plates section.)

Reaction Mechanism

Propose a reaction mechanism for the action of aconitase.

- By contrast with the majority of iron–sulfur proteins that function as electron carriers, the Fe–S cluster of aconitase reacts directly with the substrate.
- The structure of the citrate shows that the four substituent groups are attached to the central carbon atom (Scheme 5-14).
- Three Cys residues have been shown to be ligands of the [Fe$_4$S$_4$]$^{2+}$ center. In the active states, the labile iron ion of [Fe$_4$S$_4$]$^{2+}$ is coordinated not by Cys but by water molecules.

Conjugated Fe–S Proteins

What do conjugated Fe–S proteins mean?

- Iron–sulfur proteins can be classified as simple and complex proteins (Schemes 5-9 and 5–10).
- Complex iron–sulfur clusters perform other enzymatic functions besides electron transfer.

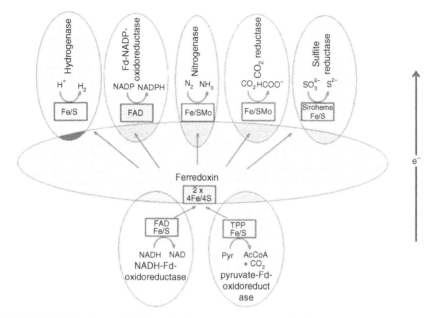

SCHEME 5-15 The soluble ferredoxin acts as electron carrier between reactants (bottom) and terminal enzymes (top). (Modified from Thauer and Schönheit, 1982.) (See the color version of this scheme in Color Plates section.)

- The conjugated Fe–S proteins contain other additional prosthetic groups, such as iron, molybdenum, nickel ion, flavin, or heme (Scheme 5-15).

HYDROXYLASES

Role, Chemical Features, and Reaction Mechanism

How can aromatic compounds be metabolized in living organisms? Give examples of iron-sulfur enzymes involved in metabolism of drugs by hydroxylation, what are the distinguishable chemical properties of this enzyme?

- Hydroxylases, Rieske nonheme iron oxygenase (ROs): Hydroxylases have a 2Fe–2S site that binds FMN:

- They are *bacterial aromatic ring hydroxylases* that incorporate two oxygen atoms into substrates, that is, a *dihydroxylation* reaction.

For example:

1. Benzoate and toluate-1,2-dihydroxylases
2. Phthalate-4,5-dioxygenase reductase (PDR)
3. Naphthalene-1,2-dioxygenase

- A large family of multicomponent mononuclear (nonheme) iron hydroxylases (oxygenases) has been identified.
- Components of bacterial aromatic hydroxylases (oxygenases) constitute two different functional proteins (Scheme 5-16):
  1. Hydroxylase components
  2. Electron transfer components
- Hydroxylase components are either $(\alpha\beta)_n$ or $(\alpha)_n$ oligomers.
- The α-subunit is associated with two prosthetic groups:
  (a) Rieske-type [(Fe$_2$S$_2$) (S$_{Cys}$)$_2$ (N$_{His}$)$_2$] center
  (b) Mononuclear iron [Fe(N$_{His}$)$_2$(O$_{Asp}$) $\cdot$ H$_2$O] center (Fig. 5-39)
- Electron transfer (reductase) components:
  (a) NADH/NDA$^+$ and FAD/FADH$_2$ (FMN and NADH$_2$ are cofactors)
  (b) Rieske–type [(Fe$_2$S$_2$) (S$_{Cys}$)$_2$ (N$_{His}$)$_2$] center.

- Despite significant progress in structural and physicochemical analyses, there is no agreement on the chemical steps in the catalytic cycles of ROs.
- Proposed mechanism:
  1. In the electron transfer component, the NADPH is oxidized to NADP$^+$.
  2. The produced electrons are stored on the flavin until the reductase completes the reduction of the Rieske cluster of the electron transfer component.

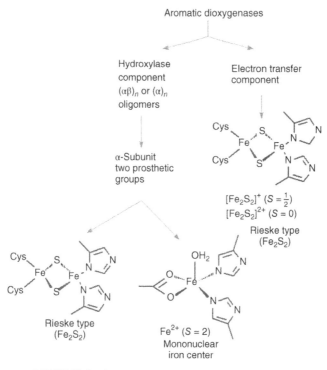

SCHEME 5-16 Chemical components of hydroxylase.

3. This shuttles the electron received from the reductase to the Rieske cluster of the hydroxylase component.

4. Reduction of the Rieske cluster of the hydroxylase is followed by binding the substrate to the nonheme iron center.

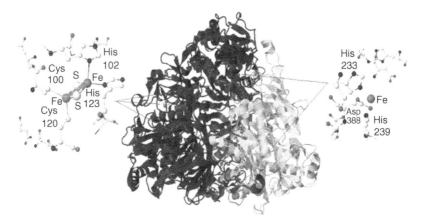

FIGURE 5-39 Crystal structure of hydroxylase component, PDB 2YFI (Mohammadi et al., 2011).

5. Binding of the substrate induces several conformational changes and makes room for oxygen binding.

6. Dioxygen bound in a side-on manner onto nonheme iron is activated by reduction to the peroxo state [Fe^{3+}-(Hydro)peroxo].

7. This state may react directly with the bound substrate; O–O bond cleavage may occur to generate a Fe-oxo-hydroxo species before the reaction.

8. After a *cis*-dihydrodiol is produced, the product is released by reducing the nonheme iron.

HYDROGENASES

Roles and Types

Define the role and main types of hydrogenases.

- Hydrogenases catalyze the reversible oxidation of molecular hydrogen (H_2) and play a vital role in anaerobic metabolism.

- Hydrogenase activity occurs in bacterial and algal species.
- Hydrogen utilization is involved:
  - In nitrogen fixation.
  - In photoproduction of hydrogen.
  - In fermentation of biomass to methane.

$$H_2 + A_{ox} \rightarrow 2H^+ + A_{red}$$

where:

  (a) H_2 is the reductant.

  (b) A is the electron acceptor, such as oxygen, nitrate, sulfate, carbon dioxide, and fumarate.

  (c) The consumption of hydrogen provides the organism with a supply of reductants that may be used for energy generation. This is the *respiratory* role of hydrogenase.

  - The reaction takes place in aerobic and anaerobic organisms:

$$2H^+ + D_{red} \rightarrow H_2 + D_{ox}$$

where:

  (a) Hydrogen production results when the proton is used as a terminal electron acceptor.

(b) Proton reduction (H_2 evolution) is essential in pyruvate fermentation or in the disposal of excess electrons.

(c) The reaction occurs under strictly anaerobic conditions.

- Metal-containing hydrogenases are subdivided into three classes:
  - Fe hydrogenase (iron only)
  - Ni–Fe hydrogenases
  - Ni–Fe–Se hydrogenases
- All contain iron–sulfur clusters.

Sources, Reactions, and ESR Investigations

Give an example of Ni containing enzymes, examples of the reactions that can be catalyzed by these enzymes, and the chemical role of Ni ion.

- The Ni–Fe hydrogenases, when isolated, are found to catalyze both H_2 evolution and uptake with low-potential cytochromes such as cytochrome c_3 acting as
  - electron donors (D) or
  - electron acceptors (A) depending on their oxidation state.
- For example (see also Table 5-5):

TABLE 5-5 Properties of Hydrogenases Containing Nickel Ion

Overall reaction catalyzed	$4H_2 + SO_4^{2-} \rightarrow S^{2-} + 4H_2O$
Source/strain on genus	*Desulfovibrio gigas*
Molecular weight	89000
ESR characteristics	Ni^{3+}: 2.31, 2.23, 2.02
	After H_2 reduction 2.19, 2.16, 2.02
	Fe–S: 2.02 (oxid.)
Source/strain on genus	*Desulfovibrio desulfuricans* II
Molecular weight	75500
ESR characteristics	Ni^{3+}: 2.3, 2.2, 2.0
	After H_2 reduction Ni: $g = 2.28$
	Fe–S: $g = 2.02$ (oxid.), $g = 1.94$ (red.)
Overall reaction catalyzed	$4H_2 + CO_2 \rightarrow CH_4 + 2H_2O$
Source/strain on genus	*Methanobacterium bryantii*
ESR characteristics	Ni^{3+}: 2.30, 2.23, 2.02
Source/strain on genus	*Methanobacterium Thermoautotrophicum* (strain ΔH), 2 types, F_{420} reducing
Molecular weight	170000
ESR characteristics	Ni^{3+}: 2.309, 2.237, 2.017
	After H_2 reduction, Ni: $g = 2.196, 2.140$
	Complex Fe–S signals in reduced state
Source/strain on genus	*Methanobacterium methylviologen*
ESR characteristics	Ni^{3+}: 2.309, 2.237, 2.017
	Complex Fe–S signals in reduced state

- *Desulfovibrio*
  - Overall reaction catalyzed:

$$4\,H_2 + SO_4{}^{2-} \rightarrow S^{2-} + 4\,H_2O$$

  - Molecular weight 8900
  - ESR: after H_2 reduction, Ni^{3+}: $g_z = 2.31$, $g_x = 2.23$, $g_y = 2.02$; oxidized, Fe–S: $g = 2.02$.
- *Methanobacterium*
  - Overall reaction catalyzed:

$$4\,H_2 + CO_2 \rightarrow CH_4 + 2\,H_2O$$

  - Molecular weight 170,000
  - ESR: after H_2 reduction, Ni^{3+}: $g_z = 2.30$, $g_x = 2.23$, $g_y = 2.02$.
- *Vibrio succinogenes*
  - Overall reaction catalyzed:

$$H_2 + fumarate \rightarrow succinate$$

  - Molecular weight 100,000
- The ESR spectra of the Ni–Fe hydrogenases show signals at $g = 2.30$, 2.23, 2.02, which were assigned to Ni^{3+}, a low-spin d^7 metal ion with one unpaired electron.
- Growth of the source bacteria of hydrogenase on a medium isotopically enriched in Ni^{61} that processes a nuclear spin of $\frac{3}{2}$ gives a well-resolved hyperfine splitting of four lines on at least one g value.
- The midpoint potential for the reduction process

$$Ni^{2+} \leftrightarrows Ni^{3+} + e-$$

  in hydrogenase has been measured by ESR spectroscopy.
  - The potential is pH dependent with a value of –145 mV at pH 7.2.
  - The pH dependence shows that one proton is taken up during reduction.
  - This redox potential is much less than those of inorganic Ni^{3+} complexes, although the g values obtained were similar to those observed for hydrogenases.
- It is clear that the protein must supply a set of ligands with suitable coordination geometry to lower Ni^{3+}/Ni^{2+} potential.

Chemical Structures

What are the main structural features of Ni–Fe–hydrogenases?

- The 3D structures of the Ni–Fe hydrogenases from *Desulfovibrio gigas* and *Desulfovibrio vulgaris* have been determined (Fig. 5-40, Scheme 5-17).

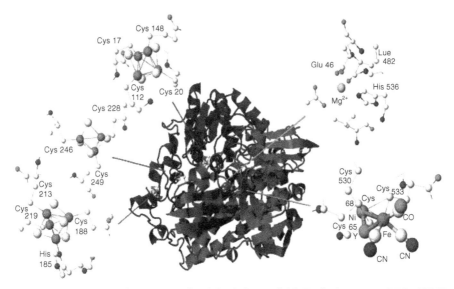

FIGURE 5-40 Crystal structure of oxidized form of Ni–Fe hydrogenase, PDB 2FRV (Volbeda et al., 1996). (See the color version of this figure in Color Plates section.)

- The Ni–Fe hydrogenases are heterodimeric proteins consisting of small (S) and large (L) subunits (Scheme 5-17).
- The large subunit contains a nickel–iron center and Mg center, while the small subunit contains three iron–sulfur clusters, two $[Fe_4S_4]^{2+/1+}$ and one $[Fe_3S_4]^{1+/0}$.
- The Ni is pentacoordinated (square pyramidal) with four equatorial S atoms of Cys residues and bridging the S or O atom as the axial ligand.
- The coordination geometry of the Fe is a slightly distorted octahedron, with three bridging ligands between Ni and Fe (two S of Cys residues and one S or O atom) and three terminal ligands.

Species	Ni-Terminal Ligands	Ni–Fe Bridging Ligands
D. gigas	S (Cys^{65}, Cys^{533})	S (Cys^{68}, O, H)
D. vulgaris	S (Cys^{81}, Cys^{546})	S (Cys^{84}, S or O, H)

- The N of His^{552}, O of Glu^{62}, carbonyl oxygen of Leu^{498}, and three H_2O molecules octahedrally coordinate the Mg ion.
- The Fe–S clusters of the small subunit are distributed along a straight line, with the Fe$_3$–S$_4$ cluster located halfway between the two Fe$_4$–S$_4$ clusters.
- The two Fe$_4$–S$_4$ clusters have been termed proximal (prox.) and distal (dist.) based on their distance to the Ni atom (Table 5-6).

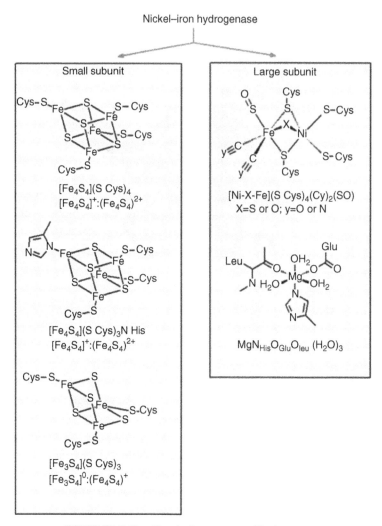

SCHEME 5-17 Chemical structures of hydrogenase.

TABLE 5-6 Redox Potentials and Amino Acid Ligands of Three Fe–S Clusters

Cluster	Redox Potential (mV)	Amino Acid Ligands, *D. gigas*
$[Fe_4\text{-}S_4]_{prox}$	-340	Cys[17], Cys[20], Cys[112], Cys[148]
$[Fe_4\text{-}S_4]_{dist}$	-290	His[185], Cys[188], Cys[213], Cys[219]
$[Fe_3\text{-}S_4]$	-35	Cys[228], Cys[246], Cys[249]

Dihydrogen and Dihydride Complexes

Characterize the binding of dihydrogen molecule to transition metal and the expected binding in the active site of the hydrogenase.

First, consider the hydrogen with the metal site:

- The hydrogen molecule reacts with the transition metal and forms either a dihydrogen or dihydride complex.
- The dihydrogen complex is formed via the σ orbital of the H–H bond, which acts as a σ donor, while the empty σ^* orbital of H_2 acts as a weak acceptor. When the back donation is strong enough, the electron density will build up in the σ^* orbital and cause cleavage of the H–H bond, leading to the formation of a dihydride complex.

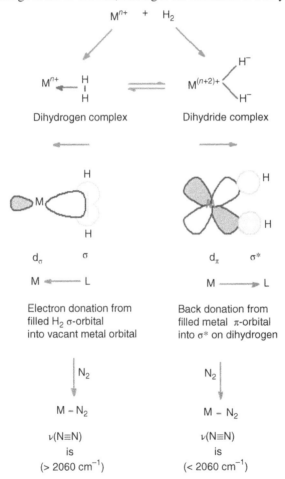

- H_2 and N_2 react in the same way with a variety of transition metals and form complexes. N_2 also displaces H_2 from the H_2 complexes and forms a N_2 complex without altering the stereochemistry of the coordination sphere. The

H_2 complexes can be judged by the stretching frequency of the corresponding N_2 complexes, $v(N \equiv N)$.

- In H_2 complexes, the formed N_2 complexes upon replacement of the H_2 by N_2 have $v(N \equiv N)$ between 2160 and 2060 cm^{-1}.
- The electron back donation from the metal ion and the formation of the dihydride are indicated when $v(N \equiv N)$ is less than 2060 cm^{-1}. For example:

$MoH_2(PCY_3)_5 + N_2$ $\rightarrow MoN_2(PCY_3)_5$
   Dihydride complex $v(N \equiv N) = 1950$ cm^{-1} (<2060 cm^{-1})

$Mo(Et_2PCH_2CH_2PEt_2)H_2(CO)_2 + N_2$ $\rightarrow Mo(Ph_2PCH_2CH_2PPh_2)N_2(CO)_2$
   Dihydride complex $v(N \equiv N) = 2050$ cm^{-1} (<2060 cm^{-1})

$Mo(Ph_2PCH_2CH_2PPh_2)H_2(CO)_2 + N_2 \rightarrow Mo(Ph_2PCH_2CH_2PPh_2)N_2(CO)_2$
   Dihydride complex $v(N \equiv N) = 2090$ cm^{-1} (>2060 cm^{-1})

- In the active site of hydrogenase, H_2 can react with the metal sulfide system at the metal site as well as at the S atom:

$R_2Mo_2S_4 + 2H_2$ $\rightarrow R_2Mo_2(SH)_4$ Bridging SH group (thiol)

$\{RhS[P(C_6H_5)_2$ $\rightarrow \{Rh(H)(SH)$ Bridging SH group +
   $CH_2CH_2]_3CH\}_2 + 2H_2$ $[P(C_6H_5)_2CH_2CH_2]_3CH\}_2$ hydride (thiol-hydride)

- Therefore, the formation of dihydrogen-, dihydride-, thio-hydride-, and/or dothiol complex must be considered as possibilities for the hydrogen-hydrogenase interaction.

Catalytic Mechanism

What is the possible catalytic mechanism of Ni–Fe–hydrogenases? What are the possible oxidation states, the forms that participate in the catalytic cycle, also those forms that are involved in activation and inactivation of Ni–Fe–hydrogenases?

- The resting state of the dinuclear clusters is Ni^{2+}–Fe^{3+}, H_2 first binds to Fe in the form of molecular hydrogen complex and then undergoes heterolytic splitting $(H_2 \rightarrow H^- + H^+)$.
- *Hydride* transfers to Fe and *proton* transfer to the adjacent Cys thiolate ligand.
- This is accompanied by decoordination of the protonated Cys thiol from Ni while remaining bound to Fe.
- Simultaneously, the cyanide ligand on Fe binds the Ni atom in a bridging binding mode.
- The hydride bound to Fe transfers to Ni, which is necessary for H^+ or electron transport.

Cys E

O=S S

O=C··· Fe³⁺ —X— Ni²⁺ — SH Cys

C≡N S SH–Cys

Cys

H⁻ + H⁺ ⇌ H₂ Net H₂-ase

Cys ES

O=S S Cys

O=C··· Fe³⁺ —X— Ni²⁺ — SH

HS CN H SH–Cys

Cys

Cys ES

O=S S Cys

O=C··· Fe³⁺ —X— Ni²⁺ — SH

C H S SH–Cys

N H Cys

Cys ES

O=S S Cys

O=C··· Fe³⁺ —X— Ni²⁺ — SH

HS H CN SH–Cys

Cys

- A mechanism of electron transfer from the Ni active site through the Fe–S clusters to cytochrome c_3 has been suggested.
- The ESR studies of Ni–Fe hydrogenases have shown the existence of three paramagnetic states, A, B, and C, and three other forms of the hydrogenases are ESR silent: X, Y, and Z.

Ni–A:	2.32	2.24	2.01 → low spin Ni^{3+}
Ni–B:	2.35	2.16	2.01 → low spin Ni^{3+}
Ni–C:	2.19	2.15	2.01 → Ni^+ or Ni^{3+} hydride have been suggested, but Ni^+ is favored because of the absence of the strong proton hyperfine coupling expected for Ni^{3+}
Ni–C$_{fully\ reduced}$:	ESR silent		→ Ni^0, Ni^{2+} or Ni^{2+} hydride

- Forms A, B, and C participate in the catalytic cycle of hydrogenases, while X, Y, and Z are involved in the activation and inactivation of the enzyme.
- The ESR signal of Ni–A disappears on hydrogen reduction, and the enzyme is converted to a higher activity form. Reoxidation of this form yields the Ni–B form. Further reduction of the enzyme shows the ESR signals of Ni–C, which also finally disappears.

NITROGENASES

Reactions Catalyzed by Nitrogenase

Give examples of reactions that are catalyzed by nitrogenase. Define the biological role of nitrogenase.

- Nitrogenase enzyme catalyzes the fixation of biological nitrogen as well as the following:

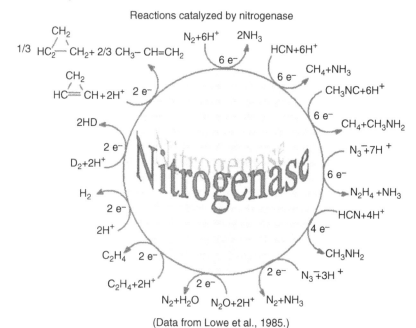

Reactions catalyzed by nitrogenase

(Data from Lowe et al., 1985.)

- Biological nitrogen fixation is the process whereby some bacteria and blue green algae convert atmospheric nitrogen into ammonia. The nitrogen cycle ($N_2 \rightarrow NH_3 \rightarrow NO_3^- \rightarrow$ plant) is of vital importance in agriculture in supplying soluble nitrates, for example, *Azobacter vinelandii* N_2-ase. The overall reaction is

$$N_2 + 6H^+ + 6e^- \rightarrow 2NH_3$$

- Nitrogenase binds and hydrolyzes 2 MgATP, yielding 2 MgADP and 2 P_i for each electron that transferred from the Fe protein to the Mo–Fe protein of the nitrogenase (Scheme 5-18):

$$16\,Mg\;ATP + 8\,H^+ + 8\,e^- \rightarrow 16\,Mg\;ADP + H_2 + 16\,P_i$$

Nitrogenase and Kinetically Inert Reactions

The reduction of N_2 to $2NH_3$ is thermodynamically favorable:

$$N_2 + 6H^+ + 6e^- \rightarrow 2NH_3 \qquad \Delta G^\circ = -3.97\,\text{kcal/mol}$$

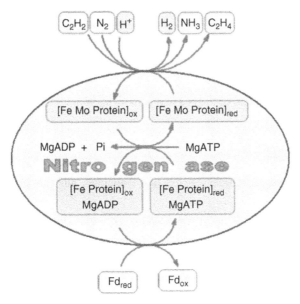

SCHEME 5-18 Nitrogenase reactions. (See the color version of this scheme in Color Plates section.)

This reduction is kinetically inert. How does nitrogenase overcome these barriers?

- The thermodynamic reduction of N_2 by H_2 is a favorable process ($\Delta G° < 0$). However, all the intermediates (N_2H_2, $N_2H_5^+$, . . .) on the pathway between N_2 and NH_3 are higher in energy than either the reactants or the products:

$$N_2 + 2H^+ + 2e^- \rightarrow N_2H_2 \quad E° \sim -1000 \text{ to } -1500 \text{ mV}$$
$$N_2 + 5H^+ + 4e^- \rightarrow N_2H_5^+ \quad E° = -695 \text{ mV}$$

Normal biological reductants cannot achieve these reactions.

- Nitrogenase overcomes these barriers:
  1. Through avoiding the unfavorable intermediates: The six-electron reduction could happen in a concerted or close to concerted way:

$$N_2 + 8H^+ + 6e^- \rightarrow 2NH_4^+ \quad E° = -280 \text{ mV}$$

  which is accessible to biological reductants such as ferredoxins.
  2. Through stabilizing the favorable intermediates: The intermediates could be complexed to the metal center to stabilize them more than either the reactants or the products.

3. Through coupling with a favorable process: The formation of the unfavorable intermediate could be coupled with a favorable process such as the hydrolysis of ATP or the evolution of dihydrogen so that the overall process is favorable.

A combination of the last two is most likely used to accelerate the reduction of N_2 to NH_3.

Structural Features

What are the main structural features of nitrogenase, and the chemical structure of the active sites?

- Nitrogenase is made up of two proteins:
    (i) Component I, Mo–Fe protein (cofactor)
    (ii) Component II, Fe protein (Fig. 5-41, Scheme 5-19)
- Sephedex column separates the Fe protein and Mo–Fe protein.
- Mo-Fe protein:
    brown,
    Tetrameric, containing two each of two subunits,
    Arranged at a corner of square
    Appears to have 7Fe & 8 S, for 1 Mo,
    Fe(1) is T_d, but other are trigonal, and may have H^- ligands?
    ○ Molecular weight 200,000–220,000

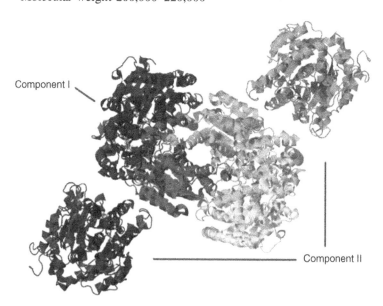

FIGURE 5-41 Structural characterization of cross-linked complex of nitrogenase, PDB 1M1Y (Schmid et al., 2002).

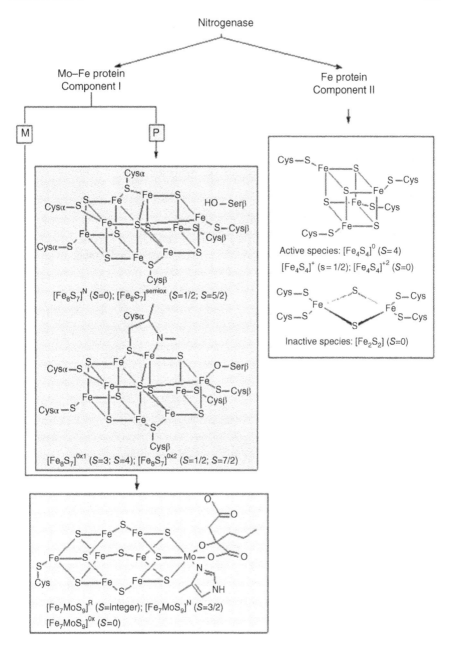

SCHEME 5-19 Chemical structures of the nitrogenase components.

- ○ Mo has $3\,\mu\text{-}S^{2-}$, plus His-N and two sites occupied by citrate cofactor, MoS_3NO_2, in a pseudo-octahedral arrangement
- ○ The Mo–Fe protein contains two types of metals centers, called M-cluster and P-cluster
- The M-cluster can adopt at least three oxidation states:
  - ○ Native, semireduced, M^N
  - ○ Oxidized, M^{Ox}
  - ○ Reduced, M^R
- The *P-cluster* can reversibly achieve four oxidation states, from native reduced P^N to P^{Ox2}, Scheme 5-19,
  - ○   Fe: $\sim 34\text{--}38$
            S^{2-}: $\sim 26\text{--}28$ in *A. vinelandii*
- The M-cluster, FeMoco, consists of two cuboidal fragments, Fe_4S_3 and Fe_3MoS_3, which are bridged by three sulfur atoms, and homecitrate, *N*-[3-(dimethylamino)propyl]N'-ethylcarbodiimide:

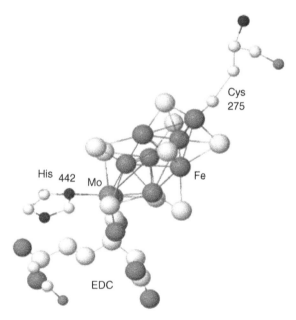

(EDC) = *N*-[3-(dimathylamino)propyl]-
N'-athylearbodimide

- The Mo is octahedrally coordinated, whereas Fe atoms at the interface of two cuboidal fragments have open coordination sites and supposedly bind N_2.
- FeMoco attaches to the α-subunit by two residues (Cys^{275} and His^{442} in the *A. vinelandii* Mo–Fe protein.
- However, the crystals of the Mo–Fe protein were characterized with respect to the oxidation state, and it was found that the P-cluster in both oxidation states exist as an [Fe_8S_7] cluster:

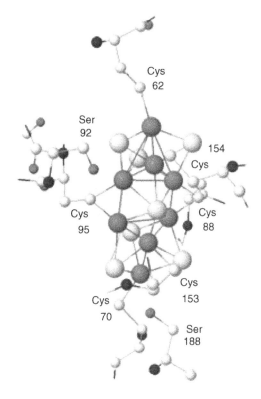

- The P-cluster may be described as containing bridged [Fe$_4$S$_4$] and [Fe$_4$S$_3$] clusters. Two bridging Cys residues originate from each of the α- and β-subunits (Cys$_\alpha$ ^{88} and Cys$_\beta$ ^{95} in the *A. vinelandii* Mo–Fe protein).

- Apart from the two bridging Cys residues, each cube of the P-cluster is attached to either the α- or β-subunit through two Cys residues (Cys$_\alpha$ ^{62}, Cys$_\alpha$ ^{154}, Cys$_\beta$ ^{70}, Cys$_\beta$ ^{153} in the *A. vinelandii* Mo–Fe protein).

- In the oxidized state (P^{ox}), two additional protein ligands to the P-cluster are present, backbone amide nitrogen of Cys$_\alpha$^{88} and O of Ser$_\beta$ ^{188}.

- In the reduced state (P^{N}), these two non-cysteinyl ligands are replaced by interactions with the central sulfur atom, which now adopts distorted octahedral coordination geometry.

- Component II (Scheme 5-19)

- Fe protein: yellow; dimer, two equal subunits; molecular weight ∼65,000, 30,000 for each subunit.

- Fe, S^{2-}

   ○ 4 Fe and 4 S^{2-}, as Fe$_4$S$_4$ cubane clusters, act as specific electron carriers to the Mo–Fe protein

   ○ Cys–Cys bridged a pair of Fe$_4$S$_4$ clusters ∼14 Å away from Mo–Fe cofactor

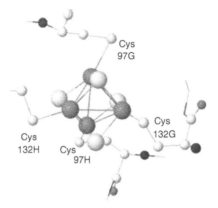

- Component II has ATP binding site(s) and one [Fe₄S₄] cluster per homodimer. It supplies energy by ATP hydrolysis and transfers electrons from reduced ferredoxin or flavodoxin to component I.
- Each subunit folds as a single α/β-type domain, which together symmetrically ligates the surface exposed [Fe₄S₄] cluster through two cysteines from each subunit. A single bound ADP molecule is located in the interface region between the two subunits:

Role of each Component in Nitrogenase

What is the role of each component of nitrogenase? The electron transfers only occur in the presence of MgATP, what is the main function of MgATP?

- The M-cluster is also termed the iron–molybdenum cofactor (FeMoco) and provides the substrate binding site:

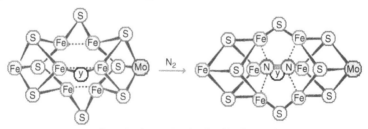

N₂ molecule attaches itself to Mo–Fe protein.

- In the crystal structure of the Mo–Fe protein, the location of the P-cluster is equidistant (~ 14 Å) between the $[Fe_4S_4]$ of the Fe protein and the FeMoco. This arrangement strongly suggests that electrons are transferred from the Fe protein to FeMoco via the P-cluster.

- The function of the P-cluster may be the electron transfer between the $[Fe_4S_4]$ center of the Fe protein and the FeMoco:

$$\begin{Bmatrix} \text{Ferredoxin} \\ \text{or} \\ \text{flavodoxin} \end{Bmatrix} \overset{e}{\rightarrow} \begin{Bmatrix} [Fe_4S_4] \\ \text{Fe protein} \end{Bmatrix} \overset{e}{\rightarrow} \begin{Bmatrix} [\text{P-cluster}] \overset{e}{\rightarrow} [\text{M-cluster}] \\ \text{Mo protein} \end{Bmatrix} \overset{e}{\rightarrow} \text{substrate}$$

- Electrons flow from the ferredoxin (a reducing agent) to the Fe protein. Then, the electrons move to the Mo–Fe protein and finally to the substrate, such as N_2, H^+, or C_2H_2. The electron transfers only occur in the presence of MgATP supplies -30.5 kJ/mol to the endergonic e-transfer.

- The main function of ATP is electron activation to form a strong nucleophile, probably a metal hydride, which evolved H_2 and/or reduced N_2-ase substrate:

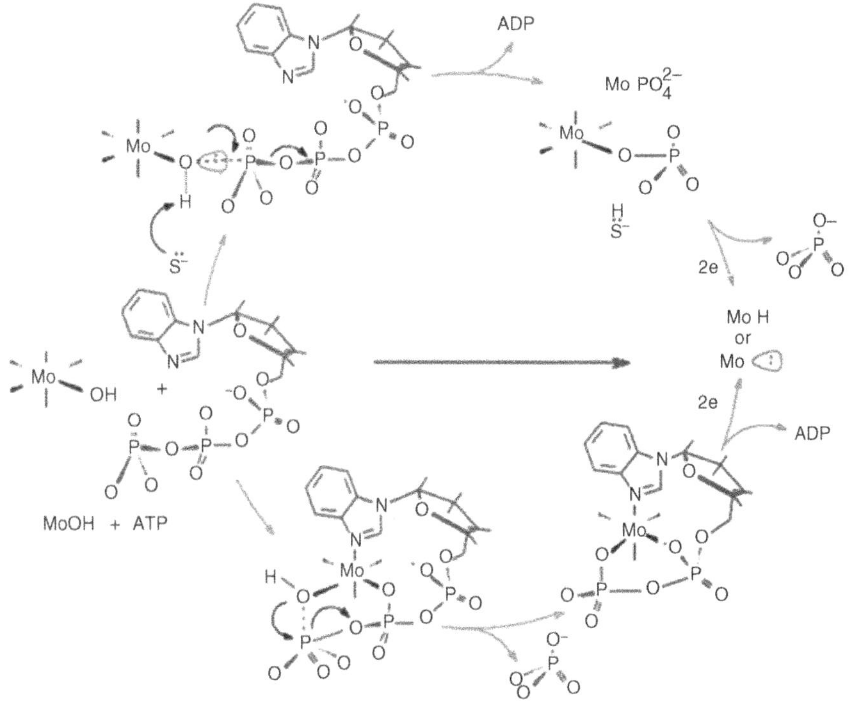

ATP facilitates formation of MoH or Mo (Modified from Hardy et al., 1973.)

Dinitrogen Complexes

How can you explain the interaction of N_2 molecule with transition metal? Give examples of dinitrogen complexes.

- N_2 at room temperature is *inert* due to the large difference between its filled (≤ -15.6 eV) and vacant (≥ -7 eV) molecular orbitals (Fig. 5-42).
- Dinitrogen complexes
  - The ligand-to-metal σ bond arises from overlap of a filled σ orbital (σ_4) of N_2 with a vacant σ orbital of the metal.
  - The ligand–to–metal π bond involves a filled metal d orbital and a vacant π_g orbital of dinitrogen.

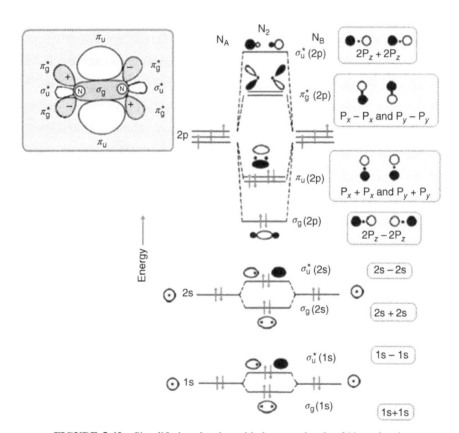

FIGURE 5-42   Simplified molecular orbital energy levels of N_2 molecule.

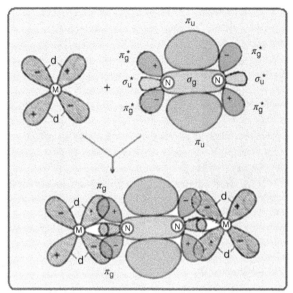

Binding possibility in *xz* or *yz* planes

- Examples of dinitrogen–metal complexes:

$$[Ru^{3+}(NH_3)_5Cl]^{2+} + OH^- \longrightarrow [Ru^{3+}(NH_3)_5(OH)_2]^+$$

$$\downarrow e^-$$

$$[Ru^{2+}(NH_3)_5N_2] \xleftarrow{\;N_2\;} [Ru^{2+}(NH_3)_5(OH)_2]$$

$$Mo^{3+}Cl_3(THF)_3 + Ph_2{-}CH_2{-}CH_2\;Ph_2 \xrightarrow{\;N_2,\,e^-\;} [Mo(N_2)(Ph_2{-}CH_2{-}CH_2{-}Ph_2)_2]$$

$$\downarrow H^+$$

$$N_2 + 2NH_4^+ + [Mo^{4+}(Ph_2{-}CH_2{-}CH_2{-}Ph_2)_2]$$

Catalytic Mechanism

What are the proposed mechanisms for biological N_2 fixation?

- Two basic mechanisms are proposed:
  1. Through metal nitrides as intermediates (Scheme 5-20):

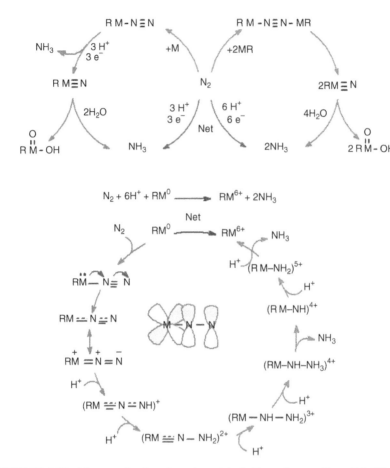

SCHEME 5-20 Nitrogen fixation through metal nitride as intermediate. (Modified from Owsley and Helmkamp, 1967, and Allen and Senoff, 1965.) (See the color version of this scheme in Color Plates section.)

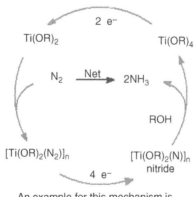

An example for this mechanism is
given by Tamelen et al. (1968)

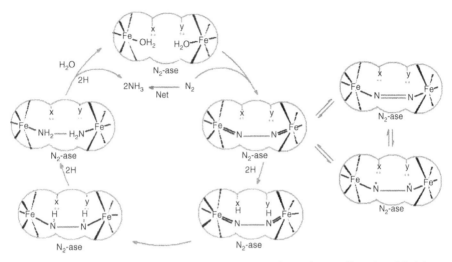

SCHEME 5-21 Nitrogen fixation through metal hydrazine as intermediate. (Modified from Hardy et al., 1968, and Winfield, 1955.)

2. Through diazene and/or hydrazine:
   (i) Based on the thermodynamics of N_2 chemisorption (Scheme 5-21) on iron
      o The attractive concepts are:
         (a) Dinuclear site
         (b) Stepwise reduction
         (c) Explanation of H_2 inhibition
         (d) Avoidance of energetically unfavorable diazene (48.7 kcal/mol)
      o The limitations are:
         (a) Incapability to accommodate many of the known substrates of N_2-ase
         (b) Absence of a predictable hydrogenase activity in purified N_2-ase
   (ii) Nitrogen fixation by avoiding the unfavorable thermodynamics of diazene intermediate (Scheme 5-22) through a four-electron addition
   (iii) Nitrogen fixation by avoiding the diazene through dinitrogen insertions (Scheme 5-23)
   (iv) Finally invoking separate sites on N_2-ase for electron activation and substrate complexation (Scheme 5-24)
- Electron activation occurs at one site while substrate binding occurs at a separate site.

BINUCLEAR IRON PROTEINS

Examples and Role

Give examples of binuclear iron proteins, and specify the biological role of each.

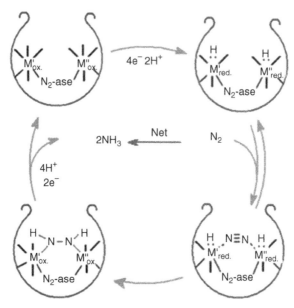

SCHEME 5-22 Nitrogen fixation through metal diazene as intermediate. (Modified from Shilov, 1970, and Borodko and Shilov, 1969.)

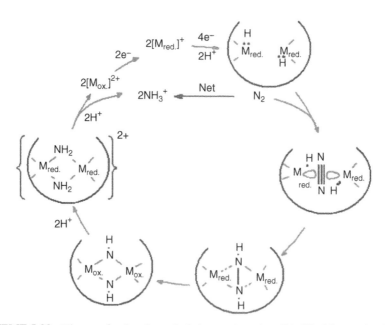

SCHEME 5-23 Nitrogen fixation through dinitrogen insertion. (Modified from Brintzinger, 1966.)

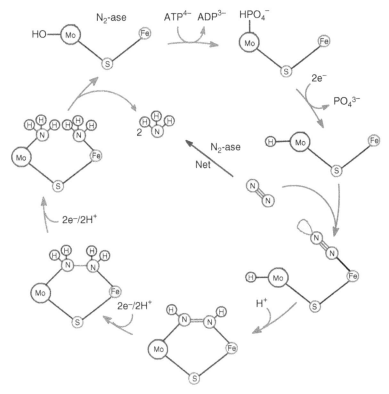

SCHEME 5-24 Nitrogen fixation through substrate complexation. (Modified from Hardy et al., 1971.) (See the color version of this scheme in Color Plates section.)

- Hemerythrin: O_2 transport protein
- Ribonucleotide reductase: Reduction enzyme
- Purple acid phosphatase: Hydrolysis enzyme
- Methane monooxygenase: Oxidation enzyme

HEMERYTHRIN

Biological Sources

What are the main sources of hemerytherins?

Hemerythrin has been found in four different invertebrate phyla:

1. *Sipunculid worms:* Marine worm, peanutworm, spoonworm, a group of elongated, often spindle-shaped, unsegmented marine worms. The head usually has one or more rings of tentacles. Peanutworms vary in length from a few to 500 mm (1.6 ft) or more in length.

2. *Polychaetes:* About 5,400 living species are known. Polychaetes, which include rag worms, lugworms, bloodworms, sea mice, and others, are marine worms

notable for well-defined segmentation of the body. They vary in size from a few millimeters to about 3 m (10 ft). In addition, they are divided informally into two groups: *errantia*, or free moving forms, and *sedentaria*, or tube-dwelling forms.

3. *Priapulids:* Marine, mud inhabiting, unsegmented. The largest of the priapulids are 10–15 cm (4–6 in.) long and inhabit the colder seas, while the smallest, several millimeters long, inhabit warmer seas.

4. *Brachiopods:* Marine invertebrates, having bivalve dorsal and ventral shells enclosing a pair of tentacled, armlike structures that are used to sweep minute food particles into the mouth. Also called *lampshell*.

Oxygen-Carrying Proteins

Give examples of oxygen–carrying proteins, and summarize the contrasts among these proteins.

Examples of oxygen-carrying proteins and the contrasts among these proteins are summarized in Fig. 5-43.

Structural Features

What are the main biological characterizations and the possible chemical forms of hemerythrin?

- Hemerythrin (Hr)
  - Molecular weight: 13.5 kDa/subunit
  - Role: Reversibly combines with molecular oxygen:

$$Hr\,(colorless) + O_2 \leftrightarrows HrO_2\,(pink)$$

  - Source: Generally found in the main body cavity of marine invertebrates and serve mainly in oxygen storage, whereas hemoglobins are found in the circulatory system and serve chiefly in oxygen transport.

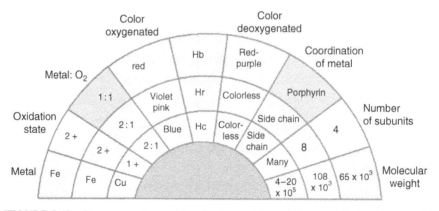

FIGURE 5-43 Oxygen-carrying proteins: Hb, hemoglobin; Hr, hemerythrin; Hc, hemocyanin.

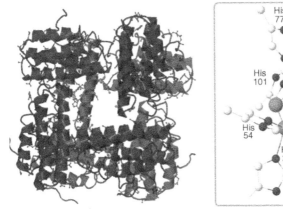

FIGURE 5-44 Structure of deoxy hemerythrin at 2.0 Å resolution, PDB 1HMD (Holmes et al., 1991).

- ○ Two Fe ligand per subunit (Binuclear iron protein).
- ○ Fe ligands: His^{25}, His^{54}, His^{73}, His^{77}, His^{101}, Glu^{58}, Asp^{106}, (Tyr^{109}) (Fig. 5-44)
- • Oxygen binding
  - ○ In oxygen transport, the sigmoidal shape and pH dependence of the oxygen equilibrium curves are functionally valuable characteristics.
  - ○ The oxygen binding curve of *Sipunculus* hemerythrin shows, in contrast to hemoglobin, only slight sigmoidal character and lacks the Bohr effect (i.e., oxygenation is independent of pH).
  - ○ In contrast, hemoglobin curves are strongly sigmoidal, especially at lower pH and exhibit a marked Bohr effect.
  - ○ The difference between oxygen equilibria in these two proteins can be rationalized in terms of their physiological function.
- • Coelomic fluid, usually octamer, $Hr_8 = 107{,}000$:

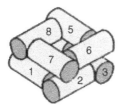

Representation of subunit
arrangement hemerythrin (Hughes, 1981)

Unusual trimers are also found, $Hr_3 = 40{,}000$.

- • Muscle (storage) myohemerythrins are monomers from *Sipuncula* (worms) *Phascolopsis* (*golfingia*) *gouldii*, *Themiste zostericola*, *Phoscalosoma spp*, *priapulids*, lampshells (*brachiopods Lingula and Glottidia*), and *ancient phyla*.
- • Native hemerythrin dissociates into its monomeric form by addition reagents that react with SH groups (Scheme 5-25).

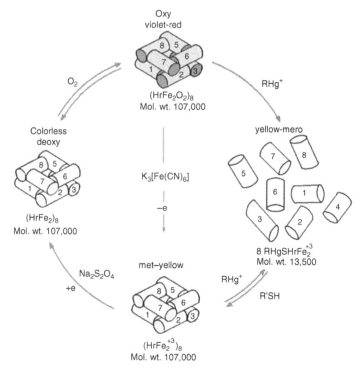

SCHEME 5-25 Relationships between deoxy-, oxy-, and methemerythrin and their monomeric subunit. (Modified from Okamura and Klotz, 1973, and Kurtz et al.,1977. Reproduced
by permission of Elsevier.)

Chemical Combination Reactions

**Methemerythrin combines with a variety of small ligands; NCO^-, N_3^-, NCS^-,
Cl^-, F^-, etc.**

Fill in the spaces:

(a)

$$Fe_2^? + O_2 \rightleftharpoons Fe_2^?O_2^? \xrightarrow{K_3[Fe(CN)_6]} (----)$$

(---Hr)   (---Hr)

$\downarrow e^-$

$Fe^?Fe^?$ (-----Hr)

$\downarrow e^-$

$Fe^?Fe^? \xrightarrow{X^-} X^-.Fe_2^?$

(---Hr)

$X^- = ----$

(b)

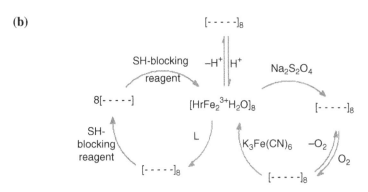

$$[- - - - -]_8$$

SH-blocking reagent $-H^+$ H^+ $Na_2S_2O_4$

$8[- - - - -]$ $[HrFe_2^{3+}H_2O]_8$ $[- - - - -]_8$

SH-blocking reagent   L $K_3Fe(CN)_6$ $-O_2$ O_2

$[- - - - -]_8$ $[- - - - -]_8$

(c) $HL + HrFe_2O_2 \rightarrow (- - - - -) + (- - - - -)$

$L = - - - - - -$

• The answers are:

(a) The addition of oxidizing agent such as $K_3Fe(CN)_6$ to oxygenated hemerythrin produces aquamethemerythrin.

$$Fe_2^{2+} + O_2 \rightleftharpoons Fe_2^{3+}O_2^{2-} \xrightarrow{K_3[Fe(CN)_6]} Fe_2^{3+}H_2O$$

Deoxy-Hr   Oxy-Hr

$\downarrow e^-$

$Fe_2^{2+}Fe^{3+}$

$\downarrow e^-$

$Fe_2^{3+} \xrightarrow{X^-} X^-.Fe_2^{3+} \xrightarrow{X^-} X^-_2.Fe_2^{3+}$

Met-Hr

$X^- = NCO^-, N_3^-, NCS^-, Cl^-, F^-, etc.$

(b) Interconversions of chemical forms of hemerytherin:

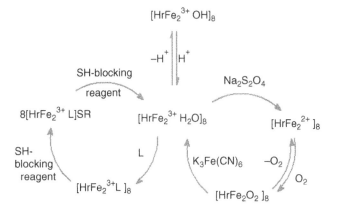

$$[HrFe_2^{3+} OH]_8$$

$-H^+$ H^+

SH-blocking reagent $Na_2S_2O_4$

$8[HrFe_2^{3+} L]SR$ $[HrFe_2^{3+} H_2O]_8$ $[HrFe_2^{2+}]_8$

SH-blocking reagent   L $K_3Fe(CN)_6$ $-O_2$ O_2

$[HrFe_2^{3+}L]_8$ $[HrFe_2O_2]_8$

- Oxyhemerythrin at 4 °C changes slowly to methemerythrin and loses its capacity to bind to oxygen. Hydrogen peroxide is detected as one of the reaction products:

  (c) $HL + HrFe_2O_2 \rightarrow HrFe_2^{3+}L + HO_2^-$

  $L = F^-, NO_2^-, N_3^-, HCO_3^-$

- Each subunit of the met-protein binds ligands such as Cl^-, Br^-, and N_3^- in a $1:2\ Fe^{3+}$ complex. The visible peak positions of these complexes vary slightly.

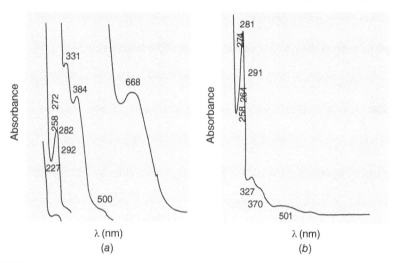

$X = Cl^-, Br^-, F^-$ $X = H_2O, HO^-, F^-$

$= \,>\!N\!\equiv\!C\!\equiv\!S,\ \ >\!N\!=\!N\!\equiv\!N,\ \ >\!C\!\equiv\!N,\ \ >\!N\!\equiv\!C\!\equiv\!O$

Electronic Spectra

Figure 5-45 represents the electronic absorption spectra of metchlorohemerythrin and oxyhemerythrin. Explain the main spectroscopic properties of hemerythrin.

- The electronic absorption spectrum of this form is relatively clear down to 300 nm. In the UV, the peak at about 280 nm points to the tyrosine residue.
- The electronic absorption spectrum of metchlorohemerythrin shows that bands at 668 and 500 nm may be assigned as low-field bands in a ^6A-type Fe^{3+} complex $[^6A \rightarrow (^4A_1, {}^4E)$, and $^6A \rightarrow {}^4T_2]$ (Fig. 5-45, see p. 153).
- The bands at 384 and 331 nm in the metchlorohemerythrin are interpreted as *simultaneous pair electronic transitions* $[(^6A \rightarrow {}^4T_1)_{Fe1} + (^6A \rightarrow {}^4T_2)_{Fe2}]$ or ligand $\rightarrow Fe^{3+}$ charge transfer band transitions are also a possible explanation.

FIGURE 5-45 Electronic absorption spectra of (*a*) metchlorohemerythrin, and (*b*) oxyhemerythrin. (From Gray and Schurar, 1973. Reproduced by permission of Elsevier.)

○ The electronic spectrum of oxyhemerythrin closely resembles that of metchlorohemerythrin (Fe^{3+}), except for the much greater intensity in the band at about 500 nm.

○ In the oxy form, both irons have been oxidized to Fe^{3+} and the O_2 has been reduced to O_2^{2-}:

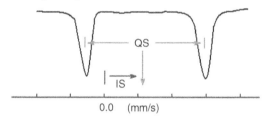

* The larger intensity at 501 nm is attributed to $O_2^{2-} \rightarrow Fe^{3+}$.

* The bands at 370 and 327 nm in oxyhemerythrin are similar to the 387- and 331- nm bands in the metchloro derivative.

Mössbauer Data and Molecular Characterization

What do you conclude from the Mössbauer data of hemerythrin derivatives given in Table 5–7?

* The Mössbauer spectrum of deoxyhemerythrin consists of a single quadrupole doublet (Table 5-7).

○ The large quadrupole splitting (QS = 2.86 mm/s) and the high isomer shift (IS = 1.15 mm/s) are characteristic of high-spin Fe(II), in agreement with the magnetic susceptibility results:

○ The similarity in the Mössbauer spectra of the two Fe nuclei in each subunit indicates that they are in a similar environment.

TABLE 5-7   Mössbauer Parameters of Methemerythrin, Oxyhemerythrin, and Deoxyhemerythrin

Complex	Ligand	Mössbauer Parameters	
		Isomer Shift (mm/s)	Quadruple splitting (mm/s)
Methemerythrin	H_2O	0.46	1.57
	N_3	0.50	1.91
	F	0.55	1.93
	Cl	0.50	2.04
	NCS	0.55	−1.92
Oxyhemerythrin	O_2	0.46	1.87
		0.47	0.94
Deoxyhemerythrin	H_2O	1.15	2.86

- In methemerythrin (Table 5-7):
  - The mössbauer spectra of the aquo, thiocyanate, fluoride, chloride, and azide derivatives of hemerythrin exhibit a single quadrupole doublet.
  - The isomer shifts of all these derivatives are in the range of 0.45–0.545 mm/s, characteristic of high-spin Fe(III).
  - The Mössbauer parameters are consistent with antiferromagnetically coupled high-spin Fe(III):
- In oxyhemerythrin (Table 5-7):
  - The Mössbauer spectrum consists of two quadrupole doublets:

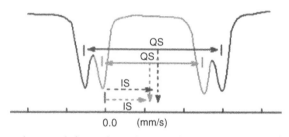

  - Both methemerythrin and oxyhemerythrin are clearly different from the spectrum of Fe (II) in deoxyhemerythrin (Table 5–7).
  - This indicates that:

    (a) Both iron atoms are involved in oxygen binding.

    (b) The environments of the two iron nuclei are different.

  - Both doublets have isomer shifts very similar to those of the met derivatives of hemerythrin.
  - The similarity between the isomer shifts suggests that both iron atoms in oxyhemerythrin are in the +3 valence state.
  - Magnetic moment data in the 2–200 K range for the oxyforms establish antiferromagnetic behavior.

RIBOTIDE REDUCTASE, PURPLE ACID PHOSPHATE, AND METHANE MONOOXYGENASE

Determine the biological role and possible reaction mechanism of:

  (i) **Ribotide reductase**

  (ii) **Purple acid phosphate**

  (iii) **Methane monooxygenase**

Ribotide Reductase

Tetramer from two different subunits

B_1 subunit, molecular weight ~80,000

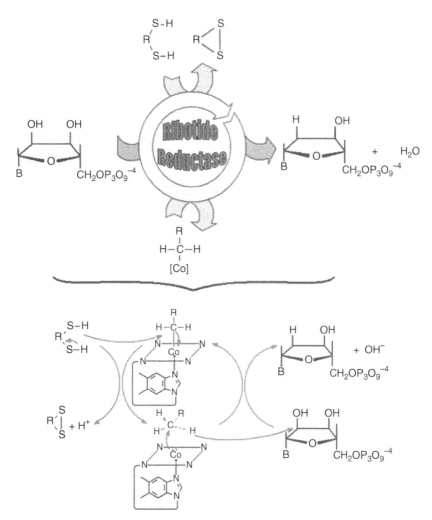

SCHEME 5-26 Proposed mechanism for ribonucleotide reductase. (Modified from Beck, 1968.)

B_2 subunit, molecular weight $\sim$39,000

$(B_1)_2(B_2 \cdot Fe)_2$, molecular weight $\sim$238,000

- Ribotide reductase converts (reduces) ribonucleotide to 2′-deoxyribonucleotide in *Lactobacillus leichmannii* (Scheme 5-26).
  - The addition of hydrogen is only the C-2′ position with retention of configuration.
  - Ribonucleotide triphosphates such as GTP, ATP, CTP, and UTP (G = guanine, A = adenine, C = cytosine, U = uracil) are suitable substrates for this enzyme.

- Requirements: α–(5, 6–Dimethylbenzimidazolyl)-5′-deoxyadenosylcobalamin coenzyme (B_{12} coenzyme) and dithiol reductant must be capable of intra-molecular cyclization on oxidation. This dithiol serves as a hydrogen donor. Dithiol reduces co-5′-deoxyadenosylcobalamin (B_{12} coenzyme) to give 5′-deoxyadenosine as intermediate, which converts ribonucleotide to 2′-deoxyribonucleotide (Scheme 5–26).

Purple Acid Phosphatase in Beef Spleen

- Role

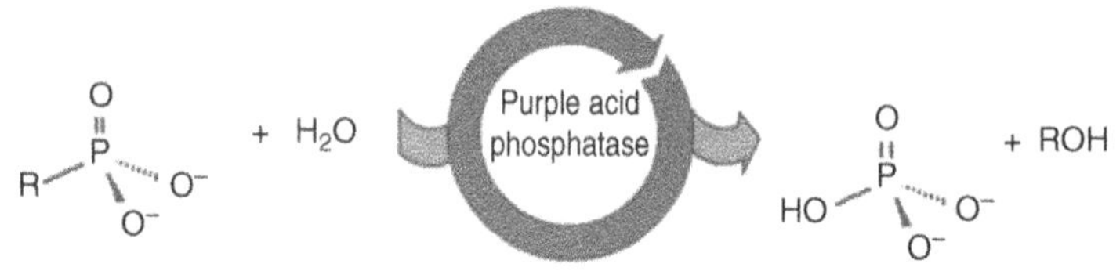

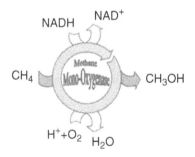

- This hydrolysis is possibly activated via the polarization effect (Scheme 5-27).
- Molecular weight 35,000
- Isolated as Fe_2^{3+} (purple), but active form is $Fe^{2+}Fe^{3+}$ (pink)

Methane Monooxygenase

- Source: *Methylococcus capsulatus*
- Role: oxidation:

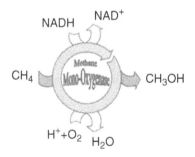

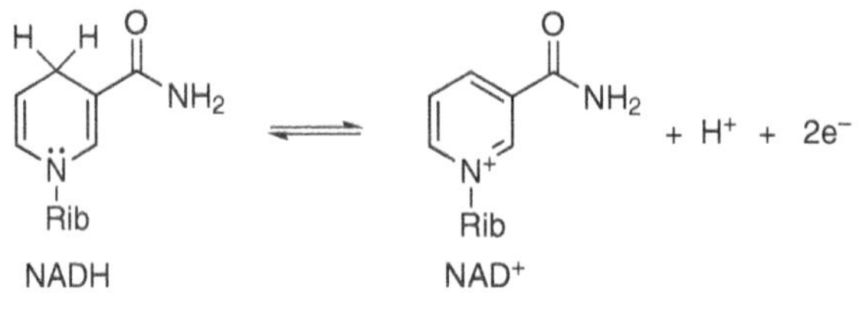

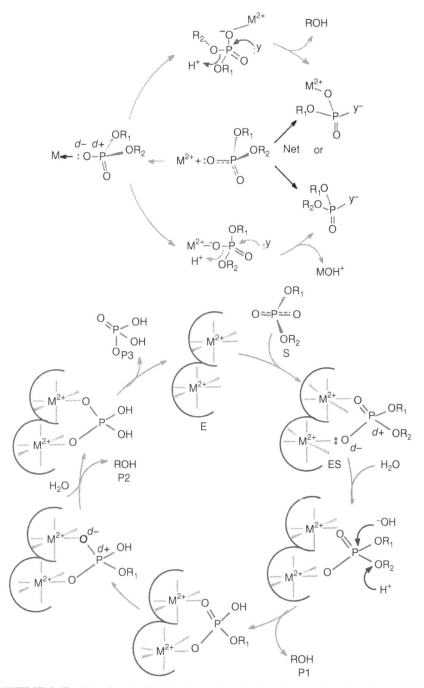

SCHEME 5-27 Phosphate hydrolysis by purple acid phosphatase. (Based on Spiro, 1973. Reproduced by permission of Elsevier.)

Chemistry of Diiron Enzymes

Discuss the general chemical properties of the diiron enzymes.

- Diiron enzymes
  - Hemerythrin: Hr
  - Ribonucleotide reductase: R_2
  - Purple acid phosphatase: PAP
  - Methane monooxygenase: MMO
- General properties
- Oxidized forms
  - Fe_2^{3+} are antiferromagnetic (AFC), e.g., Hr Fe_2^{3+} has $-2J = 270\,cm^{-1}$ and no ESR signals.
  - Optical spectra: All are colored.
    - (a) PAP/uteroferrin has charge transfer band (CT) at $\lambda \sim 580\,nm$, $\varepsilon \sim 4000$.
    - (b) Others have such CT only with certain added ligands bound to Fe. For example, Hr Fe_2O_2 has $\lambda = 500\,nm$, $\varepsilon = 2000$ suggests O_2 as a source of CT in Hr Fe_2O_2, but PAP has strong bands, $300-350\,cm^{-1}$, apparently due to Fe_2^{3+} unit.
- Reduced forms
  - Optical bands decrease
  - Native form is Fe_2^{2+}, with $-2J < 20\,cm^{-1}$ (weak AFC)
  - Hr Fe_2^{3+} is ESR silent
  - Hr Fe_2^{2+} is ESR silent
- Mixed-valence bridged diiron units
  - PAP and R reductase: $Fe_2^{3+} + e^- \leftrightarrows Fe^{3+}Fe^{2+}$
  - Hr can be reduced by $1e^-$ too
  - Hr $Fe^{2+}Fe^{3+}$ has ESR signal at $g = 1.95$, 1.7:

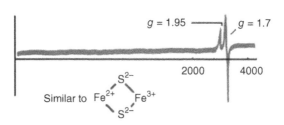

- Model compounds
  - Dinuclear-tribridged compounds, such as $[Fe_2(CH_3COO)_2(HBz_3)_2O]$ and $[Fe_2(CH_3COO)_2(TACN)_2O]^{2+}$, have been used as models for diiron enzymes:

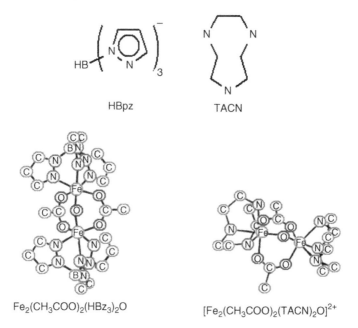

HBpz   TACN

Fe₂(CH₃COO)₂(HBz₃)₂O   [Fe₂(CH₃COO)₂(TACN)₂O]²⁺

○ The crystal structures of these models are comparable with those of dirion enzymes (Fig. 5-46), that is, 4:9.

• These models are self-assembly of active-site structural analogues with large −2J.

• In ribonucleotide reductase, R2 (Fig. 5-47): The ferric center is coordinated by two histidine, three glutamic acid, one aspartic acid, and two water molecules and a

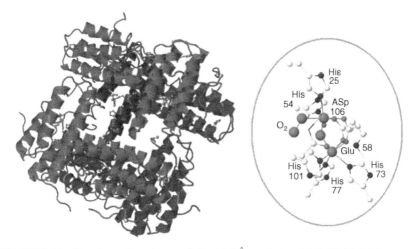

FIGURE 5-46 Structures of oxy hemerythrin at 2.0 Å resolution, PDB 1HMO (Holmes et al., 1991).

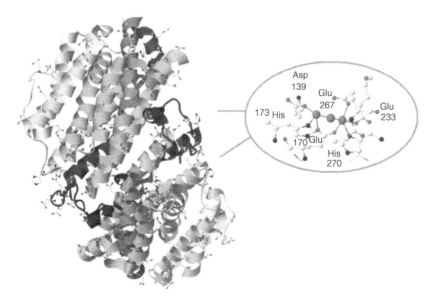

FIGURE 5-47 Crystal structural of ribonucleotide reductase protein R2 from mouse, PDB 1W68 (Strand et al., 2004).

bridging μ–oxo group. In between the radical sites Tyr^{122} and the iron ions is a narrow hydrophobic pocket formed by the conserved Phe^{212}, Phe^{206}, and Ile^{234}.

- Similar ligation in methane monooxygenase (MMO) to that of R2 is presented in Figs. 5-48 and 5-49. However, each histidine ligand is linked by an aspartic acid.

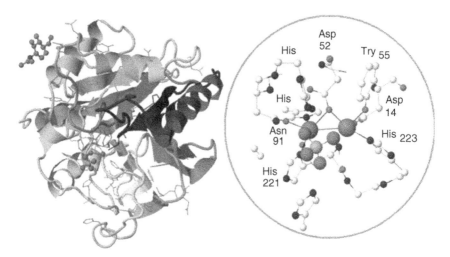

FIGURE 5-48 Crystal structure of mammalian purple acid phosphatase, PDB 1QFC (Uppenberg et al., 1999).

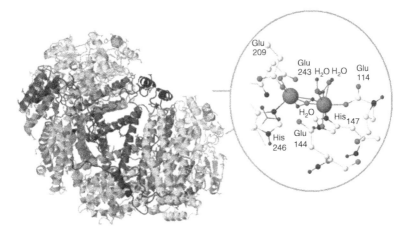

FIGURE 5-49 Crystal structures of the methane monooxygenase from *M. capsulatus,* PDB 1MTY (Rosenzweig et al., 1997).

HEMOPROTEINS: CLASSIFICATION AND BEHAVIOR OF HEME IN ABSENCE OF GLOBINS

Prosthetic Group

What is the prosthetic group of: myoglobin, hemoglobin, cytochromes, chlorophyll a, nitrite- and sulphite-reductase, and B_{12}? Suggest model compounds for these prosthetics.

- The prosthetic group of each—myoglobins, hemoglobins, cytochromes, chlorophylls, and bacterochlorophylls—are summarized in the following:

Porphyrin   Chlorin   Bacterochlorin

Fe^{2+}-Protoporphryin(XI)
In heme *b*, hemoglobins,
myoglobins, cytochromes

In chlorophyll *a*

In bacterochlorophyll

- For the *siroheme* prosthetic group in nitrite reductase of spinach and in sulfite reductase, both reductases involve six-electron oxidations:

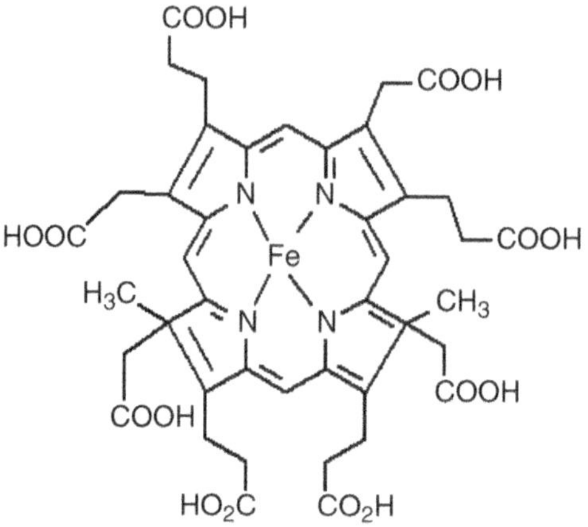

Siroheme

- Vitamin B_{12} contains a 15-membered *corrin* ring.
- The cobalt atom is in the Co^{3+} oxidation state, and the axial two positions are occupied by a variety of ligands:

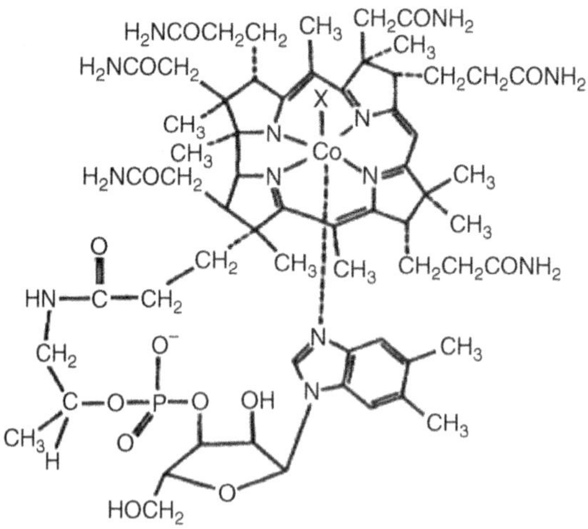

B_{12}, X = CN⁻, Co^{3+}
B_{12r}, Co^{2+}
B_{12s}, Co⁺

- In B_{12} coenzyme:

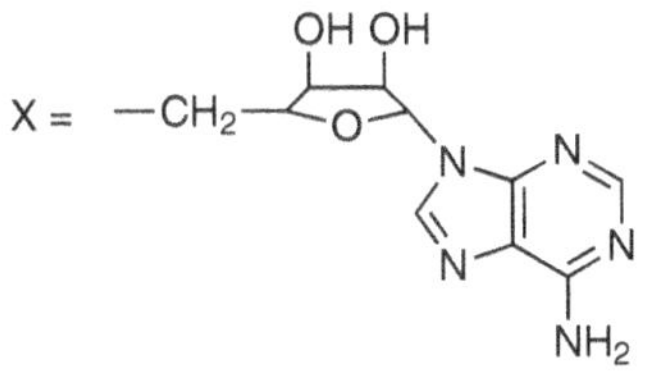

$$X = \ -CH_2-$$

OH OH

NH_2

- Phthalocyanine and *meso*-tetraphenylporphyrin are customarily used as synthetic models of the heme prosthetic group:

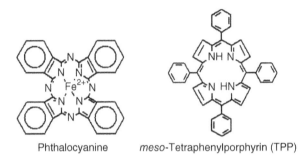

Phthalocyanine   *meso*-Tetraphenylporphyrin (TPP)

Hemoproteins

Give examples of hemoproteins, and identify the biological role of each.

- In hemoprotein:
  - ○ Hemes: the iron porphyrins, Fe-PPIX, serve as prosthetic groups
    - (a) Fe-PPIX: (proto)heme (IX)
    - (b) At neutral pH: $Fe^{II}(PPIX)$ is dianion due to propionate
    - (c) Metal porphyrin: shown as 22 π–e^- ($4n+2$ aromatic), M^{II}-Porp^{2-} is a neutral molecule
  - ○ Proteins: polypeptide chain, globins (apoproteins)
- Examples of hemoproteins
  - ○ Myoglobin: storage of O_2, and O_2 carrier
  - ○ Hemoglobin: transport of O_2, O_2 carrier
  - ○ Cytochrome c: redox protein

$$\text{Cytochrome } c \xrightarrow{\ e^-\ } \text{cytochrome oxidase} \xrightarrow{\ 4e^-\ } O_2$$

  - ○ Catalase: decompose of H_2O_2.

$$H_2O_2 + H_2O_2 \xrightarrow{\ \text{Catalase}\ } 2\,H_2O + O_2$$

  - ○ Chlorperoxidase: halogenates organic substrates

- ○ Lactoperoxidase: antibacterial, oxidation of NCS^-
- ○ Peroxidase: $H_2O_2 + SH_2 \xrightarrow{\text{peroxidase}} 2\,H_2O + S$
- ○ Cytochrome P-450: hydroxylates organic substrates
- ○ Cytochrome c peroxidase: reduction of H_2O_2

Conformations in Hemoproteins

What are the general conformations for hemoproteins?

Three general conformations have been assigned to hemoproteins:

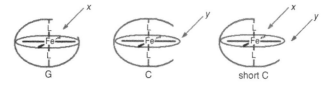

- • *G-conformation* for globins (Hb, Mb): an easy approach to an axial position, while the porphyrin periphery is sterically blocked.
- • *C-conformation* for cytochromes: an exposed periphery and blocked or difficult to access axial ligands.
- • *Short C-conformation* for certain cytochromes, catalase, and peroxidases: both an axial position on iron and the porphyrin periphery are exposed.

Diverse Functions of Heme Group

Why are there differences in the physiological functions when all of the hemoproteins have Fe-porphyrin in the active site?

The physiological functions and reaction pathway in heme proteins are controlled by:

1. Variation in Fe chemistry via axial fifth and sixth ligands, L_5 and L_6, in proximal site
2. Architecture of distal active site cavity (heme pocket)
3. Conformation of heme proteins

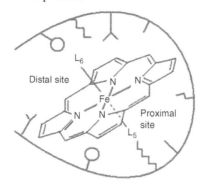

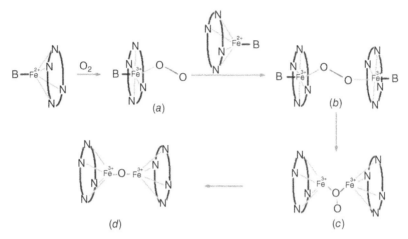

SCHEME 5-28 Direct reaction of heme with O_2 yields irreversible formation of stable μ-oxo-diiron(III)heme.

Heme chemistry

What prevents simple iron porphyrins from functioning as O_2 carriers? What approaches have been used to overcome this problem?

- "Free heme" chemistry
  - The initial and final states of the physiological processes that involve hemoproteins are revisable.
  - However, the direct reaction of heme with O_2 yields irreversible formation of the stable μ-oxo-diiron(III)heme (Scheme 5-28):

$$Fe^{2+} \cdot Porp + O_2 \xrightarrow{\text{room temp.}} Porp \cdot Fe^{3+} - O - Fe^{3+} \cdot Porp$$

In benzene   μ-oxo-diiron(III)hemes

(Porp = porphyrin)

where

$$Fe^{2+} \rightarrow Fe^{3+} - O_2^- \xrightarrow{Fe^{3+}} Fe_2^{3+} \cdot O_2^{2-} \rightarrow Fe^{3+} - O - Fe^{3+} \quad \text{(Scheme 5-28)}$$

  - When B = N-donor unsaturated ligands, such as imidazole (Im):
    High-spin Fe^{2+} Porp(Im) give a clean modeling of

$$ImFe^{2+} Porp + L \leftrightarrows Im Porp Fe^{2+} \cdot L$$

where

L = CO, CNR, NO, N-donors, O-donors and, with O_2, give Fe^{3+}–O–Fe^{3+} or Fe^{3+}–O_2^{2-}–Fe^{3+}.

However,

$$Fe^{2+} Porp + Im \leftrightarrows \tfrac{1}{2} Fe^{2+} Porp(Im)_2 + \tfrac{1}{2} Fe^{2+} Porp$$

High spin   Low spin

o Three approaches have been used to promote the formation of stable $Fe^{2+} Porp(B)$ and inhibit the formation of μ-oxo-diiron(III)hemes:

1. The use of low temperature so that the reactions leading to dimerization are very slow, at $-79\,°C$, in CH_2Cl_2:

$$Fe\,(TPP)(B)_2 + O_2 \leftrightarrows Fe(TPP)(B)(O_2) + B$$

(a) The rate-determining step is the dissociative process

$$Fe(TPP)(B)_2 \leftrightarrows Fe(TPP)(B) + B$$

(b) Followed by rapid reaction of the five-coordinate intermediate, $Fe(TPP)(B)$, with either O_2 or CO:

$$Fe(TPP)(B) + O_2 \leftrightarrows Fe(TPP)(B)(O_2)$$

$$Fe(TPP)(B) + CO \leftrightarrows Fe(TPP)(B)(CO)$$

2. Using steric constrains in such a way that the dimerization is inhibited:

(a) The stability of the O_2 ligation is mainly dependent upon the nature and the concentration of the coordinating base B:

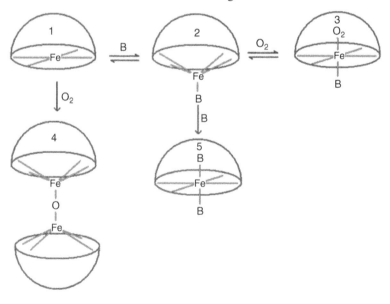

(b) In the absence of B, with O_2, there is a rapid formation of μ-oxo-dimer, 4, Fe^{3+}–O–Fe^{3+} or Fe^{3+}–O_2^{2-}–Fe^{3+} at best.

(c) A large excess of B shifts the equilibrium towards the five–coordinate complex, 2, with simultaneous reduction in the rate of autooxidation.

(d) The oxidative pathway 4, and octahedral species 5, are avoided by adding satirically hindering superstructure to the heme:

  (i) Useing Dmim, CH_3 steric effect destabilized Fe.Porp(Dmim)$_2$, and promoting the formation of the five-coordinate Fe.Porp(Dmim), complex 2.

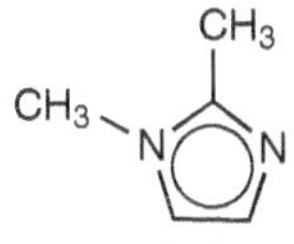

Dmim

$$Fe^{2+} Porp + Dmim \leftrightharpoons Fe^{2+} Porp(Dmim)$$

  (ii) Tailbase, "tailed" hemes:

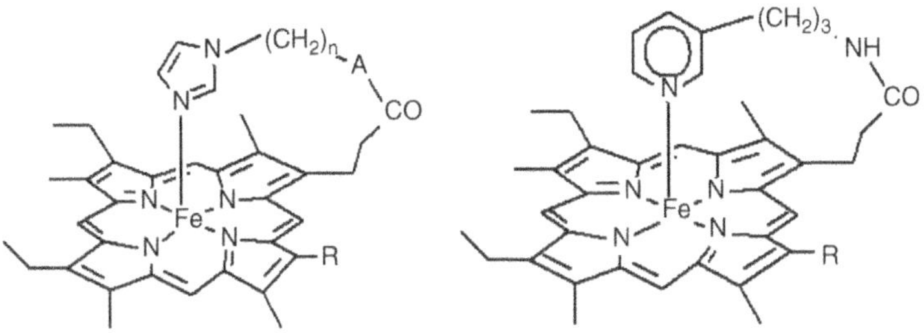

This tailbase is capable of forming a μ–oxo–dimer and reversibly oxygenates at $-45\,°C$.

  (iii) Picket fence porphyrins: N–Alkyl imidazole, a π-donor, is coordinated on the unhindered side of the porphyrins, leaving a hydrophobic pocket for reaction with oxygen.

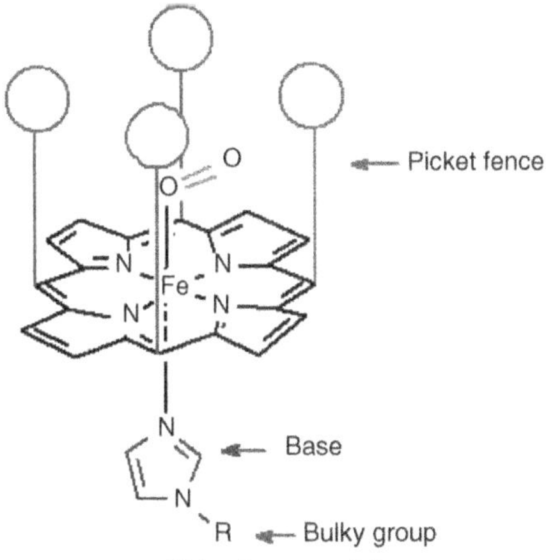

Picket fence model

(iv) Picket fence tailbase

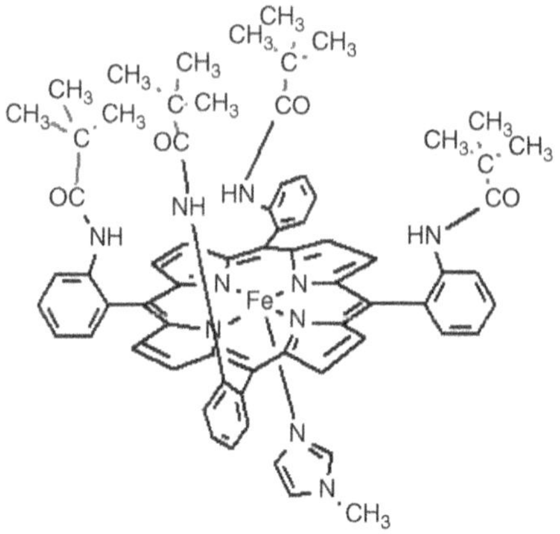

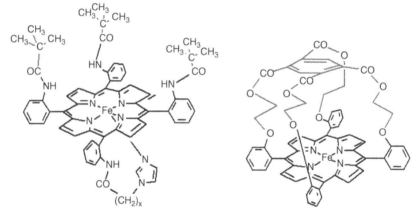

Fe(TpivPP)(1-MeIm)

(v) Capped: The iron complex of capped porphyrin reacts rapidly with O_2 in pyridine. The lifetime of O_2 adduct in pyridine is $\sim$20 h, after which complete oxidation to the Fe(III) complex occurs:

Picket fence tailbase

x = 3: FePiv$_3$(4CimP)Por

x = 4: FePiv$_3$(5CimP)Por

From Hughes (1981)

Capped Fe porphyrin

3. Rigid surface attachment of the iron complex to a surface (e.g., silica gel) so that the dimerization is prevented.

- Treatment of polystyrene–Fe(II)(TPP)Im complex with O_2 in benzene led to oxidation and formation of the μ-oxo dimer:

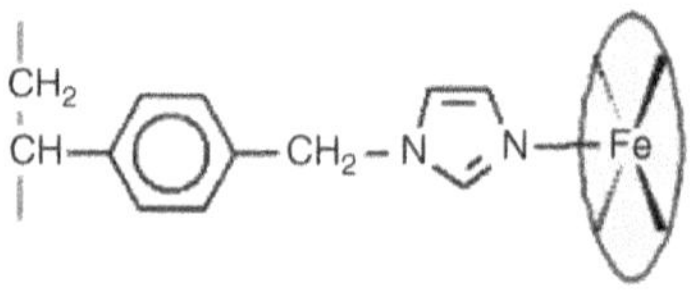

Polystyrene–Fe(II)(TPP)Im complex

Therefore, the cross-linked polystyrene ligand was not sufficiently rigid to prevent dimerization on treatment with oxygen.

- However, Fe (II)(TPP)Im–silica gel complex reversibly binds O_2:

Attached to rigid modified silica gel (Hughes, 1981)

MYOGLOBIN AND HEMOGLOBIN

Biological Oxygen Carriers

Why do biological systems develop oxygen carriers?

- The solubility of O_2 in H_2O is low: $6.59\,cm^3/L$ at 1 atm and 20°C, giving $3 \times 10^{-4}\,M$.
- Therefore, O_2 delivery by the circulatory system is limited.
- Biological systems have developed O_2 carriers that reversibly coordinate oxygen to transition metal (Fe, Cu, V) bound to protein.
- Blood carries 30 times as much oxygen as pure water can hold.
- In human blood, $200\,cm^3/L$ is the solubility of O_2 in equilibrium with air at 20 °C, giving $9 \times 10^{-3}\,M$.

MYOGLOBIN

Biological Roles and Properties

What are the biological roles and the main characters of myoglobin?

- The form of hemoglobin (Hb) found in muscle fibers has a higher affinity for oxygen than the hemoglobin in the blood (Fig. 5-50):

$$Mb + O_2 \leftrightarrows MbO_2 \qquad K \simeq 10^3\,atm^{-1}$$

$$Mb + HbO_2 \leftrightarrows MbO_2 + Hb$$

- Role of Mb: storage and possible transport of O_2 across membranes.
- Molecular weight 17,000, 1 heme per Mb: Consist of 153-residue peptide and Fe^{2+} -protoporphyrin IX as a prosthetic group.

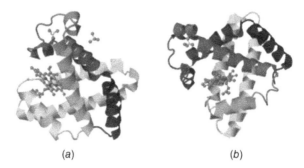

(a)   *(b)*

FIGURE 5-50 (*a*) Crystal structures of deoxymyoglobin, PDB 1A6N (Vojtechovsky et al., 1999). (*b*) Structure of oxymyoglobin at 1.6 Å, PDB 1MBO (Phillips, 1980).

- The polypeptide chain is conformed to eight α-helices, lettered from A nearest the amino end to H nearest the carboxyl end of the chain:

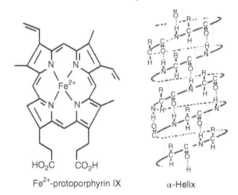

Fe^{2+}-protoporphyrin IX   α-Helix

- Special attention should be given to the unique chemistry of the active site (Fig. 5-51).

Diamagnetic Oxymyoglobin

Explain the spin-coupling in oxymyoglobin. List the possible spin states of the iron ion in deoxymyoglobin and oxygen molecule. Why is oxymyoglobin

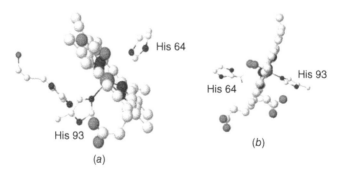

His 64

His 93

His 64

His 93

(b)

(a)

FIGURE 5-51 (*a*) Active site of deoxymyoglobin, PDB 1A6N (Vojtechovsky et al., 1999). (*b*) Active site of oxymyoglobin at 1.6 Å, PDB 1MBO (Phillips, 1980).

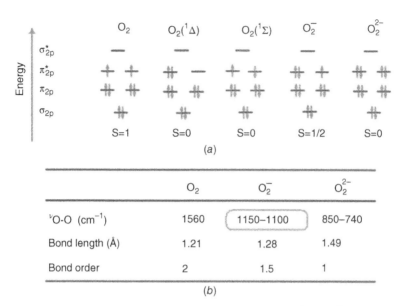

FIGURE 5-52 (*a*) Molecular orbital schemes for O_2, O_2^-, O_2^{2-}, $O_2(^1\Delta)$, and $O_2(^1\Sigma)$. (*b*) Bond order, bond length, and stretching frequency of O_2, O_2^-, and O_2^{2-}. (From Hughes, 1981. Reproduced by permission of John Wiley & Sons.)

diamagnetic at room temperature? Support your answer using observable evidence.

- The spin angular momentum of iron in deoxymyoglobin is $S_{Fe^{2+}} = 2$, while that of oxygen is $S_{O_2} = 1$ (Fig. 5-52, all possible electron spin arrangements of oxygen molecules). However, the formed oxymyoglobin has $S_{Fe^{2+}-O_2} = 0$ (Scheme 5-29) and therefore the reaction is more than just coupling.
- The oxidation states of Fe^{2+} and O_2^0 cannot explain the value of $S_{Fe^{2+}-O_2} = 0$ for any ligand L (Scheme 5-29).
- From the previous approach for hemerythrin (H_R) and hemocyanin (H_C), this would correspond to the O_2^- level of charge transfer:

$$Fe^{2+} + O_2 \quad \rightarrow \quad Fe^{3+} - O_2^-$$
$$\text{Deoxy-Mb} \qquad \quad \text{Oxy-Mb, Fig. 5-52}$$

  In $Fe^{3+}-O_2^-$

$$Fe^{3+} \rightarrow S = \tfrac{1}{2}$$
$$O_2^- \rightarrow S = \tfrac{1}{2} \quad \text{coupled to } S = 0 \quad \text{at the ground state}$$

- The charge transfer in $Fe^{3+}-O_2^-$ is supported by the following:
  - The electronic spectra of oxymyoglobin (or oxyhemoglobin) and alkaline metmyoglobin (low-spin, hydroxymetmyoglobin) are quite similar.

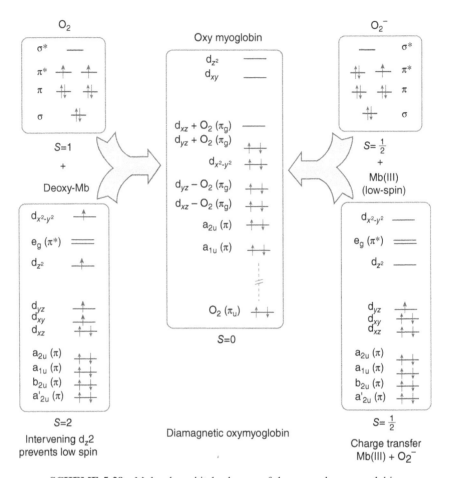

SCHEME 5-29 Molecular orbital schemes of deoxy- and oxymyoglobin.

- Various molecular orbital calculations show charge transfer bands both ways, MLCT and LMCT, supporting the formation of Fe^{3+}–O_2^-.
- The binding of O_2^- is evident in the infrared and in the crystal structure by its characteristic stretching frequency and bond length (Fig. 5-52).
- The abnormal large quadrupole splitting in the Mössbauer spectra of oxymyoglobin is comparable to that of the model compounds.
- The formation constant, K_{O_2}, is directly proportional to the iron oxidation potential of various Hb's and Mb's:

$$MbFe^{3+} + e^- \rightarrow MbFe^{2+} \qquad E^{0'} = 0.090 \text{ V}$$

- The linear relationship between K_{O_2} and $E^{0'}$ supports the mutual charge transfer between the two centers:

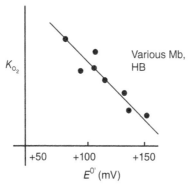

Binding constant versus various hemoproteins

Ligand Binding and Myoglobin

How can you monitor the binding process of small molecules to myoglobin?

- Beside O_2, other ligands (L's) bind well to the iron ion of myoglobin:

  Fe^{2+}: $\quad L = CN^-, O_2, CO, -CN-R, NO$

  Fe^{3+}: $\quad L = OH_2, OH^-, F^-, N_3^-, NCS^-, CN^-$, azoles such as imidazole, etc.

- If y = fraction of Mb_{total} as MbL, then

$$Mb + L \rightleftarrows MbL$$
$$ {}_{1-y} {}_{y}$$

$$K \quad = \frac{[MbL]}{[L][Mb]}$$

$$[MbL] \quad = y[Mb_{total}]$$

$$[Mb] \quad = [Mb_{total}] - [MbL] = [Mb_{total}] - Y[Mb_{total}] = [Mb_{total}](1 - Y)$$

$$\frac{[MbL]}{[Mb]} = \frac{y[Mb_{total}]}{[Mb_{total}](1 - y)} = \frac{y}{1 - y}$$

$$K \quad = \frac{y}{[L](1 - y)}$$

$$Y \quad = \frac{K[L]}{1 + K[L]}$$

$$\frac{y}{1 - y} = K[L]$$

- A plot of the fraction of oxymyoglobin versus oxygen concentration exhibits a hyperbolic curve:

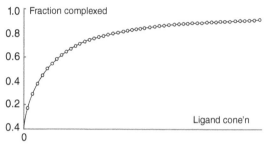

Hyperbolic saturation curve of binding oxygen to myoglobin

- The plot of $\log[y/(1-y)]$ versus $\log[L]$ would have a straight line:

$$\log \frac{y}{1-y} = \log K + \log[L]$$

with slope 1 and intercept $\log K$.

HEMOGLOBIN

Role and Properties

What are the properties and the biological roles of hemoglobin?

- There are 5 billion red cells/Ml, 280 million Hb molecules/red cell, and 4 heme groups/Hb molecule.
- Similar to Mb, each heme is tightly bound to a protein (globin) through about 80 hydrophobic interactions and a single coordinate bond between imidazole of the proximal histidine and Fe(II).
- Molecular weight 68,000, tetramer (Fig. 5-53).
- Human Hb has four chains, $\alpha_2\,\beta_2$, two identical chains labeled α, with 141 amino acids, and two identical β-chains, with 146 amino acids.

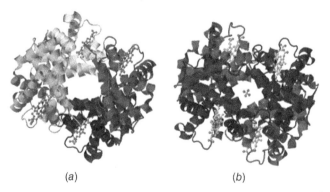

(a)   (b)

FIGURE 5-53 (*a*) The crystal structure of human deoxyhemoglobin at 1.74 Å resolution, PDB 2HHB (Fermi et al., 1984). (*b*) Structure of human oxyhemoglobin at 2.1 Å resolution, PDB 1HHO (Shaanan, 1983).

- Roles
    - (i) Hemoglobin binds O_2 to its heme iron at the lungs and delivers O_2 to myoglobin, which stores the O_2 until it is required for metabolic oxidation.
    - (ii) A second task of Hb is to bring the CO_2 by-product of oxidation back to the lungs to get rid of it:

$$R-NH_2 + CO_2 \rightarrow R-NH-COO^- + H^+$$

CO_2, also buffering, as a result of the reversible formation of HCO_3^-.

$$CO_2 + H_2O \rightarrow HCO_3^- + H^+$$

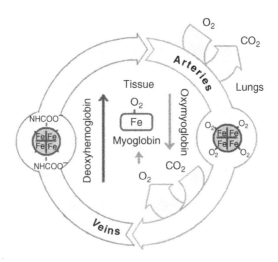

Oxygen Binding

What are the differences between binding of oxygen either to Mb or to Hb?

How can the statistical O_2 binding constants K_1, K_2, K_3, and K_4 be calculated in the absence of interactions between the four sites of Hb and considering these sites are identical?

$$Hb + O_2 \overset{K_1}{\Leftrightarrow} Hb\,O_2 + O_2 \overset{K_2}{\Leftrightarrow} Hb(O_2)_2 + O_2 \overset{K_3}{\Leftrightarrow} Hb(O_2)_3 + O_2 \overset{K_4}{\Leftrightarrow} Hb(O_2)_4$$

For myoglobin

$$Mb + O_2 \rightleftharpoons MbO_2$$

$$K_M = \frac{[MbO_2]}{[Mb][O_2]}$$

- The fractional saturation y is given as

$$y = \frac{[\text{Occupied sites}]}{[\text{Possible sites}]} = \frac{[MbO_2]}{[Mb] + [MbO_2]} = \frac{K_M[Mb][O_2]}{[Mb] + K_M[Mb][O_2]} = \frac{K_M[O_2]}{1 + K_M[O_2]}$$

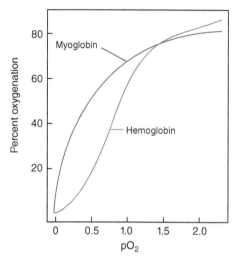

FIGURE 5-54 Hyperbolic and sigmoid saturation curves for oxygen binding of myoglobin and hemoglobin. (Cotton and Wilkinson, 1980. Reproduced by permission of John Wiley & Sons.)

If $[O_2] = pO_2$, the partial pressure of oxygen, the following equation is useful:

$$H = \frac{y}{1-y} = K_M \cdot pO_2$$

This form is known as the *Hill equation* (Fig. 5–54, Hill plot).

- This equation describes the saturation curve of Mb. In the case of Hb, is Hb one protein or four?
  - If each subunit is acting independently, we expect an Mb-like–O_2 binding curve such as the one for four Mb molecules in solution (hyperbolic curve) (Fig. 5-54).
  - If Hb is acting as a single molecule,

$$\underset{1-y}{Hb} + 4\,O_2 \;\rightleftharpoons\; \underset{y}{Hb(O_2)_4}$$

where y = fraction of $Hb(O_2)_4$, $K = K_f = \beta_4$. Then

$$K = \beta_4 = \frac{[Hb(O_2)_4]}{[Hb][O_2]^4} = \frac{y}{(1-y)[O_2]^4}$$

where $K_1 \geq K_2 \geq K_3 \geq K_4$ (the variation of K's is due to the electronic and steric effects):

$$K = \frac{y}{(1-y)[O_2]^4}$$

$$y = \frac{K[O_2]^4}{1+K[O_2]^4}$$

$$\frac{y}{1-y} = K[O_2]^4$$

where y becomes a function of $[L]^4$ and the oxygen saturation curve will have a sigmoid shape.

- The plot of $\log[y/(1-y)]$ versus $\log[L]$ would have a straight line with slope of 4:

$$\log\frac{y}{1-y} = \log K + 4\log[L]$$

- Hb has a log-log plot with slope 2–3. The relative K_i then follows

$$K_4 > K_1 - K_3 \quad (\text{or } K_4, K_3 > K_1, K_2) \text{ depends on source.}$$

- The saturation curve for hemoglobin is more complicated, as it has a sigmoid shape (Fig. 5-54) and cannot be derived from any simple equilibrium.
- Before discussing the association of O_2 to Hb, it is useful to differentiate between *macroscopic* and *microscopic* equilibrium constants.
- Hb contains four sites for O_2, and the *macroscopic equilibria* are

$$\text{Hb} + O_2 \quad \overset{K_1}{\rightleftharpoons} \quad \text{HbO}_2$$

$$\text{HbO}_2 + O_2 \quad \overset{K_2}{\rightleftharpoons} \quad \text{Hb}(O_2)_2$$

$$\text{Hb}(O_2)_2 + O_2 \quad \overset{K_3}{\rightleftharpoons} \quad \text{Hb}(O_2)_3$$

$$\text{Hb}(O_2)_3 + O_2 \quad \overset{K_4}{\rightleftharpoons} \quad \text{Hb}(O_2)_4$$

and

$$[\text{HbO}_2] = K_1[\text{Hb}][O_2]$$
$$[\text{Hb}(O_2)_2] = K_1 K_2[\text{Hb}][O_2]^2$$
$$[\text{Hb}(O_2)_3] = K_1 K_2 K_3[\text{Hb}][O_2]^3$$
$$[\text{Hb}(O_2)_4] = K_1 K_2 K_3 K_4[\text{Hb}][O_2]^4$$

- The saturation curve is given by

$$y = \frac{[\text{occupied sites}]}{[\text{possible sites}]}$$

$$= \frac{[\text{HbO}_2] + 2[\text{Hb}(O_2)_2] + 3[\text{Hb}(O_2)_3] + 4[\text{Hb}(O_2)_4]}{4([\text{Hb}] + [\text{HbO}_2] + [\text{Hb}(O_2)_2] + [\text{Hb}(O_2)_3] + [\text{Hb}(O_2)_4])}$$

- Therefore:

$$y = \frac{K_1[O_2] + 2K_1K_2[O_2]^2 + 3K_1K_2K_3[O_2]^3 + 4K_1K_2K_3K_4[O_2]^4}{4\left(1 + K_1[O_2] + K_1K_2[O_2]^2 + K_1K_2K_3[O_2]^3 + K_1K_2K_3K_4[O_2]^4\right)}$$

- We now examine the microscopic states of Hb. Hb contains four sites:

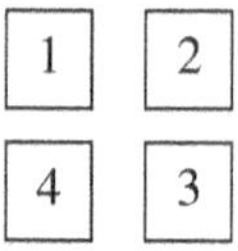

$$\begin{array}{cc} \boxed{1} & \boxed{2} \\ \boxed{4} & \boxed{3} \end{array}$$

Hb(O$_2$)$_i$ is the total set of microscopic species that have i bound molecules of O$_2$. For example:

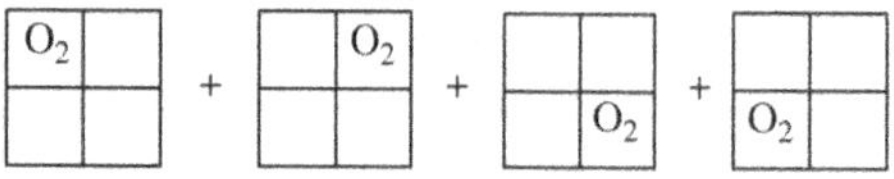

$[HbO_2] = [^1HbO_2] + [^2HbO_2] + [^3HbO_2] + [^4HbO_2]$

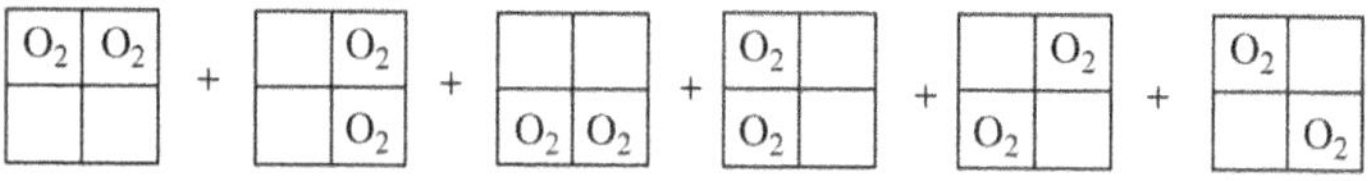

$[Hb(O_2)_2] = [^{1,2}Hb(O_2)_2] + [^{2,3}Hb(O_2)_2] + [^{3,4}Hb(O_2)_2]$
$\quad\quad\quad + [^{4,1}Hb(O_2)_2] + [^{2,4}Hb(O_2)_2] + [^{1,3}Hb(O_2)_2]$

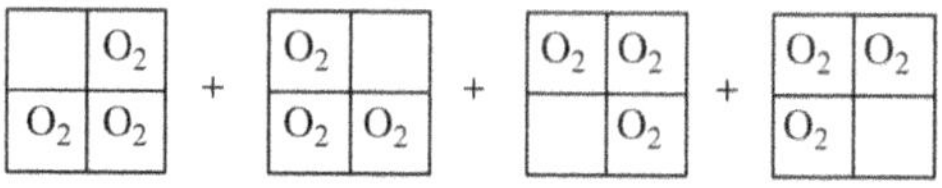

$[Hb(O_2)_3] = [^{2,3,4}Hb(O_2)_3] + [^{1,3,4}Hb(O_2)_3] + [^{1,2,3}Hb(O_2)_3] + [^{1,2,4}Hb(O_2)_3]$

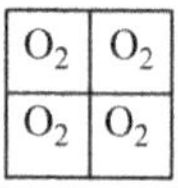

$[Hb(O_2)_4] = [Hb(O_2)_4]$

- Assume:
  - Each site has the same microscopic association constant k.
  - The sites are independent.
  - The microscopic association constant k is the same regardless of the state of occupancy of the other sites:

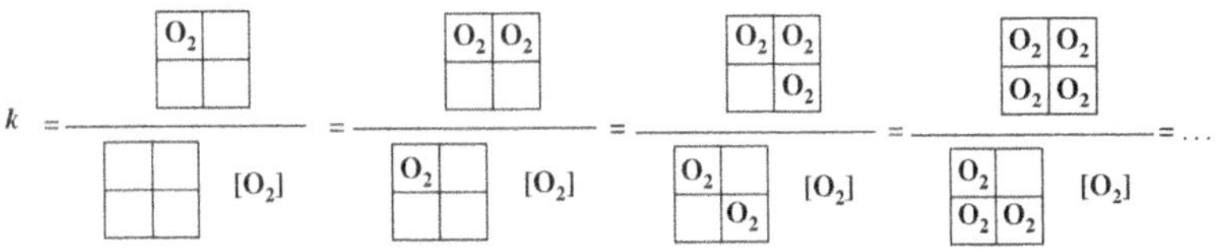

$$k = \frac{\boxed{O_2}}{\boxed{}\,[O_2]} = \frac{\boxed{O_2\,O_2}}{\boxed{O_2}\,[O_2]} = \frac{\boxed{O_2\,O_2 \atop O_2}}{\boxed{O_2 \atop O_2}\,[O_2]} = \frac{\boxed{O_2\,O_2 \atop O_2\,O_2}}{\boxed{O_2 \atop O_2\,O_2}\,[O_2]} = \dots$$

- Therefore:

$$\sum[HbO_2] = 4[HbO_2]$$
$$\sum[Hb(O_2)_2] = 6[Hb(O_2)_2]$$
$$\sum[Hb(O_2)_3] = 4[Hb(O_2)_3]$$
$$\sum[Hb(O_2)_4] = [Hb(O_2)_4]$$

- Consequently:

$$K_1 = \frac{[HbO_2]}{[Hb][O_2]} = \frac{[^1HbO_2] + [^2HbO_2] + [^3HbO_2] + [^4HbO_2]}{[Hb][O_2]}$$

$$K_1 = \frac{[HbO_2]}{[Hb][O_2]} = \frac{k[Hb][O_2] + k[Hb][O_2] + k[Hb][O_2] + k[Hb][O_2]}{[Hb][O_2]}$$

$$\boxed{K_1 = 4k}$$

$$K_2 = \frac{[Hb(O_2)_2]}{[HbO_2][O_2]}$$

$$K_2 = \frac{[^{1,2}Hb(O_2)_2] + [^{2,3}Hb(O_2)_2] + [^{3,4}Hb(O_2)_2] + [^{4,1}Hb(O_2)_2] + [^{2,4}Hb(O_2)_2] + [^{1,3}Hb(O_2)_2]}{[[^1HbO_2] + [^2HbO_2] + [^3HbO_2] + [^4HbO_2]][O_2]}$$

$$K_2 = \frac{6[Hb(O_2)_2]}{4[HbO_2][O_2]} = \frac{6k[HbO_2][O_2]}{4[HbO_2][O_2]}$$

$$\boxed{K_2 = \tfrac{3}{2}k}$$

$$K_3 = \frac{[Hb(O_2)_3]}{[Hb(O_2)_2][O_2]}$$

$$K_3 = \frac{4[Hb(O_2)_3]}{6[Hb(O_2)_2][O_2]} = \frac{4k[Hb(O_2)_2][O_2]}{6[Hb(O_2)_2][O_2]}$$

$$K_3 = \tfrac{2}{3}k$$

$$K_4 = \frac{[Hb(O_2)_4]}{[Hb(O_2)_3][O_2]} = \frac{[Hb(O_2)_4]}{4[Hb(O_2)_3][O_2]} = \frac{k[Hb(O_2)_3][O_2]}{4[Hb(O_2)_3][O_2]}$$

$$K_4 = \tfrac{1}{4}k$$

- Therefore:

$$y = \frac{K_1[O_2] + 2K_1K_2[O_2]^2 + 3K_1K_2K_3[O_2]^3 + 4K_1K_2K_3K_4[O_2]^4}{4\left(1 + K_1[O_2] + K_1K_2[O_2]^2 + K_1K_2K_3[O_2]^3 + K_1K_2K_3K_4[O_2]^4\right)}$$

$$y = \frac{k[O_2] + 3k^2[O_2]^2 + 3k^3[O_2]^3 + k^4[O_2]^4}{\left(1 + 4k[O_2] + 6k^2[O_2]^2 + 4k^3[O_2]^3 + k^4[O_2]^4\right)}$$

$$y = \frac{k[O_2](1 + 3k[O_2] + 3k^2[O_2]^2 + k^3[O_2]^3)}{\left(1 + 4k[O_2] + 6k^2[O_2]^2 + 4k^3[O_2]^3 + k^4[O_2]^4\right)}$$

- Note the binomial expansion p:

$$p = (1+x)^n = 1 + nx + \frac{n(n-1)}{2!}x^2 + \frac{n(n-1)(n-2)}{3!}x^3 + \cdots$$

$$p = (1 + k[O_2])^4 = 1 + 4k[O_2] + 6k^2[O_2]^2 + 4k^3[O_2]^3 + k^4[O_2]^4$$

$$dp = 4 + 12k[O_2] + 12k^2[O_2]^2 + 4k^3[O_2]^3 = 4(1 + k[O_2])^3$$

Therefore:

$$y = \frac{k[O_2]dp}{4p}$$

$$y = \frac{k[O_2](1 + k[O_2])^3}{(1 + k[O_2])^4}$$

$$y = \frac{k[O_2]}{(1 + k[O_2])}$$

which is identical to the obtained expression of O_2 binding to Mb.

- Assume y can be calculated spectroscopicaly from absorbance changes. Let A_{oxy} and A_{deoxy} be the absorbances of the fully oxygenated and fully deoxygenated

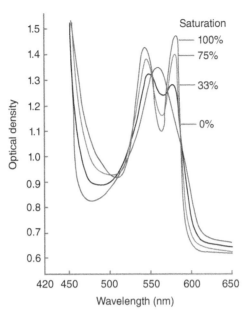

FIGURE 5-55   Electronic spectra of human hemoglobin at different oxygen concentrations. (Giardnia and Amiconi, 1981. Reproduced by permission of Academic Press.)

samples, respectively, and A the absorbance at a given pO_2 (Fig. 5-55). Then, when $l=1$ cm, (light path),

$$A_{\text{deoxy}} \quad = \varepsilon[\text{Hb}]_0 \quad \varepsilon = \text{molar absorptivity of fully deoxygenated sample}$$

$$A_{\text{oxy}} \quad = \varepsilon_{O_2}[\text{HbO}_2]_\infty \quad \varepsilon_{O_2} = \text{molar absorptivity of fully oxygenated sample}$$

$$A \quad = \varepsilon[\text{Hb}] + \varepsilon_{O_2}[\text{HbO}_2]$$

$$A \quad = \varepsilon\{[\text{Hb}]_0 - [\text{HbO}_2]\} + \varepsilon_{O_2}[\text{HbO}_2]$$

$$A \quad = \varepsilon[\text{Hb}]_0 - \varepsilon[\text{HbO}_2] + \varepsilon_{O_2}[\text{HbO}_2]$$

$$A \quad = A_{\text{deoxy}} - [\text{HbO}_2]\{\varepsilon - \varepsilon_{O_2}\}$$

$$A_{\text{deoxy}} - A = [\text{HbO}_2]\{\varepsilon - \varepsilon_{O_2}\}$$

$$[\text{HbO}_2] \quad = \frac{A_{\text{deoxy}} - A}{\varepsilon - \varepsilon_{O_2}}$$

$$[\text{Hb}]_{\text{total}} \quad = [\text{Hb}]_0 = [\text{HbO}_2]_\infty$$

Multiply $[\text{Hb}]_{\text{total}}$ by $\{\varepsilon - \varepsilon_{O_2}\}$:

$$[Hb]_{total}\{\varepsilon - \varepsilon_{O_2}\} = \varepsilon[Hb]_{total} - \varepsilon_{O_2}[HbO_2]_{total}$$

$$\varepsilon[Hb]_0 - \varepsilon_{O_2}[Hb_{O_2}]_\infty = A_{deoxy} - A_{oxy}$$

$$[Hb]_{total}\{\varepsilon - \varepsilon_{O_2}\} = A_{deoxy} - A_{oxy}$$

$$[Hb]_{total} = \frac{A_{deoxy} - A_{oxy}}{\varepsilon - \varepsilon_{O_2}}$$

$$y = \frac{[HbO_2]}{[Hb]_{total}} = \frac{A_{deoxy} - A}{A_{deoxy} - A_{oxy}}$$

or

$$y = \frac{A - A_{deoxy}}{A_{oxy} - A_{deoxy}}$$

Then

$$\boxed{y = \frac{k[O_2]}{1 + k[O_2]} = \frac{A_{deoxy} - A}{A_{deoxy} - A_{oxy}}}$$

- The oxygen electrode provides a direct means for measuring the pO_2.
- The microscopic equilibrium parameter (k) can be evaluated from these curves. Consequently, in the absence of interactions between the identical sites, the four macroscopic constants can be evaluated:

$$K_1 = 4k \quad K_2 = \tfrac{3}{2}k \quad K_3 = \tfrac{2}{3}k \quad K_4 = \tfrac{1}{4}k$$

Statistic Analysis and Difficulties

Why cannot the behavior of oxygen binding to hemoglobin be described by the simple statistical values?

- The oxygen binding of hemoglobin cannot be described by simple statistical values because:
  - The sites are naturally not equivalent; there are α- and β-subunits.
  - Differences in the active sites may be associated with functional interactions.
  - Positive or negative interactions may occur.
- Therefore, the microscopic association constant k may not be the same for the different states of occupancy.

Bohr Effect

What is the effect of pH change on the oxygen affinity of hemoglobin?
How can you explain this effect? Define the Bohr effect.

- The *oxygen Bohr effect* is the dependence of the affinity of hemoglobin for oxygen on pH.

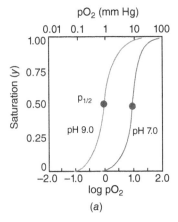
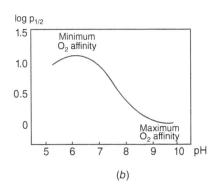

FIGURE 5-56 (*a*) Displacement of saturation curve with pH. (*b*) Maximum and minimum limits of oxygen affinity.

- The effect of pH on the oxygen equilibrium curves (Fig. 5-56) indicates that:
  - ○ The two curves are very similar in shape but are displaced from each other along the pressure axis. Therefore, they have different values for $p_{1/2}$, e.g., the oxygen affinity changes with pH.
  - ○ The saturation curve has been displaced to higher pressures and the affinity has been reduced.
- The effect of pH on log $p_{1/2}$ indicates that:
  - ○ The curve has a maximum at pH 6.5. Therefore, this pH corresponds to minimum O_2 affinity:

$$\text{Maximum-affinity conformation} \rightleftharpoons \text{minimum-affinity conformation}$$
$$\text{High pH} \qquad\qquad\qquad \text{Low pH}$$

  - ○ The physiological pH is 7.4, which lies on the alkaline side of maximum affinity.
- The alkaline Bohr effect associates the binding of O_2 with the release of H^+. Oxyhemoglobin is a stronger acid (proton donor) than deoxyhemoglobin.

Oxygenation and Structural Alternations

What are the structural variations that follow the oxygenation and deoxygenation of hemoglobin?

- Consider one of hemoglobin's four hemes:
  - ○ Deoxy-Hb is high spin and a penta-coordinated Fe^{2+}.
  - ○ Oxy-Hb is low spin and a six-coordinated Fe^{2+} (Fig. 5-57).
- Oxygenation pulls proximal His and Fe together; and induces a shift in the F-helix. The helix's motion is transmitted through subunit-to-subunit interface.
- As a consequence, the oxygen binding also initiates the breakup of the net hydrogen bonds and salt links at the end of the molecule. This leads to change in the conformation from "tense" to "relaxed."

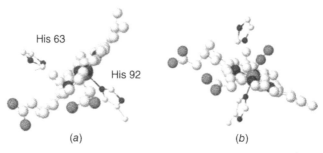

	(a)		(b)

FIGURE 5-57 (*a*) Crystal structure of human deoxyhemoglobin at 1.74 Å resolution, PDB 2HHB (Fermi et al., 1984). (*b*) Structure of human oxyhemoglobin at 2.1 Å resolution, PDB 1HHO (Shaanan, 1983).

	Deoxy-Hb	$\leftrightarrows$	$Hb(O_2)_4$
• Quaternary structure: (4° structure)	Tightens up subunits twisted together		Loosens
	T-conformation		R-conformation
• Heme geometry	CN = 5		CN = 6
• O_2 affinity	Lower		Higher
	$K_{p(HbA,T)} = 9$		$K_{p(HbA,R)} = 950 \text{ atm}^{-1}$

$$Hb + O_2 \overset{K_1}{\rightleftarrows} HbO_2 + O_2 \overset{K_2}{\rightleftarrows} Hb(O_2)_2 + O_2 \overset{K_3}{\rightleftarrows} Hb(O_2)_3 + O_2 \overset{K_4}{\rightleftarrows} Hb(O_2)_4$$

• T → R is discontinuous vs. oxygenation.

Monod–Wyman–Changeux (MWC) model

Derive the expressions for:

   (i) The fraction of saturation of hemoglobin with oxygen, *y*, when:

      (a) [R] ≠ [T]

      (b) [R] = [T]

      (c) [R] and [T] act independently

  (ii) The oxygen binding constants of hemoglobin (Adair constants) in terms of the allosteric parameters using the Monod–Wyman–Changeux model

 (iii) Write your conclusions.

• The basic assumptions of the MWC model are:

   1. No distinction is made between α- and β-subunits.

   2. There are two quaternary conformations: R and T.

   3. Both forms can bind up to four molecules of O_2.

   4. Subunits in R and T have different affinities for O_2.

5. In each conformation, there is a series of stepwise equilibria:

$$R_{n-1} + O_2 \overset{K_{R_n}}{\Leftrightarrow} R_{n-1}O_2 = R_n$$

$$T_{n-1} + O_2 \overset{K_{T_n}}{\Leftrightarrow} T_{n-1}O_2 = T_n$$

where $n = 0,1,2,3,4$.

- The constant $K_{R1}, \ldots, K_{R4}$ can all be expressed in terms of a single microscopic constant k_R.
  - This implies that in each conformation the binding of oxygen is noncooperative.
  - The probability of binding oxygen at a particular subunit is independent of the state of oxygenation in the other subunits of the same molecule.
- There is an equilibrium between R and T forms:

$$R_0 \overset{L}{\Leftrightarrow} T_0 \quad L = \frac{[T_0]}{[R_0]}$$

- Under these assumptions, if

$$y = \frac{[Hb](K_1[O_2] + 2K_1K_2[O_2]^2 + 3K_1K_2K_3[O_2]^3 + 4K_1K_2K_3K_4[O_2]^4)}{4[Hb]\left(1 + K_1[O_2] + K_1K_2[O_2]^2 + K_1K_2K_3[O_2]^3 + K_1K_2K_3K_4[O_2]^4\right)}$$

$$[Hb] = [R_0] + [T_0]$$

then

$$4y = \frac{[R_0]\left(K_{R1}[O_2] + \cdots + 4K_{R1}K_{R2}K_{R3}K_{R4}[O_2]^4\right) + [T_0]\left(K_{T1}[O_2] + \cdots + 4K_{T1}K_{T2}K_{T3}K_{T4}[O_2]^4\right)}{[R_0]\left(1 + K_{R1}[O_2] + \cdots + K_{R1}K_{R2}K_{R3}K_{R4}[O_2]^4\right) + [T_0]\left(1 + K_{T1}[O_2] + \cdots + K_{T1}K_{T2}K_{T3}K_{T4}[O_2]^4\right)}$$

where

$$K_{R1} - 4k_R \quad K_{T1} = 4k_T$$

$$K_{R2} = \tfrac{3}{2}k_R \quad K_{R1}K_{R2} = 6k_R^2 \quad K_{T2} = \tfrac{3}{2}k_T \quad K_{T1}K_{T2} = 6k_T^2$$

$$K_{R3} = \tfrac{2}{3}k_R \quad K_{R1}K_{R2}K_{R3} = 4k_R^3 \quad K_{T3} = \tfrac{2}{3}k_T \quad K_{T1}K_{T2}K_{T3} = 4k_T^3$$

$$K_{R4} = \tfrac{1}{4}k_R \quad K_{R1}K_{R2}K_{R3}K_{R4} = k_R^4 \quad K_{T4} = \tfrac{1}{4}k_T \quad K_{T1}K_{T2}K_{T3}K_{T4} = k_T^4$$

- If

$$c = \frac{k_T}{k_R}$$

$$K_{T1} = 4c\,k_R$$

$$K_{T1}K_{T2} = 6c^2\,k_R^2$$

$$K_{T1}K_{T2}K_{T3} = 4c^3\,k_R^3$$

$$K_{T1}K_{T2}K_{T3}K_{T4} = c^4\,k_R^4$$

then

$$4y = \frac{4[R_O]\, k_R[O_2](1 + k_R[O_2])^3 + 4[T_O]k_T[O_2](1 + k_T[O_2])^3}{[R_O](1 + k_R[O_2])^4 + [T_O](1 + k_T[O_2])^4}$$

$$y = \frac{[R_O]k_R[O_2](1 + k_R[O_2])^3 + cL[R_O]k_R[O_2](1 + ck_R[O_2])^3}{[R_O](1 + k_R[O_2])^4 + L[R_O](1 + ck_R[O_2])^4}$$

if $\rho = k_R[O_2]$. The fraction of saturation with oxygen is given by

$$y = \frac{\rho(1 + \rho)^3 + cL\rho(1 + c\rho)^3}{(1 + \rho)^4 + L(1 + c\rho)^4}$$

- The fraction of saturation with oxygen, y, when $[R] = [T]$ can be simply calculated when $L = 1$:

$$y = \frac{\rho(1 + \rho)^3 + c\rho(1 + c\rho)^3}{(1 + \rho)^4 + (1 + c\rho)^4}$$

- When the two states act independently, $L = 0$:

$$y = \frac{\rho}{1 + \rho}$$

which is the expression for a hyperbolic saturation curve.

- The following are the derivations of the oxygen binding constants in terms of the allosteric parameters L and c:

$$L = \frac{[Hb_T]}{[Hb_R]}$$

$$c = \frac{k_T}{k_R}$$

$$K_1 = \frac{[HbO_2]}{[Hb][O_2]} = \frac{[Hb_RO_2] + [Hb_TO_2]}{([Hb_R] + [Hb_T])[O_2]} = \frac{[Hb_RO_2] + [Hb_TO_2]}{([Hb_R] + L[Hb_R])[O_2]}$$

$$[Hb_TO_2] = k_T[Hb_T][O_2] = ck_RL[Hb_R][O_2] = cL[Hb_RO_2]$$

$$K_1 = \frac{[Hb_RO_2] + cL[Hb_RO_2]}{([Hb_R] + L[Hb_R])[O_2]} = \frac{[Hb_RO_2](1 + cL)}{[Hb_R][O_2](1 + L)}$$

$$K_1 = \frac{1 + cL}{1 + L}K_R$$

$$K_2 = \frac{[Hb(O_2)_2]}{[HbO_2][O_2]} = \frac{[Hb_R(O_2)_2] + [Hb_T(O_2)_2]}{([Hb_RO_2] + [Hb_TO_2])[O_2]} = \frac{k_R[Hb_RO_2][O_2] + k_T[Hb_TO_2][O_2]}{(k_R[Hb_R][O_2] + k_T[Hb_T][O_2])[O_2]}$$

$$K_2 = \frac{k_R[Hb_RO_2] + c^2Lk_R[Hb_RO_2]}{k_R[Hb_R][O_2] + ck_RL[Hb_R][O_2]}$$

$$K_2 = \frac{1 + c^2 L}{1 + cL} K_R$$

$$K_3 = \frac{[Hb(O_2)_3]}{[Hb(O_2)_2][O_2]} = \frac{k_R^2[Hb_R O_2] + c^3 Lk_R^2[Hb_R O_2]}{k_R^2[Hb_R][O_2] + c^2 Lk_R^2[Hb_R][O_2]}$$

$$K_3 = \frac{1 + c^3 L}{1 + c^2 L} K_R$$

$$K_4 = \frac{[Hb(O_2)_4]}{[Hb(O_2)_3][O_2]} = \frac{k_R^3[Hb_R O_2] + c^4 Lk_R^3[Hb_R O_2]}{k_R^3[Hb_R][O_2] + c^3 Lk_R^3[Hb_R][O_2])}$$

$$K_4 = \frac{1 + c^4 L}{1 + c^3 L} K_R$$

- Assume the range $L \sim 10^3 - 3 \times 10^5$, with corresponding range $c \sim 0.04 - 0.001$. Consequently:
  - A high value of L implies that deoxygenated hemoglobin mainly exists in T-form.
  - A small value of c means that $k_T \ll k_R$, or oxygen binds more readily to the R-form.
  - As R-subunits become oxygenated, many more subunits must transform from T to R in order to maintain L constant.
  - This means that more R molecules become available for oxygenation. Therefore, information about conditions of the ligated hemes is transmitted to heme receiving next ligand.
- The *first* O_2 attaches itself *very weakly* to the heme (mostly in T-form, $k_T \ll k_R$). It has a low association constant, K_1 of 5–60 atm^{-1}, depending on pH, [Cl$^-$], [CO$_2$], and [DPG],(DPG : 2,3-diphosphoglycerate)

$$Hb + O_2 \leftrightarrows HbO_2$$
$$K_1 = \frac{[HbO_2]}{[Hb][O_2]} = 5 - 6 \, atm^{-1}$$

- The *second and third* O_2 then bind *more strongly*, and the *last* O_2 has two to three orders of magnitude *greater affinity* than the first O_2:

$$Hb(O_2)_3 + O_2 \leftrightarrows Hb(O_2)_4$$
$$K_4 = \frac{[Hb(O_2)_4]}{[Hb(O_2)_3][O_2]} = 3000 - 6000 \, atm^{-1}$$

At 25 °C, the hemoglobin oxygenation is exothermic, so the oxygen affinity decreases with rising temperature.

- When there are three ligated $(Hb(O_2)_3)$:

$K_4 > K_3 > K_2 > K_1$,   and $K_4 \approx K_{Mb} \approx K_{Hb\alpha} \approx K_{Hb\beta}$,

where K is oxygen binding to monomeric hemes. Therefore, increase of oxygen affinity is not due to just the $\alpha_2\beta_2$ structure.

$$K_4 \approx K_{Mb} = 1500 \text{ atm}^{-1} \qquad K_\alpha = 1500 \text{ atm}^{-1} \quad K_\beta = 2600 \text{ atm}^{-1}$$

- The four-chain Hb tetramer functions not by enhancing the binding of the last oxygen but by diminishing the tendency to bind the first one.

Oxygen Dissociation

What are the expected dissociation behaviors of $Hb(O_2)_4$ based on MWC model?

- When a fully loaded Hb arrives at the tissues, its tendency to keep or lose the first oxygen is about the same as that of myoglobin. But with each oxygen lost, the next one falls off more easily (*the all-or-nothing, Matthew effect*).

Heterotropic Allosteric Cooperativity

Define: 1. Cooperatives
          2. Positive cooperativity
          3. Negative coopertivity
          4. Homotropic allosteric cooperativity
          5. Heterotropic allosteric cooperativity

- *Cooperativity*: Bound ligand affects binding of another ligand.
- *Allostery*: Bound ligand affects binding by changing a protein's conformation.
- *Positive cooperativity*: Bound ligand *enhances* another binding.
- *Negative cooerativity*: Bound ligand *inhibits* another binding.
- *Homotropic cooperativity*: Both ligands are the same species. Therefore O_2 binding by Hb is homotropic allosteric cooperativity.
- *Heterotropic cooperativity*: The two ligands are different.

Heterotropic Allosteric Cooperativity

Muscle	Lungs
$HbO_2 \rightarrow Hb$	$Hb \rightarrow HbO_2$
CO_2 taken up	CO_2 released

- At muscle, higher $[CO_2]$:

$$CO_2 + H_2O \leftrightarrows H^+ + HCO_3^-$$

Therefore pH decreases and

$$Hb[O_2]_4 \rightarrow Hb + 4O_2 \qquad \text{(Bohr effect)}$$

Takes on 0.5–3.0 H^+

$\therefore$ H^+ stabilized deoxy-Hb

FIGURE 5-58 Positive residues of amino acid side chains around DPG binding site in human deoxyhemoglobin-2,3-diphosphoglycerate complex, PDP 1B86 (Richard et al., 1993).

$\therefore$ CO_2 stabilized deoxy-Hb

Also via carbamylation

that is, binding of O_2, CO_2, and H^+ are all linked.

- Diphosphoglycerate (DPG) and some other drugs bind in the central cavity and stabilize the T-state of Hb (Fig. 5-58).

Homotopic Effect

Draw a scheme to represent the homotropic effect of oxygen binding to the four sites of the hemoglobin, indicating the microscopic equilibrium constants. Find an expresion for the fractional oxygen saturation y. Make some assumtions about and express the macroscopic equilibrium constants in terms of the microscopic values.
How can you estimate the following from the Hill plot $[\log(Y/1 - Y)$ vs. $\log pO_2]$:

(i) $\log p_{50}$

(ii) K_1 and K_4

(iii) Hill coeffiecent, n_{max}

(iv) ΔG_{41}

- Scheme 5-30 presents oxygen binding by hemoglobin and contains 16 microscopic equilibrium constants donating κ_1 to κ_{16}.
- By the law of energy conservation only nine of the κ's are *independent*:
  (a) Oxygenation through κ_5 and κ_7 excludes the oxygenation through κ_9 and κ_{11}, also though κ_3 and κ_{15} excludes that of κ_6, κ_8, κ_{10}, and κ_{12}.
  (b) The set κ_1, κ_{13}, κ_2, κ_5, κ_7, κ_{14}, κ_3, κ_{15}, κ_4, and κ_{16} can be selected as a possible choice.
- Then, by taking the statistical factors into account, the concentrations of the molecular species combined with oxygen molecules of different numbers are given as

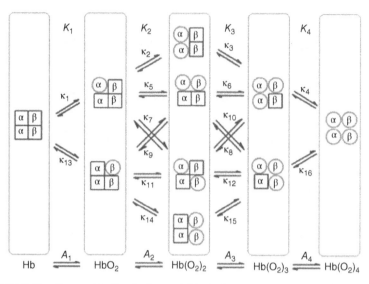

SCHEME 5-30 Oxygen binding by hemoglobin (red are oxygenated and blue squares are deoxygenated). (See the color version of this scheme in Color Plates section.)

$$[HbO_2] = 2(\kappa_1 + \kappa_{13})[O_2][Hb]$$

$$[Hb(O_2)_2] = (\kappa_1\kappa_2 + 2\kappa_1\kappa_5 + 2\kappa_1\kappa_7 + \kappa_{13}\kappa_{14})[O_2]^2[Hb]$$

$$[Hb(O_2)_3] = 2(\kappa_1\kappa_2\kappa_3 + \kappa_{13}\kappa_{14}\kappa_{15})[O_2]^3[Hb]$$

$$[Hb(O_2)_4] = \kappa_1\kappa_2\kappa_3\kappa_4[O_2]^4[Hb]$$

The fractional oxygen saturation y is given by

$$y = \frac{[HbO_2] + 2[Hb(O_2)_2] + 3[Hb(O_2)_3] + 4[Hb(O_2)_4]}{4([Hb] + [HbO_2] + [Hb(O_2)_2] + [Hb(O_2)_3] + [Hb(O_2)_4])}$$

- Assume α- and β-subunits are totally equivalent in both
  - their intrinsic affinity for oxygen and
  - their manner of interaction.

  Then all the molecular species are combining with the same number of oxygen molecules, forming one species.
- In consequence:
  - $\kappa_1 = \kappa_{13} = K_1$
  - $\kappa_2 = \kappa_5 = \kappa_7 = \kappa_9 = \kappa_{11} = \kappa_{14} = K_2$
  - $\kappa_3 = \kappa_6 = \kappa_{10} = \kappa_8 = \kappa_{12} = \kappa_{15} = K_3$
  - $\kappa_4 = \kappa_{16} = K_4$

Consequently:

$$[HbO_2] \quad = 2(\kappa_1 + \kappa_{13})[O_2][Hb] = 4K_1[O_2][Hb]$$
$$[Hb(O_2)_2] = (\kappa_1\kappa_2 + 2\kappa_1\kappa_5 + 2\kappa_1\kappa_7 + \kappa_{13}\kappa_{14})[O_2]^2[Hb] = 6K_1K_2[O_2]^2[Hb]$$
$$[Hb(O_2)_3] = 2(\kappa_1\kappa_2\kappa_3 + \kappa_{13}\kappa_{14}\kappa_{15})[O_2]^3[Hb] = 4K_1K_2K_3[O_2]^3[Hb]$$
$$[Hb(O_2)_4] = 2(\kappa_1\kappa_2\kappa_3\kappa_4)[O_2]^4[Hb] = 2K_1K_2K_3K_4[O_2]^4[Hb]$$

and the oxygen binding is described by only four equilbria:

$$Hb + O_2 \overset{A_1}{\Leftrightarrow} HbO_2 + O_2 \overset{A_2}{\Leftrightarrow} Hb(O_2)_2 + O_2 \overset{A_3}{\Leftrightarrow} Hb(O_2)_3 + O_2 \overset{A_4}{\Leftrightarrow} Hb(O_2)_4$$

Thus

$$y = \frac{A_1[O_2] + 2A_2[O_2]^2 + 3A_3[O_2]^3 + 4A_4[O_2]^4}{4\left(1 + A_1[O_2] + A_2[O_2]^2 + A_3[O_2]^3 + A_4[O_2]^4\right)}$$

Substituting

$$A_1 = 4K_1$$
$$A_2 = 6K_1K_2$$
$$A_3 = 4K_1K_2K_3$$
$$A_4 = K_1K_2K_3K_4$$

yields

$$y = \frac{K_1[O_2] + 3K_1K_2[O_2]^2 + 3K_1K_2K_3[O_2]^3 + K_1K_2K_3K_4[O_2]^4}{1 + K_1[O_2] + 6K_1K_2[O_2]^2 + 4K_1K_2K_3[O_2]^3 + K_1K_2K_3K_4[O_2]^4}$$

and

$$\boxed{\frac{y}{1-y} = \frac{K_1[O_2]\left(1 + 3K_2[O_2]^2 + 3K_2K_3[O_2]^3 + K_2K_3K_4[O_2]^4\right)}{1 + K_1[O_2] + 3K_1K_2[O_2]^2 + 3K_1K_2K_3[O_2]^3}}$$

- The ligand curve is often represened by $\log[y/(1-y)]$ versus $\log pO_2$ (Hill plot) (Fig. 5-59).
- $y/(1-y)$ is the fraction of liganded sites to unliganded sites.
- The intersection of this plot with the $\log pO_2$ axis gives the $\log p_{50}$ value.
- The slop of this plot, n, is given by

$$n = \frac{d\log[y/(1-y)]}{d\log pO_2}$$

(a) In the absence of interaction among sites, i.e., $K_1 = K_2 = K_3 = K_4$, $n = 1$ for any $\log pO_2$, giving a straight line with a slope of unity.

(b) If $K_1 < K_2 < K_3 < K_4$, n is larger than unity in the middle of the plot.

(c) The slope n approches unity at the bottom and top extremes.

(d) The maximal slope, n_{max}, is called the *Hill coefficient* and is used as a simple measure of cooperativity.

(e) The value of n_{max} never exceeds the number of ligand binding sites.

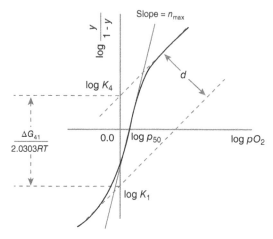

FIGURE 5-59 Hill plot. (Brunor et al., 1985. Reproduced by permission of John Wiley & Sons).

- The bottom and the top asymptotes are expressed by

$$\log \frac{y}{1-y} = \log pO_2 + \log K_1 \rightarrow \text{ has intercept } = \log K_1$$

$$\log \frac{y}{1-y} = \log pO_2 + \log K_4 \rightarrow \text{ has intercept } = \log K_4$$

- The distance between the intersections of the two asymptotes with the line $\log pO_2 = 0$ gives $\Delta G_{41}/(2.303\,RT)$, where
$\Delta G_{41} = RT \ln \frac{K_4}{K_1}$ or

$\Delta G_{41} = 2.303\sqrt{2}\,RTd$ or

ΔG_{41} is the *free energy of the cooperativity* but does not give the true free energy of interaction. It measures the enhancement of the ligand affinity at the fourth ligation step relative to that of the first step. The cooperativity exhibited at i_{th} oxygenation step with reference to the first can be expressed by the difference of the free energy, $\Delta G_i - \Delta G_1$.
$K_1 \approx K_T$ and $K_4 \approx K_R$, $\Delta G_{41} = \Delta G_T - \Delta G_R$
The free energy of cooperation can be separated into the enthalpy contribution, $\Delta H_T - \Delta H_R$, and entropy contribution $-T(\Delta S_T - \Delta S_R)$.

Evaluation of the true interaction energy involves a model defining the non-interacting system as a reference.

CYTOCHROME *C*

Identify the cytochromes. What are the advantages of such a complex system of cytochromes? Define the biological function of cytochrome–c.

- Cytochromes are heme proteins whose characteristic mode of action is electron and/or hydrogen transport by a reversible valency change of its heme.
- The cytochrome members differ from one another in reduction potentials by about 0.2 V or less, with a total potential difference of about 1 V.
- The differences in potentials for Fe^{2+}/Fe^{3+} oxidation result from change in
  - the porphyrin substituents,
  - the porphyrin environment, where the polypeptide component tunes the metal center to the required potential, and
  - the axial ligands.
- Cytochrome c (Cyt. c) accepts an electron from cytochrome c_1 and transfers this electron to cytochrome c oxidase, $E^{0\prime} = 260\,mV$.

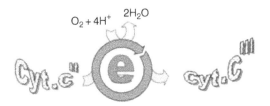

- Molecular weight ~ 37000 and subject to extensive aggregation (1.2×10^6).
- The protein is acidic with an isoelectric point near 3.6.
- Cyt. c contains a single heme group per molecule.
- The polypeptide chain has ~ 104–113 amino acids.
- The heme group is bound covalently to the protein via thioether linkages to the side chains of the porphyrin (Fig. 5-60).
- Most Cyt. c molecules are low spin.
- The fifth and sixth iron ligands are histidine (His) and methionine (Met).

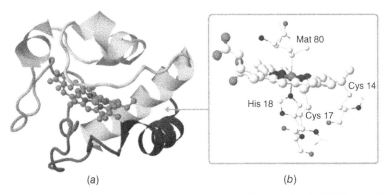

FIGURE 5-60   Three-dimensional structure of horse heart Cyt. c, PDB 1HRC (Bushnell et al., 1990).

- Cyt. *c* of human heart differs in only *one* of 104 amino acids when compared with that of the rhesus monkey and in 11 amino acids when compared with that of the dog.
- Cyt. *c* can be reduced by dithionite or by ascorbic acid.

Hemochrome Structure

Why cannot the central iron atom of cyto.c form adduct with other small ligands?

The peptide chain is arranged in such a manner that the *charged groups* are on the *outside* of the molecule, while *hydrophobic groups* are for the most part in the *interior*, clustered about the heme group.

- Cyt. *c* has a *buried* active site (the heme) to prevent its reaction with potential substrates (ligands).
- Cyt. *c* does not react with oxygen or carbon monoxide at neutral pH.

Absorbance Bands

What are the characteristic features of the electronic absorbance bands of cyto.c, (Figs. 5-61 and 5-62)?

- Cyt. *c* exhibits absorption spectra that are *typical* of ferri- and ferroporphyrin complexes, in which the fifth and sixth coordination positions about the iron atom are occupied by strong field ligands (Fig. 5-61).
- The "mammalian-type" molecules *retain* their homochrome properties to unusually *low* values of pH (Fig. 5-62).

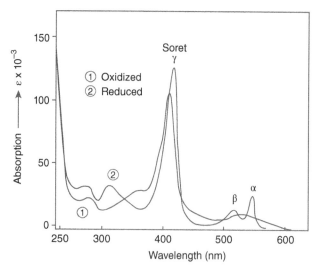

FIGURE 5-61 Absorption spectra of oxidized and reduced horseheart Cyt. *c* (From Greenwood, 1985, Stephanos and Addison, 1990. Reproduced by permission of John Wiley & Sons.)

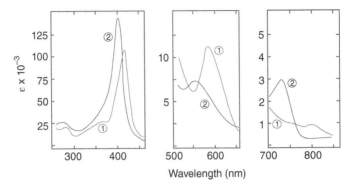

FIGURE 5-62   Absorbance spectra of oxidized horse heart Cyt. *c*: (1) at pH 7 and (2) at pH 1. (From Harbury and Marks, 1973. Reproduced by permission of Elsevier.)

- Horse ferrocytochrome *c* remains in the low-spin state even at pH 1, while the oxidized form undergoes transition to the high-spin state.
- The iron atom in Cyt. *c* alternates between the oxidized state in which iron is ferric (Fe^{3+}) and has a single unpaired electron and the reduced or ferrous state where there are no unpaired electrons.
- The reduced form is pink with absorption bands at 551, 522, and 415 nm.
- It is reoxidized by air but in vivo by cytochrome oxidase.
- Oxidized form is yellow with absorption bands at 530 and 400 nm.
- The reduction of Cyt. *c* by various inorganic reducing agents has been studied kinetically.

Examples

Show some examples of the electron transfer between cyto. c and other small inorganic redox reagents.

Electron transfer between Cyt. *c* and other small inorganic redox reagent is presented in Table 5-8.

ELECTRON TRANSFER IN PORPHYRINS AND METALLOPORPHYRINS

What are the general mechanisms for one–electron oxidation and reduction of porphyrins and metalloporphyrins?

The general routes for one-electron oxidation and reduction of porphyrins and metalloporphyrins are presented in Scheme 5-31.

Axial Electron Transfer

What are the requirements and the possible mechanisms of axial electron transfer in heme proteins? Give example for each mechanism.

TABLE 5-8   Rates of Electron Transfer between Cytochromes and Small Inorganic Redox Reagents (Cyt. c = Horse Cytochrome c) at 25 °C

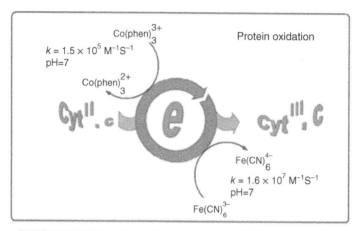

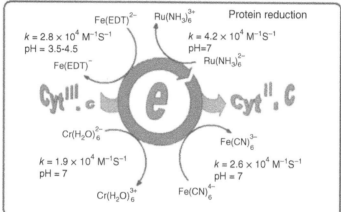

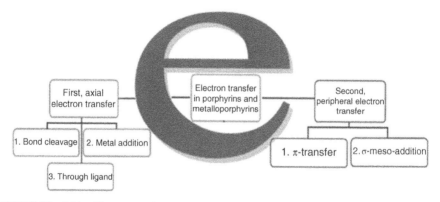

SCHEME 5-31 The general routes for one-electron oxidation and reduction in metalloporphyrins.

- The G-conformations assign these proteins as an easy approach to an axial position of the iron porphyrin.
- Axial electron transfer can proceed through three mechanisms:
  1. Bond cleavage (Scheme 5-32). This mechanism requires that the axial ligand is the oxidant, that is, the *inner sphere mechanism*. Examples
     (a) The chloride ion of chloro-Fe(III)porphyrin does not exchange with radio-labeled chloride ion, $Cl^* = {}^{36}Cl$:

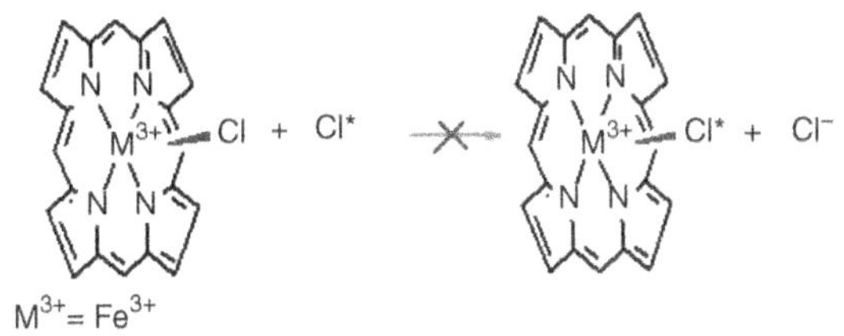

$M^{3+} = Fe^{3+}$

No Cl^- exchange → no axial bond cleavage

However, axial electron transfer proceeds by axial bond cleavage in the presence of $CrCl^+$:

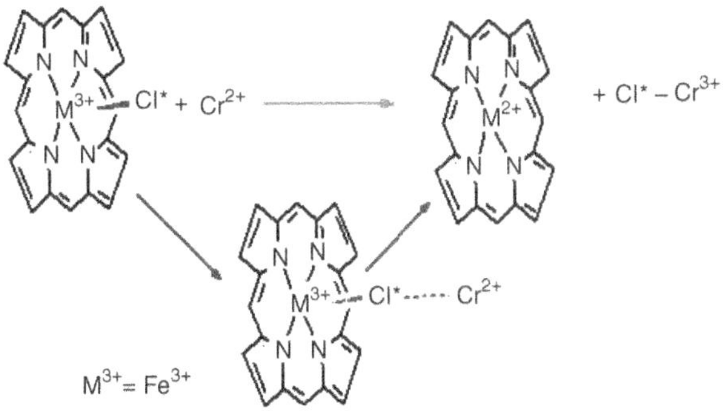

$M^{3+} = Fe^{3+}$

Electron transfer through bond cleavage (Modified from Cohen et. al., 1972.)

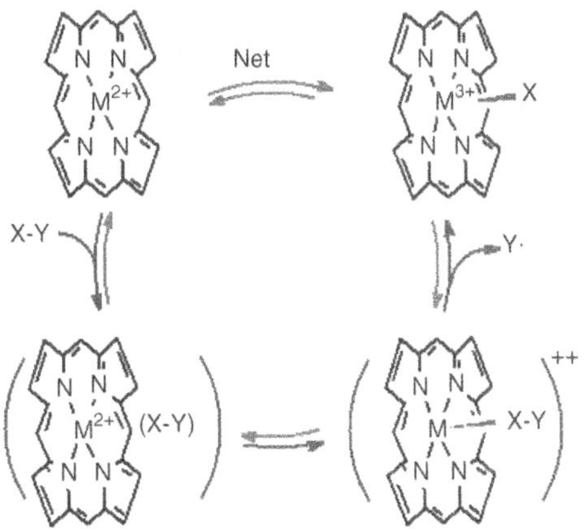

SCHEME 5-32   Electron transfer through bond cleavage.

(b) Oxidation of high-spin Fe(II)porphyrin with alkyl halides:

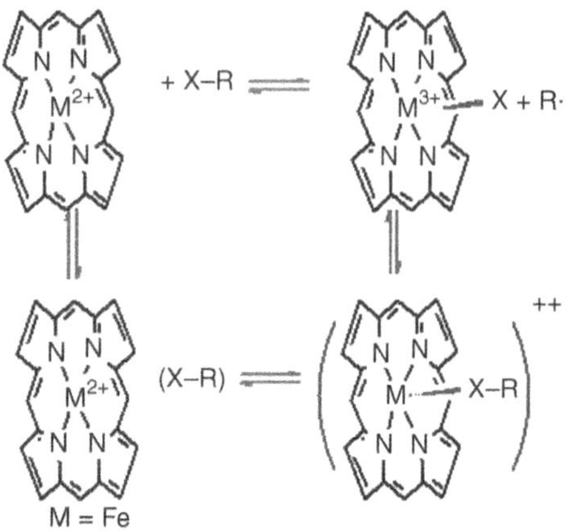

Axial electron transfer through bond cleavage
(Modified from Wade and Castro, 1973, and Castro et al., 1974.)

The radical R · may either be dimerized when R = α-phenylethyl or be reduced when R = C(CN)(CH$_3$)$_2$ to 2–cyano-propane, HC(CN)(CH$_3$)$_2$:

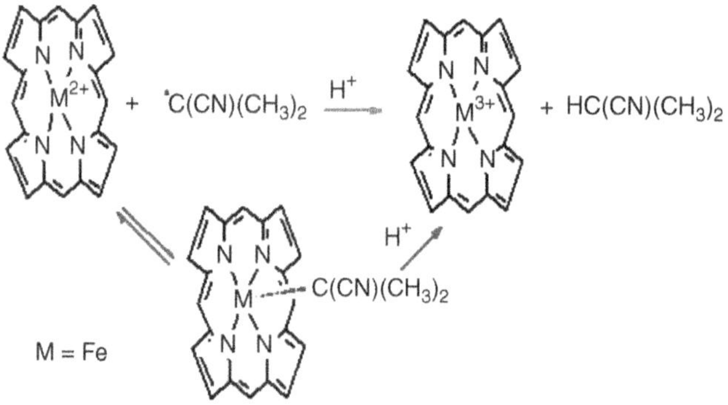

(c) Oxidation of Hb(II) and Mb(II) by alkyl halides although they are not easily oxidized by oxygen. Fe(II) Cyt. c is inert.

2. *Metal addition*: There are two possible routes for reaction of A = B and Fe(II) iron porphyrin (Scheme 5-33). This requires that the axial ligand is the oxidant. However:

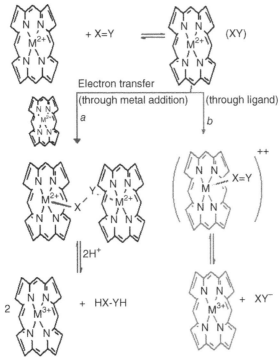

SCHEME 5-33   Axial electron transfer through metal addition, reaction (*a*) (Modified from Castro, 1978.)

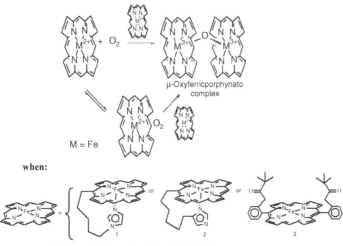

when:

- Complex 1 forms O_2 adduct at − 45 °C
- Complex 2 is rapidly oxidized by O_2
- Complex 3 forms O_2 adduct in amide solvent at room temperature

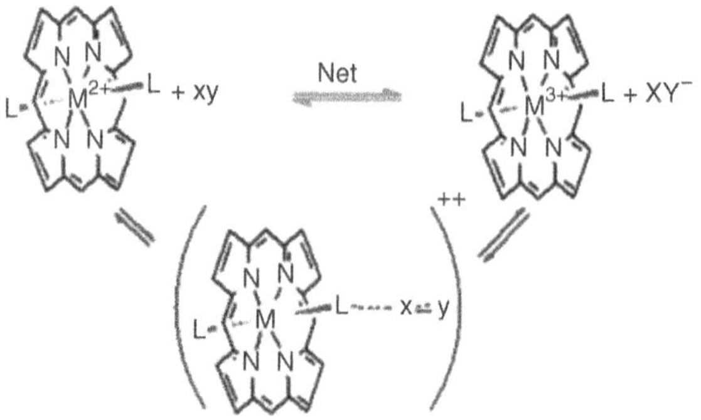

SCHEME 5-34   Axial electron transfer through ligand.

3. *Through ligand* (Scheme 5-34)

   (a) This mechanism requires orbital overlap between the *axial ligand* and the oxidant or reductant.

   (b) This is an outer sphere mechanism;

   (c) Examples:

   (i) In the reaction of tetra-*n*-butylammonium superoxide with Fe^{3+}-protoporphyrin dimethyl ester perchlorate, the spectral studies suggest the reduction of Fe^{3+} to Fe^{2+} by superoxide ions.

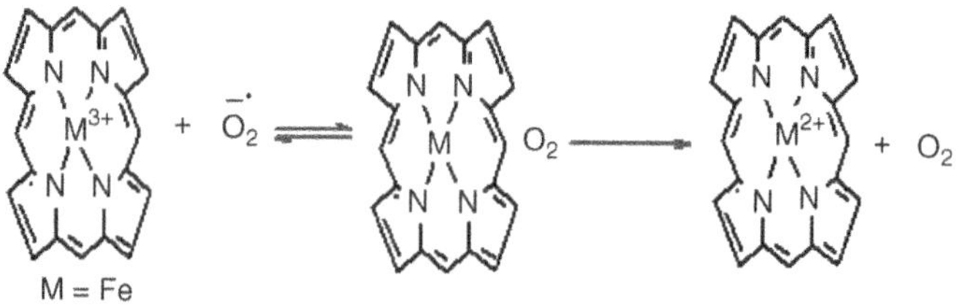

M = Fe

Reduction of Fe^{2+} through superoxide axial addition
(Modified from Hill et al., 1974.)

   (ii) The high-spin Fe(II) porphyrin derivatives are rapidly oxidized by quinines:

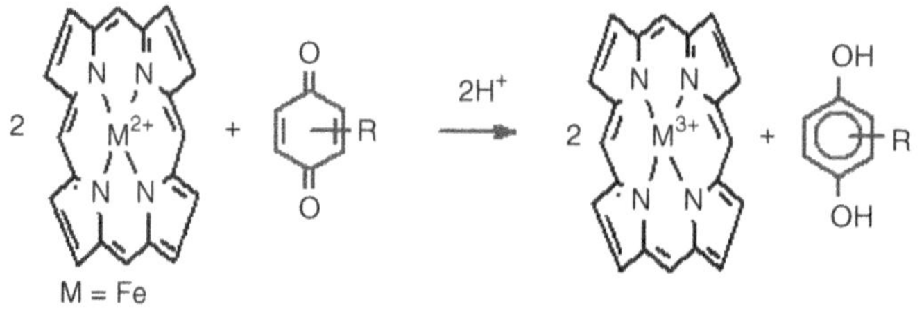

M = Fe

Reduction through ligand (Modified from Castro et al., 1975.)

(iii) Ferric hemoglobins and ferric myoglobins react with NO (nitric oxide) to give Hb(III)NO and Mb(III)NO, respectively.

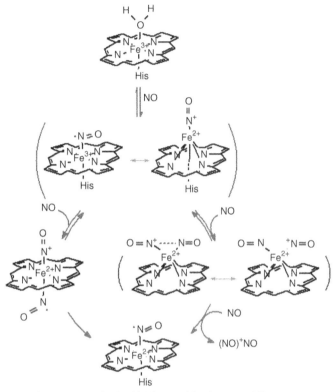

Nitrosyl autoreduction (Addison and Stephanos, 1986).

- After 2 hours, in the presence of NO, these adducts are reduced to Hb(II) NO and Mb(II)NO.
- In this autoreduction mechanism, NO is kinetically the reductant. The two possible mechanisms were suggested through axial ligand redox.

Peripheral Electron Transfer

What are the requirements and the possible mechanisms of peripheral electron transfer?

- It is a pure electron transfer "outer sphere" process similar to the axial electron transfer through the ligand.
- Electron transfer through the porphyrin requires:
  1. The C-conformation: This exposes the periphery and blocks the accessibility to the axial ligand.

2. Peripheral electron transfer requires that metal d electrons be in conjugation with the porphyrin.

   (a) Therefore, the transition metal must be at the core of the porphyrin.

   (b) If the metal is a nontransition element (Mg) or no metal is present, "π-cation radicals" or "π-anion radicals" will be formed:

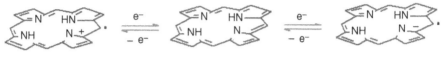

π-Cation radical   **π-Anion radical**

3. Axial ligands direct the degree to which metal d electrons are conjugated with the porphyrin π-system. Axial ligands have large effect upon the peripheral electron transfer. There are at least two modes of spin pairing in metalloporphyrin complexes:

   (a) *Strong σ-bonding axial ligand*, such as ammonia, hydrazine, and *n*-butylamine. These ligands enhance metal d–π interaction to the extent that low-spin complexes ensue.

   (b) *π-Bonding ligands*, such as CO, O_2, and thioethers. These ligands coordinate to d^6 Fe^{2+} and enhance the metal d–π interaction and yield low-spin complexes.

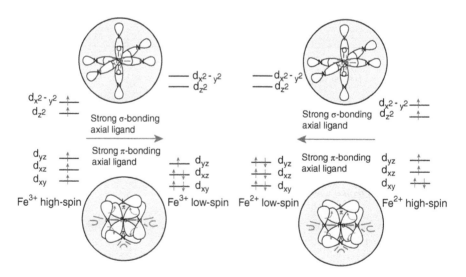

4. A sixth ligand must coordinate to the metal, a hexacoordinate arrangement.

   (a) Unless a sixth ligand is coordinated to the metal, a pentacoordinate complex will form, and the metal ion will not be in the plane of the porphyrin, and this decreases d–π interaction.

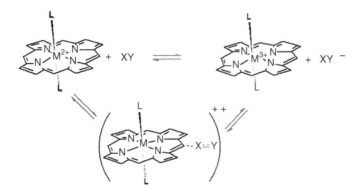

SCHEME 5-35   Peripheral π-electron transfer.

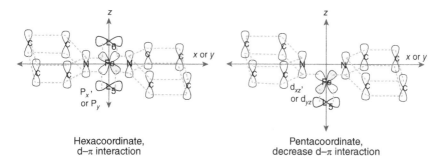

Hexacoordinate,
d–π interaction

Pentacoordinate,
decrease d–π interaction

(b) The sixth ligand of Mb, Hb, and Cyt. c is an imidazole, which has the capacity for both σ- and π–coordination, enhances

(i) d–π interaction,

(ii) stability, and

(iii) electron transfer in cytochromes.

- There are two distinct peripheral processes:

1. π-*Transfer.* There is no σ-bond formed or broken with the porphyrin (Scheme 5-35).

(a) The metalloporphyrin can be electrochemically or chemically oxidized through the formation of a positive π-cation radical:

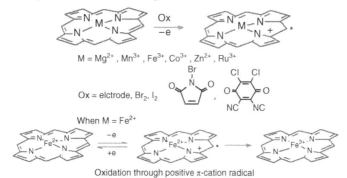

$M = Mg^{2+}$, Mn^{3+} , Fe^{3+}, Co^{3+} , Zn^{2+} , Ru^{3+}

$Ox = $ elctrode, Br_2, I_2

When $M = Fe^{2+}$

Oxidation through positive π-cation radical

(b) Also, the metalloporphyrin can be electrically or chemically reduced via π-anion radical formation:

$M = Fe^+, Zn^{2+}$
Red. = electrode, Na^+ anthranecide
When $M = Fe^{3+}$:

Oxidation through positive π-anion radical

2. σ-*Meso addition*

(a) A new σ-bond is formed at the meso position.

(b) It requires the metal d-electron to be in conjugation with the porphyrin.

(c) The σ-meso addition reduction is not likely to be reversible (Scheme 5-36).

(d) The free radical acts as an electron carrier.

(e) Example: when diacetyl peroxide is decomposed in pyridine solutions of chloro Fe(III) octaethylporphyrin:

    (i) Low-spin dipyridyl adduct is formed.

    (ii) The generated methyl radical is added to the meso position.

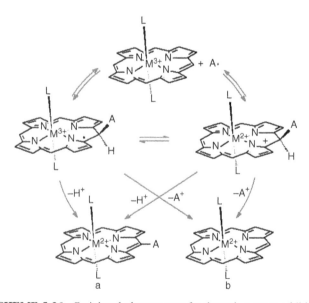

SCHEME 5-36 Peripheral electron transfer through σ-meso addition.

(iii) Fe^{3+} is reduced to Fe(II)porphyrin:

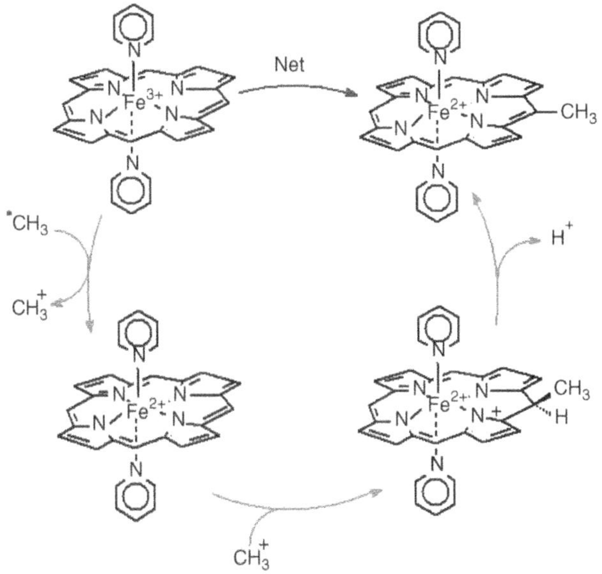

(f) On the other hand, in the *absence* of pyridine, when high-spin chloro- and acetate-Fe(III) porphyrin is mixed with methyl radical (no d–π interaction), an axial redox reaction is predominates:

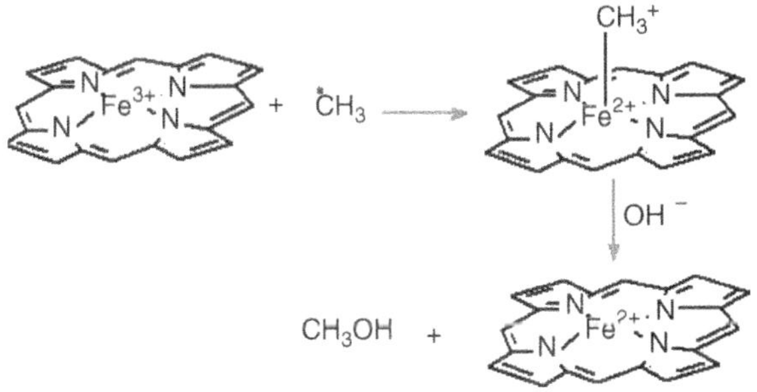

CATALASES

Role and Properties

Define the biological role and the main properties of catalases.

- Catalases are found in:
  - Erythrocyte
  - All plants
  - Most aerobic bacteria
- Catalases have the short C-conformation, in which both an axial position on iron and the porphyrin periphery are exposed.

- Catalases catalyze the decomposition of H_2O_2 into H_2O and O_2:

$$2\ H_2O_2 \quad\quad\quad 2\ H_2O_2 + O_2$$
$$k = 10^7\ M^{-1}s^{-1}$$

- Molecular weight 240,000.
- Catalases are composed of four identical subunits, each containing high-spin Fe^{3+}–porphyrin prosthetic groups, and one heme group at the active site.
- The resting state of the enzyme is Fe^{3+}.
- Catalases are very closely related to the peroxidases.
- Absorption differences between native and modified catalase in the presence of cyanogen bromide (BrCN) were barely noticeable, suggesting that no major alteration occurs in the environment of tyrosine (see spectrum; Jajczay, 1966). *Note*: BrCN oxidizes the sulfhydryl groups of the protein, with simultaneous libration of cyanide ion.

Reaction Mechanism

Suggest a possible mechanism for the action of catalase.

- Chemical modification studies have suggested that histidine and tyrosine are involved in the activity of the enzyme.
- The axial metal sites are occupied by water and an amino acid residue.
- Molecules of H_2O_2 are dealt with one at a time. If simultaneous:
  1. Rate α $[H_2O_2]^2$, and the reaction would be very slow, with $[H_2O_2] \ll 1$ mol, where $t_{1/2} = 1/kC_0$.
  2. But $k = 10^7\ M^{-1}\ s^{-1}$, and the catalase can decompose as 6×10^6 molecules of peroxide per second.
- Catalase reacts with H_2O_2 to give compound I, which is able to oxidize H_2O_2:

$$H_2O_2 + Fe^{3+} \quad\quad\quad \rightarrow H_2O + Fe\ compound\ I$$
$$Fe\ compound\ I + H_2O_2 \rightarrow H_2O + O_2 + Fe^{3+}$$

Compound I is also an intermediate in the catalytic cycle of other heme enzymes such as peroxidase.

- Kinetic and spectroscopic studies led to:

- By using ^{18}O labeling, it has been shown that both oxygens of the O_2 molecule originated from the same H_2O_2 molecule:

$$H_2O_2 \rightleftharpoons H^+ + HO_2^-$$

$$(PpIX)^{2-}Fe^{3+} + HO_2^- \underset{k_2}{\overset{k_1}{\rightleftharpoons}} (PpIX)^{2-}Fe^{3+}HO_2^-$$

$$(PpIX)^{2-}Fe^{3+}HO_2^- + H^+ \overset{k_3}{\longrightarrow} [(PpIX)^{2-}Fe^{4+}=O]^{+\bullet} + H_2O$$

$$[(PpIX)^{2-}Fe^{4+} = O]^{+\bullet} + HO_2^- \overset{k_4}{\longrightarrow} (PpIX)^{2-}Fe^{3+} + O_2 + H_2O$$

$$2H_2O_2 \xrightarrow{(PpIX)^{2-}Fe^{3+}} 2H_2O + O_2$$

Compound I is believed to be Fe^{4+} with a radical group in the protein or porphyrin: $[(PpIX)^{2-}Fe^{4+} = O]^{+\bullet}$.

- Nature of compound I:
  1. Formation of H_2O:
     (i) Requires getting O^{2-} from compound I,
     (ii) For Por^{2-} and O^{2-}, requires having Fe^{4+}
        Therefore $Fe^{3+} \rightarrow Fe^{4+}$ in compound I. Similar to $V(OH_2)_6^{+4} \rightarrow VO(H_2O)_4^{2+}$, $V^{+4}O^{-2}$, vanadyl [oxovanadium (IV) ion, $V^{+4} = O$]
        It has been proposed that O in $Fe^{4+} = O$ should be specified as $Fe^{3+} - O^{\bullet-}$, consistent with quantum chemical results.
  2. Redox of Fe–Por:

$$\begin{array}{c} \text{Unstable} \\ \text{Reactive} \end{array} \underset{+E}{\overset{-e}{\longleftarrow}} [Fe^{3+} - Por]^0 \overset{+e}{\longrightarrow} Fe^{2+} - Por$$

(a) Easier if M-por is less + ve, it forms radical cation.

$$[Zn^{2+} - Por]^{+\bullet} \underset{+E}{\overset{-e}{\longleftarrow}} [Zn^{2+} - Por]^0 \overset{+e}{\longrightarrow} [Zn^{2+} - Por]^{-\bullet}$$
$$\quad d^{10} \qquad\qquad\quad \text{ESR silent}$$

(b) The visible spectra of the π-cation radicals of Co(III) octaethylporphyrin or *meso*-tetraalkyl Zn(II) porphyrins closely resemble the spectrum of compound I of catalase.

(c) Radical cation $[Zn^{2+} - Por]^{+\bullet}$ has $S = \frac{1}{2}$, and ESR analysis shows e on the ring, not Zn (oxidation of the ring).

Oxidation Site

Show that the site of oxidation in heme proteins depends on axial ligand.

- The site of oxidation in heme proteins depends on the axial ligand as well as on the metal porphyrin:

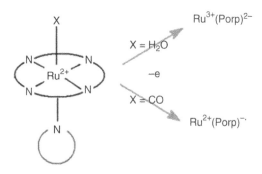

- Stabilization of $(PpIX^{+\bullet})$ $Fe^{4+}=O$: requires strongly σ-basic anionic; oxyanion is a good candidate as an axial ligand to stabilize Fe^{4+} such as tyrosine.
- The essential residues are tyrosine-357 on the proximal side and asparagine-147 and histidine-74 on the distal side (Fig. 5-63).
- Then compound I is identified as

$$[PpIX]^{2-}Fe^{4+}=O]^{+\bullet}$$

which is able to oxidize H_2O_2:

$$[(PpIX)^{2-}Fe^{4+} = O]^{+\bullet} + H_2O_2 \rightarrow (PpIX)^{2-}Fe^{3+} + O_2 + H_2O$$

Catalase Mimics

Give a model for the catalase action.

- *Triethylenetetramine* chelates as a model for catalase action (Scheme 5-37)
- The reaction seems to involve combination of the metal ion with two peroxide moieties followed by an intramolecular electron shift.

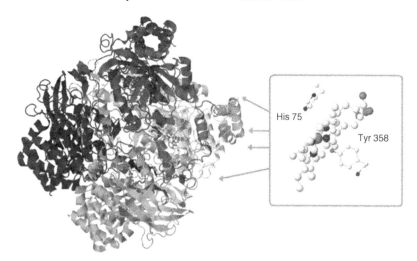

FIGURE 5-63 Structure of human erythrocyte catalase, PDB 1QQW (Ko et al., 2000).

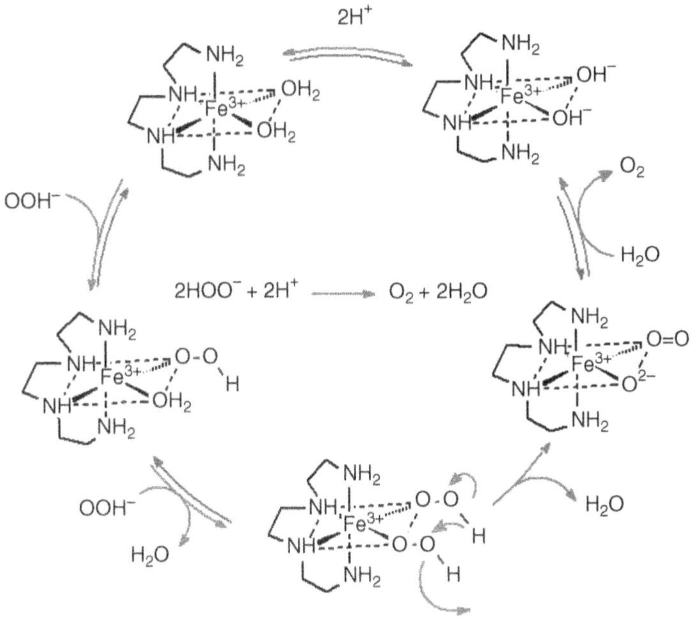

SCHEME 5-37 Triethylenetetramine chelates as a model for catalase action. (Based on Hamilton, 1969.)

PEROXIDASES

Biological Roles

Define the main biological role of peroxidases, the differences between catalases and peroxidases, and give examples for peroxidases.

- Peroxidases are enzymes catalyzing the *oxidation* of a variety of organic and inorganic compounds by H_2O_2, also acting as dehydrogenases:

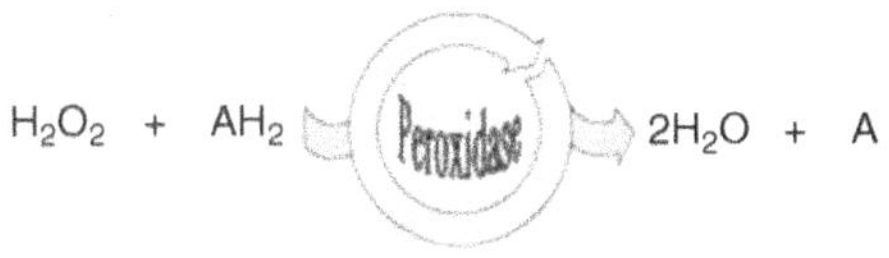

$$H_2O_2 \; + \; AH_2 \quad \text{Peroxidase} \quad 2H_2O \; + \; A$$

or

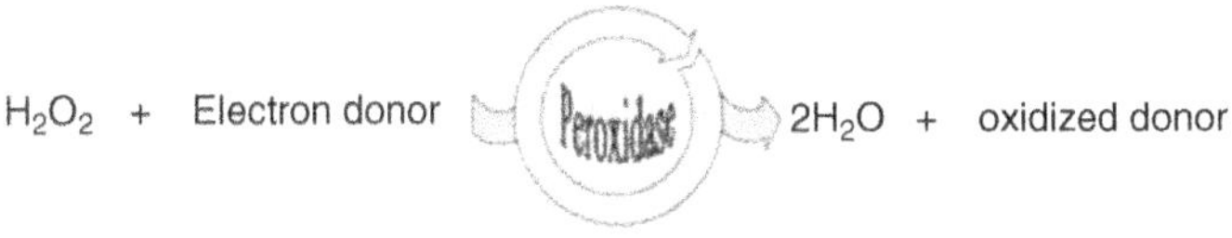

$$H_2O_2 \; + \; \text{Electron donor} \quad \text{Peroxidase} \quad 2H_2O \; + \; \text{oxidized donor}$$

- Peroxidases and catalases are related enzymes; both are capable of promoting the oxidation of H_2O_2. The mechanism of this oxidation involves a similar enzymatic intermediate.

- Catalases decompose H_2O_2:

$$H_2O_2 + H_2O_2 \xrightarrow{\text{catalase}} 2H_2O + O_2$$

or

$$H_2O_2 + Fe^{3+} \rightarrow H_2O + Fe \text{ compound I}$$
$$Fe \text{ compound I} + H_2O_2 \rightarrow H_2O + O_2 + Fe^{3+}$$

Catalases have a high degree of *specificity* and show *strong preference* for a molecule of *unsubstituted hydrogen peroxide* in the second reaction.

- Examples of peroxidases:
  - Chloroperoxidase: halogenates organic substrates.
  - Lactoperoxidase: antibacterial, oxidizes NCS^-.
  - Cytochrome P-450: hydroxylates organic substrates.
  - Cytochrome c peroxidase: reduces H_2O_2.

$$H_2O_2 + SH_2 \xrightarrow{\text{peroxidase}} 2H_2O + S \quad \text{where } S = \text{substrate}$$

- All the peroxidases purified from plants contain the heme group (Fe^{3+}–protoporphyrin IX).
- Horseradish roots and the sap of fig trees are the richest source of plant peroxidases. Cytochrome c peroxidase from baker's yeast also contains Fe^{3+}–protoporphyrin IX.
- Many of these enzymes are mixtures of *isozymes* of differing physical but similar catalytic properties.
- The isozymes A and C of horseradish peroxidase both contain Ca^{2+}, which is involved in *stabilization protein conformation*. This was evidenced by the lower thermal stability of calcium-free isozyme. These isozymes differ substantially in their ability to exchange Ca^{2+}.
- Most peroxidases are glycoproteins.

Characteristic Features

What are the main chemical properties of horseradish peroxidase?

- Color brown, molecular weight $\sim 40,000$
- Horseradish peroxidase (HRP) involves a histidine residue and an aquo group as the fifth and sixth ligands to Fe^{3+}, while a second histidine residue may be hydrogen bonded to the aquo group (Fig. 5-64).
- Ferriperoxidases react reversibly with CN^-, S^{2-}, F^-, N_3^-, and NO (Fig. 5-65).
- HRPs have some resemblance to methemoglobin. Peroxidases and methemoglobin show similarity in color and absorption spectra.
- In solutions of low pH, HRP is high spin, but low-spin species are formed at high pH (pH 11).

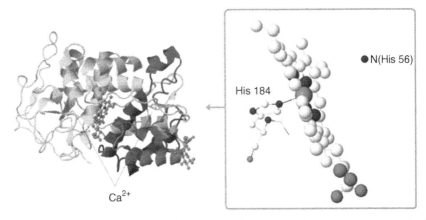

FIGURE 5-64 Crystal structure of fungal peroxidase from *Arthromyces ramosus* at 1.9 Å resolution, PDB 1ARP (Kunishima et al., 1994).

Reaction Mechanism

Suggest the reaction route for HRP actions.

- HRP + H_2O_2 → HRP-I + H_2O

  HRP-I is a green complex, contains Fe^{4+}, is a radical species, and the extra electron is removed from the protein or porphyrin, $(Por\text{-}Fe^{4+} = O)^{+\bullet}$.

- The similarity of optical spectra between compound I of catalase or HRP-I and $Co^{3+}OEP^{2+\bullet}(Br^-)_2$ or $Co^{3+}OEP^{2+\bullet}(ClO_4)_2$ suggests that compounds I are

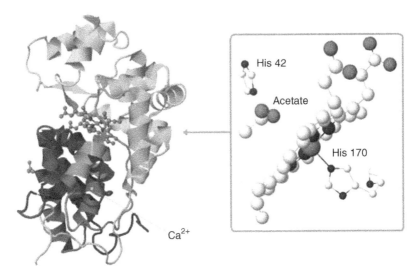

FIGURE 5-65 Structure of ferric horseradish peroxidase C1A in complex with acetate, PDB 1H5A (Berglund et al., 2002).

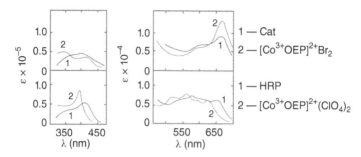

FIGURE 5-66 Comparison of optical absorption spectra of $[Co^{3+}OEP]^{2+\bullet}(ClO_4)_2$ with Cat (upper figure) and $[Co^{3+}OEP]^{2+\bullet}Br_2$ with HRP (lower figure). (Data from Dolphin et al., 1973. Reproduced by permission of John Wiley & Sons.)

Fe^{4+} porphyrin π-cation radicals (Fig. 5-66). However, the electronic distributions of $[(PpIX)^{2-}Fe^{4+}O]^{+\bullet}$ in catalase and HRP are slightly different. This difference is observed in their electronic spectra and also in the slightly different behaviors. Unlike catalase, HRP cannot decompose H_2O_2. This implies that $[(PpIX)^{2-}Fe^{4+}O]^{+\bullet}$ of HRP cannot react with H_2O_2.

- NMR investigations suggest high-spin Fe^{4+}.
- HRP-I has absorption maximum at 658 and 407 nm (Fig. 5-66).
- A weak ESR signal, the magnetic susceptibility (6600×10^{-6} cgs), and the electronic spectrum suggest $S = \frac{3}{2}$ for compound I.

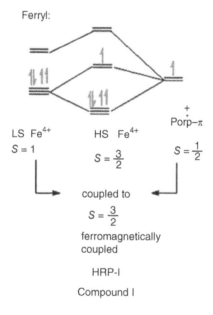

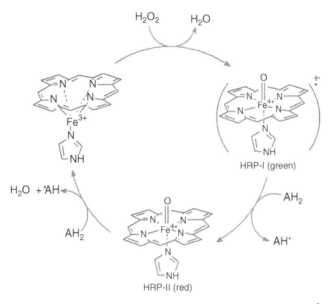

SCHEME 5-38 Catalytic cycle of horseradish peroxidase, Fe–O is 1.62 Å, → Fe = O. (Modified from Yonetani et al., 1966.)

- The fifth ligand for Fe in catalase is tyrosine's O^-, and the N^- of histidine in HRP and cytochrome c peroxidase must be sufficient to stabilize the Fe^{4+} oxidation state.
- HRP-I is unstable and is readily converted to HRP-II:

$$HRP\text{-}I + AH_2 \rightarrow HRP\text{-}II + AH$$

- Addition of one electron, 1 mol of H_2O_2, to HRP-I gives (decomposes to) the nonradical, *red* product HRP-II (Scheme 5-38),
- Absorption maxima of HRP-II: 561, 530, and 417 nm.
- NMR investigations of HRP-II suggest *low-spin* Fe^{4+}:

$$HRP\text{-}II + AH_2 \rightarrow HRP + AH$$

This leads to the formation of free radicals, which then react further, for example, with O_2 to give superoxide ion.
- Fe^{3+}–HRP is reduced by dithionite, $Na_2S_2O_4$, to give Fe^{2+}–HRP, which reacts with O_2 to give an inactive species HRP–III.

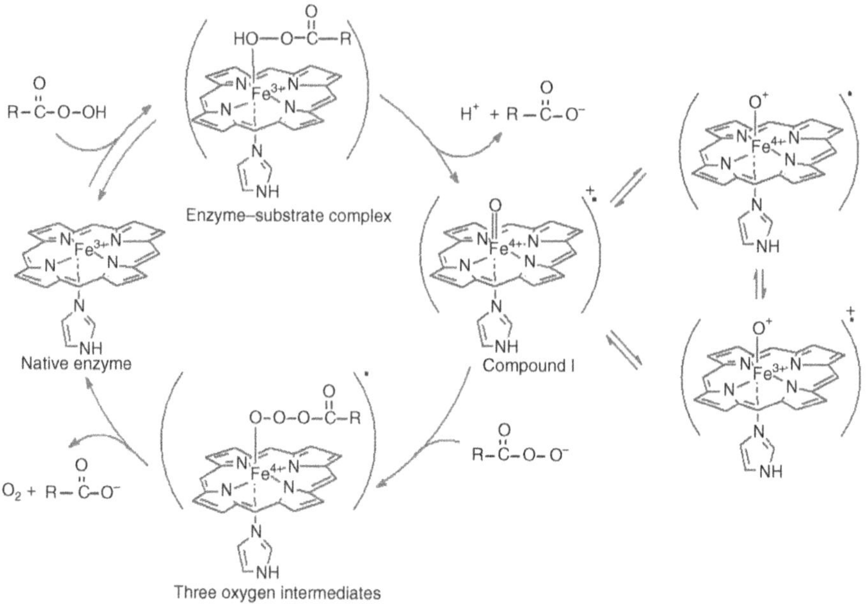

SCHEME 5-39 Proposed mechanism for compound I formation and oxygen evolution from peroxy acids. (Modified from Hager et al., 1972, and Murray et al., 1985.)

Model Example

Give examples of peroxidases reactions.

- HRP reacts with ROOH as well as HOOH but not with ROOR:

$$
\begin{aligned}
\text{Per-OH} \qquad\qquad &= \quad \text{peroxidase} \\
\text{Per-OH} + \text{HOOR} \quad &\leftrightarrows \quad \text{Per-OOR(I)} + H_2O \\
\text{Per-OOR(I)} \qquad &\rightarrow \quad \text{Per-OOR(II)} \\
\text{Per-OOR(II)} + AH_2 \quad &\rightarrow \quad \text{Per-OH} + \text{ROH} + A \\
AH_2 \qquad\qquad &= \quad \text{ascorbic acid}, \; \text{HOOR} = H_2O_2
\end{aligned}
$$

- For AH_2 = peroxyacids see Scheme 5-39.

Models for the Mechanism

Suggest some models for peroxidase actions.

- Model peroxidase systems consist of Fe^{2+} ion, H_2O_2, and a two-electrons donor, such as ascorbic acid, isoascorbic acid, or dihydroxy fumaric acid.
- Examples:
  1. The presence of a *complexing agent such as EDTA* enhances the catalytic activity of the metal ion, possibly by increasing its oxidation

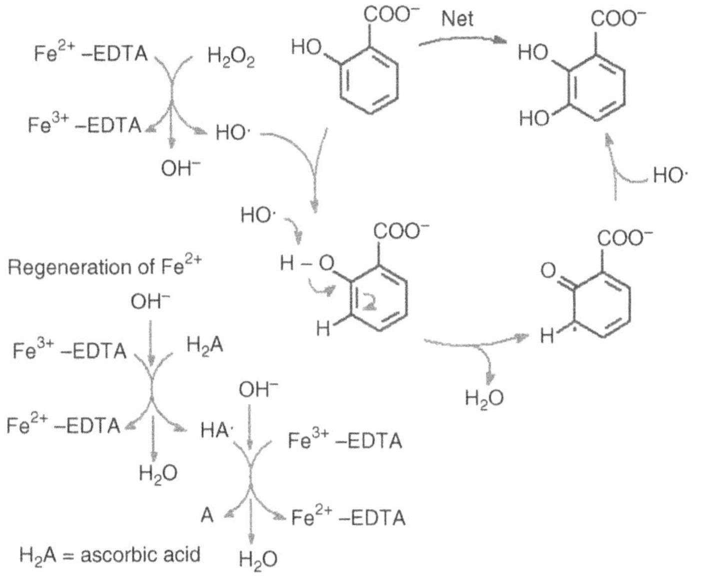

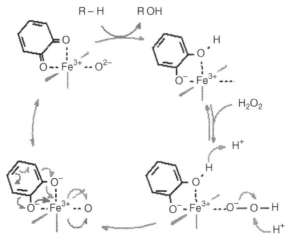

SCHEME 5-40 Free-radical mechanism for hydroxylation of salicylic acid. (Modified from Breslow and Lukens, 1960, and Grinstead, 1960.)

potential. Hydroxylation of salicylic acid requires equimolar amounts of the two-electron donor and hydrogen peroxide such as ascorbic acid (Scheme 5-40).

2. The oxidation of aromatic compounds and aliphatic alcohols by hydrogen peroxide with *Fe(III)–catechol complex as a catalyst* (Scheme 5-41).

SCHEME 5-41 Proposed mechanism for oxidation of aromatic compounds. (Modified based on Hamilton et al., 1963, 1966.)

CYTOCHROME P-450

Roles and Main Properties

What are the biological roles and the main properties of cytochrome P-450?

- These heme proteins are *monooxygenases*; they catalyze the *hydroxylation* of a substrate RH:

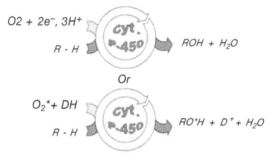

$$O2 + 2e^-, 3H^+$$
$$R - H$$

$$ROH + H_2O$$

Or

$$O_2^* + DH$$
$$R - H$$

$$RO^*H + D^+ + H_2O$$

(a) DH is a proper electron donor such as NADPH.

(b) R–H: benzpyrene, aminopyrine, aniline, morphine, and benzphetamine.

- Cytochrome P-450 is part of the body detoxification system, *hydroxylating* compounds (making them more water soluble), so that urinary excretion is favored over fat storage.

- The action of cytochrome P-450 on certain substrates results in the production of highly reactive intermediates, which can disrupt other cellular compounds.

- The carcinogenicity of polycyclic aromatic hydrocarbons has been attributed to their conversion by P-450 in vivo to arene oxides.

- Cytochrome P-450 is found in plants, animals, and bacteria.

- Molecular weight $\sim$ 50,000.

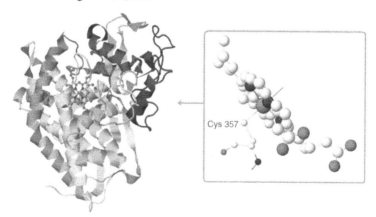

FIGURE 5-67 Crystal structure of substrate-free *Pseudomonas putida* cytochrome P-450 PDB 1PHC (Poulos et al., 1986).

- The heme group in P-450 is located in a hydrophobic environment (Fig. 5-67).
- At neutral pH and in the absence of substrate, oxidized P-450 exists mainly in the low-spin state.
- In the low–spin state, a mercaptide group of a cysteine residue is bound to the iron.
- The imidazole group may occupy the remaining coordination position of the oxidized form.
- The mercaptan tail porphyrin has been synthesized as a model:

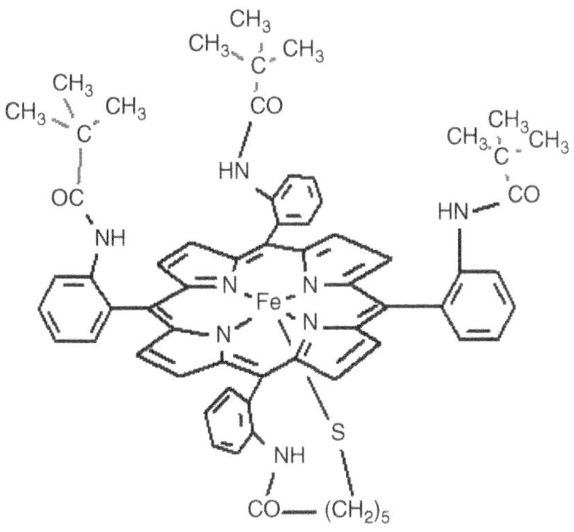

- The six-coordinate mercaptide–Fe(II)–CO complexes can be generated. This model compound compares well with the characteristic absorption and MCD spectra of (MCD: magnetic circular dichroism).

Electronic Spectra

What are the main differences between electronic spectrum of the oxidized and reduced forms of P–450?

- Cytochrome P-450 exhibits an intense Soret band, near ultraviolet, which is displaced ~30 nm from the corresponding band observed in the spectra of carbon monoxide adduct of other hemoproteins (Fig. 5-68).
- At neutral pH, the oxidized and reduced forms yield Soret bands at 415 and 412 nm, respectively.
- The carbon monoxide derivative of the reduced species has a band at 449 nm, with a shoulder at 424 nm.
- On the other hand, the oxidized form of P-450 exhibits both low- and high-spin features.

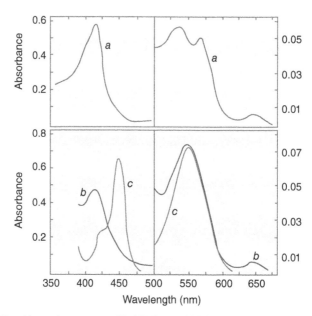

FIGURE 5-68   Absorption spectra of P-450 from rabbit liver microsomes: (a) oxidized form; (b) reduced form; (c) in presence of CO. (From Coon and white, 1980. Reproduced by permission of John Wiley & Sons.)

Substrate and Spectral Changes

How can you explain the spectral changes that follow the substrate binding to P-450?

- Addition of substrates to the oxidized form affects the absorption spectrum in one of two ways:
  - In type I, there is an *increase* in the absorbance at 390 nm and a *decrease* at 420 nm. Hexobarbital, among others, induces a type I change:

Hexobarbital

  - In type II, there is a *decrease* in the absorbance at 390 nm and an *increase* at 430 nm. Adding aniline to P-450 induces a type II change.
- It is postulated that the increase in the *high–spin character* observed upon addition of type I substances to oxidized P-450 reflects a *conformational-induced change* in which the distance between the iron and the ligand opposite the mercaptide group is stretched, with the metal out of the plane with respect to porphyrin, toward the coordinated sulfur atom.

- Type II substances, on the other hand, are thought to become bound more tightly to the iron through displacement of the coordinated group in the position opposite the cysteine residue to yield a low-spin complex.
- The oxidized form of P-450 yields an ESR spectrum with features characteristic of a low-spin ferriheme system.
- The g values obtained ($g = 1.92, 2.27, 2.47$) are similar to the ones observed in model studies of complexes of thiols with ferriheme.

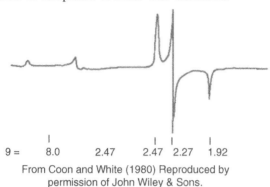

$g =$ 8.0   2.47   2.47   2.27   1.92

From Coon and White (1980) Reproduced by
permission of John Wiley & Sons.

Cytochromes P-450 and P-420

What are the differences between P-450 and P-420?

Urea, guanidine hydrochloride, a high concentration of neutral salts, and a change in pH to either side of neutrality convert the P-450 to an inactive form P-420, which yields a carbon monoxide complex with a Soret maximum at 420 nm.

Catalytic Cycles for P-450

Suggest possible catalytic cycles for P-450 actions, compare P-450 to HRP and to cytochrome c peroxidase, and give examples to demonstrate the action of P-450.

- Most P-450 follow the catalytic cycle in Scheme 5-42:

  Step 1: Association of a hydrophobic molecule, A–H, with P-450, which is the substrate binding to ferric P-450.

  Step 2: First reduction to ferrous state, allowing oxygen binding.

  Step 3: Binding dioxygen, two intermediates (compounds I and II) observed kinetically when dioxygen reacts with ferrous P-450.

  Step 4: Second reduction induces splitting of bound oxygen.

  Step 5: Splitting oxygen–oxygen bond of peroxide, one oxygen atom expelled as water, and two alternative modes have been proposed:

  (a) The "active oxygen" in superferryl intermediate $(Fe^{5+} = O)^{3+}$: The electronic configuration of this species can be described as a resonance of three structures:

$$O^{2-} - Fe^{5+} \leftrightarrow O = Fe^{5+} \leftrightarrow O^0 - Fe^{3+}$$

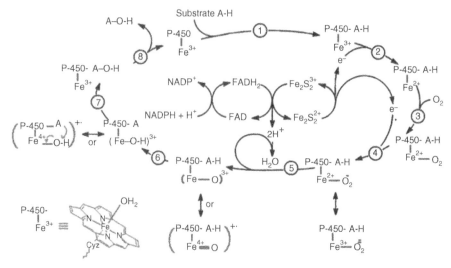

Catalytic cycle of cytochrome P-450 hydroxylase, modified from Sono et al., 1996.

NADH ———▸ Flavoprotein$_1$ ———▸ Cyt. b$_5$

NADPH ———▸ Flavoprotein$_2$ ———▸ Cyt. P-450 ———▸ Hydroxylation

SCHEME 5-42 The electron transport chain in liver P-450 does not require the iron-sulfur protein Fe$_2$-S$_2$. (See the color version of this scheme in Color Plates section.)

The active oxygen of O^0-Fe^{3+} inserts into a proximate carbon–hydrogen bond on the substrate *oxenoid mechanism*.

(b) Similar to the oxygen of $[(PpIX)^{2-}Fe^{4+}O]^{+\bullet}$, the cation radical that is found in catalase or HRP.

Step 6: Hydrogen abstraction by the iron–oxenoid species led to a transient carbon radical juxtaposed to the iron-bound hydroxyl radical. In the cation radical mechanism

$$[(PpIX)^{2-}Fe^{4+} = O]^{+\bullet} \leftrightarrow [(PpIX)^{2-}Fe^{3+}\text{-}O^{-\bullet}]^{+\bullet} + \text{R-H}$$
$$\rightarrow R + \{[(PpIX)^{2-}Fe^{4+} = O\text{-}H]^{+\bullet} \leftrightarrow [(PpIX)^{2-}Fe^{3+}\text{-}O^{-\bullet}\text{-}H]^{+\bullet}\}$$

Step 7: Either

(i) direct combination of the radicals or

(ii) Simultaneous one–electron oxidation of the carbon radical and collapse of the hydroxide–carbonium ion pair generated an alcohol functional group:

$$R + \{[(PpIX)^{2-}Fe^{4+} = O\text{-}H]^{+\bullet} \leftrightarrow [(PpIX)^{2-}Fe^{3+}\text{-}O^{-\bullet}\text{-}H]^{+\bullet}\}$$
$$\rightarrow ROH + (PpIX)^{2-}Fe^{3+}$$

Steps 6 and 7: Hydrogen abstraction–radical recombination was termed *oxygen rebound*.

Step 8: Dissociation of product.

P-450 versus HRP

- In HRP:
  (i) The substrate does not sit along the Fe axis but approaches the heme edge.

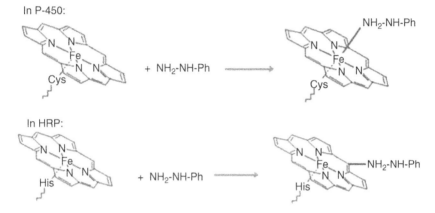

In P-450:

$+ \ NH_2\text{-NH-Ph}$

In HRP:

$+ \ NH_2\text{-NH-Ph}$

  (ii) The redox reaction is by electron transfer and not atom transfer (Scheme 5-43)

P-450 versus HRP and Cyt. c Peroxidase

- In HRP and Cyt. *c* peroxidase $\rightarrow [(PPIX)Fe^{4+} = O]^{+\bullet}$ (compound I): stabilized by the presence of aromatic residues surrounding the heme, such as His, Typ, Met, and Tyr.
- Examples of action of P-450:
  (i) Hydroxylation of norbornane catalyzed by P-450 (Scheme 5-44)
  (ii) Reaction between P-450 and arenes (Scheme 5-45)

ELECTRONIC SPECTRA OF HEMOPROTEINS

Optional Transitions of Hemoproteins

Describe and classify the optical transitions of the different hemoproteins.

- Examples of the solution absorption spectra for common hemoglobin complexes are represented in Fig. 5-69.
- It is useful to classify the optical transition according to their use in inorganic spectroscopy.
- The $\pi \rightarrow \pi^*$ transitions of the porphyrin ring:
  ○ The letters Q, B, N, and L will used for $\pi \rightarrow \pi^*$ absorption bands.
  ○ A very intense band near 400 nm is the Soret or B band.

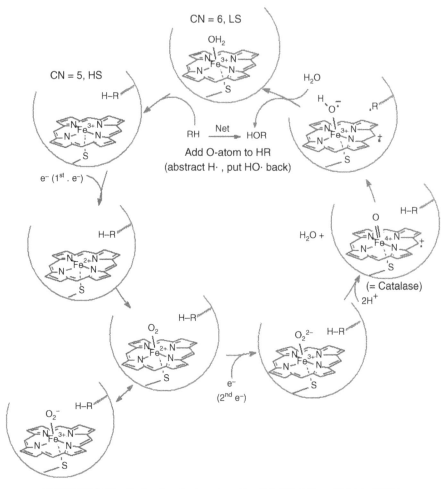

SCHEME 5-43   Redox by electron transfer. (Modified from Ochiai, 2008.)

- ○ Much weaker bands between 500 and 600 nm are called Q bands.
- ○ Two broad absorption bands between 200 and 400 nm, called the N and L bands, are much weaker than the Soret band.
- The d → d transitions of the iron atom.
- The charge transfer transitions between the iron and the porphyrin.
- The charge transfer transitions between the iron and the axial ligand. The Roman numerals will signify d → d or charge transfer transitions.

Molecular Orbital and Molecular Energy Wave Functions of Porphyrin Molecule

Derive the molecular orbital and the molecular energy wave functions of the porphyrin molecule, using free electron molecular orbital model.

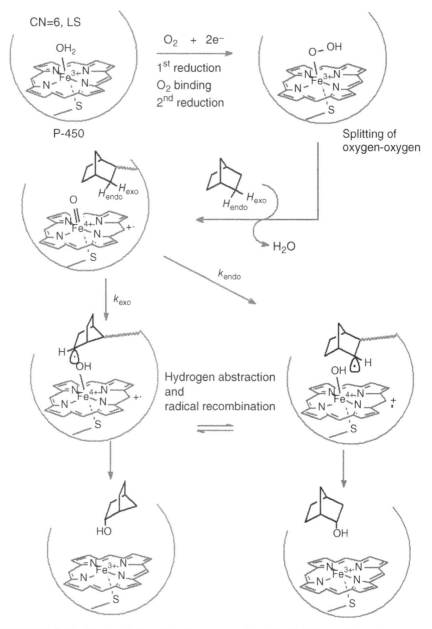

SCHEME 5-44 Hydroxylation of norbornane catalyzed by P-450$_{LM2}$, illustrating oxygen rebound and partial epimerization. (From Coon and white, 1980. Reproduced by permission of John Wiley & Sons.)

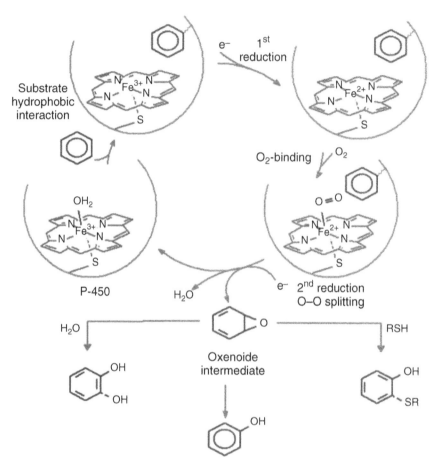

SCHEME 5-45 Formation of arene oxides via cytochrome P-450 and their further reactions in microsomes. (Modified from Estabrook et al., 1972.)

- The free-electron molecular orbital model is the simplest model that accounts for the intensity difference of $\pi \rightarrow \pi^*$ transitions.
- We start by assuming that a particle can be described by a wave equation (general wave equation as a function of distance x and time t) of the form

$$\psi(x, t) = a_o \exp\left[2\pi i\left(\frac{x}{\lambda} - \nu t\right)\right]$$

where a_o is a constant and λ and ν are the wavelength and frequency of the wave and describe the behavior of the particle.

- The wave function ψ contains all of the information about the particle that one can know: its position, momentum, energy, etc.
- We now derive the differential equation. If

$$y = ae^{bx}$$

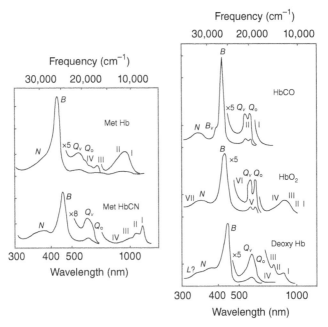

FIGURE 5-69 Solution absorption spectra of hemoglobin derivatives. (Data from Eaton and Hofichter, 1981. Reproduced by permission of Elsevier.) (See the color version of this figure in Color Plates section.)

then

$$\frac{dy}{dx} = abe^{bx}$$

Assume $i^2 = -1$; momentum, $p = mv$, and $V(x, t)$ is the potential energy. Then

$$\frac{\delta^2 \psi(x, t)}{\delta x^2} = -\left(\frac{2\pi}{\lambda}\right)^2 \psi(x, t)$$

$$\because \lambda = \frac{h}{p} \qquad \frac{1}{\lambda} = \frac{p}{h} \qquad \hbar = \frac{h}{2\pi}$$

$$\because p^2 = 2mK \qquad K = \text{kinetic energy}$$

$$\therefore \frac{\delta^2 \psi(x, t)}{\delta x^2} = -\left(\frac{p^2}{\hbar^2}\right) \psi(x, t) = -\left(\frac{2mK}{\hbar^2}\right) \psi(x, t)$$

$$\because E = K + V(x, t) \qquad \therefore K = E - V(x, t)$$

$$\frac{\delta^2 \psi(x, t)}{\delta x^2} = -\left(\frac{2mE}{\hbar^2}\right) \psi(x, t) + \left(\frac{2mV(x, t)}{\hbar^2}\right) \psi(x, t)$$

$$-\frac{\hbar^2}{2m} \frac{\delta^2 \psi(x, t)}{\delta x^2} + V(x, t) \psi(x, t) = E\psi(x, t)$$

- The wave function can be factored into functions of position and time:

$$\because \psi(x,t) = a_o \exp\left[2\pi i\left(\frac{x}{\lambda} - vt\right)\right] = a_o \exp\left(\frac{2\pi i x}{\lambda}\right)\exp\left(-\frac{itE}{\hbar^2}\right) = \psi(x)\phi(t)$$

$$-\frac{\hbar^2}{2m}\frac{\delta^2\psi(x)\ \phi(t)}{\delta x^2} + V(x)\psi(x)\phi(t) = E\psi(x)\phi(t)$$

$$-\frac{\hbar^2}{2m}\frac{\delta^2\psi(x)}{\delta x^2} + V(x)\psi(x) = E\psi(x) \quad \text{(Schrödinger equation)}$$

  - In the free-electron molecular orbital model, the porphyrin is considered as a one-dimensional ring or loop of constant potential.
  - This means that the electron cannot exist in the region outside $0 \leq x \leq d$, where d is the length of the ring.
- Therefore, the wave function must be zero for $x < 0$ and $x > d$, and the electron constrains to move along x in interval $0 \leq x \leq d$. The *Schrödinger equation* with $V = 0$ is given as

$$-\frac{\hbar^2}{2m}\frac{\delta^2\psi(x)}{\delta x^2} = E\psi(x)$$

- In general ψ as a function of x can be expressed as

$$\psi = A\cos(\alpha x) + B\sin(\alpha x)$$

  where A and B are constants.
- The condition $\psi = 0$ at $x = 0$ requires that the cosine term vanish. Therefore, $A\cos.\alpha x. = 0$, and

$$\psi = B\sin(\alpha x)$$

- The condition $\psi = 0$ at $x = d$ demands that

$$\sin(\alpha d) = 0$$

If

$$\sin(180°) = \sin(2 \times 180°) = \sin(3 \times 180°) = \cdots = 0$$

then

$$\sin(l2\pi) = 0$$

where l is an integral called the orbital angular momentum about x. Therefore, $\alpha d = 2l\pi$, and

$$\alpha = \frac{2l\pi}{d} \qquad \psi = B\sin\frac{2l\pi}{d}x$$

- The requirements that ψ be normalized give us the value of B:

$$\int_0^d \psi\psi \, \delta x = 1$$

$$\int_0^d \left(B \sin \frac{2l\pi x}{d}\right)\left(B \sin \frac{2l\pi x}{d}\right)\delta x = B^2 \int_0^d \sin^2\left(\frac{2l\pi x}{d}\right)\delta x = \frac{1}{2}B^2 d = 1$$

$$B = \sqrt{\frac{2}{d}}$$

where

$$\int \sin^2\left(\frac{2l\pi x}{d}\right)\delta x = \frac{1}{2}d + \text{const}$$

- Therefore, the wave function is

$$\psi = \sqrt{\frac{2}{d}} \sin \frac{2l\pi}{d}x$$

- The exact form of the wave function ψ depends on the integral quantum number l.
- The Schrödinger equation can now be solved for E:

$$-\frac{\hbar^2}{2m}\frac{\delta^2\psi(x)}{\delta x^2} = E\psi(x)$$

$$-\frac{\hbar^2}{2m}\frac{\delta^2\left[\sqrt{2/d}\sin(2l\pi/d)x\right]}{\delta x^2} = E_l\left[\sqrt{\frac{2}{d}}\sin\frac{2l\pi}{d}x\right]$$

$$\because \frac{\delta}{\delta x}\sin ax = a\cos ax, \quad \text{and} \quad \frac{\delta}{\delta x}\cos ax = -a\sin ax$$

$$\therefore \frac{\delta^2}{\delta x}\sin ax = \frac{\delta}{\delta x}\left(\frac{\delta}{\delta x}\sin ax\right) = \frac{\delta}{\delta x}(a\cos ax) = -a^2\sin ax$$

$$-\frac{\hbar^2}{2m}\left(\frac{2\pi l}{d}\right)\sqrt{\frac{2}{d}}\frac{d}{dx}\cos\frac{2\pi l}{d}x = \frac{\hbar^2}{2m}\left(\frac{2\pi l}{d}\right)^2\sqrt{\frac{2}{d}}\sin\frac{2\pi l}{d}x = E_l\left[\sqrt{\frac{2}{d}}\sin\frac{2l\pi}{d}x\right]$$

$$\therefore \frac{\hbar^2}{2m}\frac{4\pi^2 l^2}{d^2} = E_l$$

Therefore, the molecular energy wave function of the porphyrin molecule is given as

$$E_l = \frac{l^2 h^2}{2md^2}$$

π → π Optical Transitions in Hemoproteins

What are the origins of the π → π optical transitions in the hemoproteins, and estimate the center of gravity for these transitions?

- The carbon–nitrogen skeleton of the porphinato group without peripheral constituents is a planar tetrapyrrole macrocycle with conjugated structure of 11 double bonds and 22 π electrons.
- Hydrogenation of two of the peripheral double bonds of the pyrrole rings leads to relatively small changes in the optical spectrum.
- Accordingly, the ring of constant potential is taken as the ring formed by the inner 16 atoms of a total 18 π electrons and can be regarded as a circle:

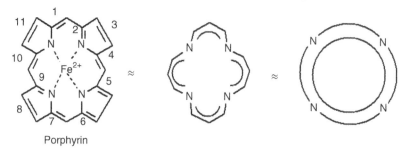

Porphyrin

- The 18 π electrons can fill the orbital levels up to $l = \pm 4$:

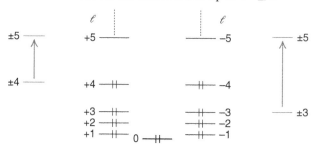

Energy levels for 18 "free" electrons in 16-membered cyclic polyene
(Malley, 1973. Reproduced by permission of John Wiley & Sons.)

- Orbitals with $L = 0$, ± 1, ± 2, ± 3, ± 4 are occupied in the ground state.
- The lowest frequency transitions arise from $l = \pm 4$ to $l = \pm 5$ orbitals.
- These promotions produce four excited states characterized by their net orbital angular momentum $L = \pm 1$, ± 9.
- Interelectronic repulsion splits the four states into two degenerate pairs.
- The $L = \pm 9$ states will be at lower energy than the $L = \pm 1$ states because of the higher angular momentum.
- Invoking Hund's selection rule, only transitions for which $L = \pm 1$ and $S = 0$ are allowed and are in agreement with the B (Soret) band.

- The transitions for which $L=\pm9$ and $S=0$ are strongly forbidden and are identified with the Q bands.
- Similarly, the excitations from $l=\pm2$ to $l=\pm5$ are identified for N and L bands:

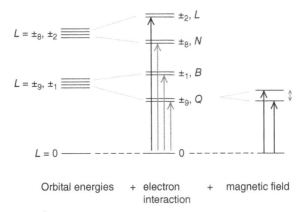

Orbital energies   + electron   + magnetic field
                     interaction

- Taking $d=22$ Å, which is the sum of the bond lengths for the inner 16-member ring, the calculated center of gravity for the Q and B bands agrees with experimental value (about $21{,}000\,\mathrm{cm}^{-1}$, $\lambda=480\,\mathrm{nm}$).
- The center of gravity for the N and L bands can be calculated, $40{,}000\,\mathrm{cm}^{-1}$ (250 nm), in excellent agreement with experimental value of about $33{,}000\,\mathrm{cm}^{-1}$ (300 nm).
- This prediction was confirmed by measuring the splitting of the Q band in the high magnetic field using circular polarized light.

Molecular Orbitals of Porphyrin Complex

How can you construct the molecular symmetry orbitals for a simplified iron porphyrin complex without axial ligands?

- Both σ and π bonds must be considered. The nitrogen atoms are taken as hybridized sp^2 σ orbitals, and each of the four nitrogen atoms forms a σ bond with the iron.
- For π bonds, it is assumed that each of the 20 carbon atoms contributes a 2π p orbital and each of the 4 nitrogen atoms contributes 2π p orbitals.
- The molecular orbitals are constructed from linear combinations of P_z orbitals of all 24 atoms of the porphyrin skeleton.
- The 24 π orbitals are subdivided into four sets of equivalent atoms:

π_{N1}, π_{N2}, π_{N3}, π_{N4}

π_{C1}, π_{C2}, π_{C3}, π_{C4}

π_{C5}, π_{C6}, π_{C7}, π_{C8}, π_{C9}, π_{C10}, π_{C11}, π_{C12}

π_{C13}, π_{C14}, π_{C15}, π_{C16}, π_{C17}, π_{C18}, π_{C19}, π_{C20}

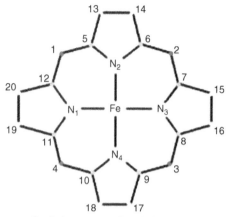

Porphyrin structure and numbering system

- Each set will be a basis set for a reducible representation of D_{4h}. The characters of these reducible representations are:

D_{4h}	E	$2C_4$	C_2	$2C_2'$	$2C_2''$	i	$2S_4$	σ_h	$2\sigma_v$	$2\sigma_d$
$\Gamma\,(\pi_{N1}\cdots{}_{N4})$	4	0	0	-2	0	0	0	-4	2	0
$\Gamma\,(\pi_{C1}\cdots{}_{C4})$	4	0	0	0	-2	0	0	-4	0	2
$\Gamma\,(\pi_{C5}\cdots{}_{C12})$	8	0	0	0	0	0	0	-8	0	0
$\Gamma\,(\pi_{C13}\cdots{}_{C20})$	8	0	0	0	0	0	0	-8	0	0
$\Gamma\,(\sigma_{N1}\cdots{}_{N4})$	4	0	0	2	0	0	0	4	2	0

- Γ_i, a reducible representation, is a sum of irreducible representations. Use

$$\Gamma_i = \frac{1}{h}\sum_R \chi(R)\chi_i(R)$$

and the following D_{4h} character table:

D_{4h}	E	$2C_4$	C_2	$2C_2'$	$2C_2''$	I	$2S_4$	σ_h	$2\sigma_v$	$2\sigma_d$		
A_{1g}	1	1	1	1	1	1	1	1	1	1		$x^2+y^2,\,z^2$
A_{2g}	1	1	1	-1	-1	1	1	1	-1	-1	R_z	
B_{1g}	1	-1	1	1	-1	1	-1	1	1	-1		x^2-y^2
B_{2g}	1	-1	1	-1	1	1	-1	1	-1	1		xy
E_g	2	0	-2	0	0	2	0	-2	0	0	(R_x, R_y)	(xz, yz)
A_{1u}	1	1	1	1	1	-1	-1	-1	-1	-1		
A_{2u}	1	1	1	-1	-1	-1	-1	1	1	z		
B_{1u}	1	-1	1	1	-1	-1	1	-1	-1	1		
B_{2u}	1	-1	1	-1	1	-1	1	-1	1	-1		
E_u	2	0	-2	0	0	-2	0	2	0	0	(x, y)	

- $\Gamma(\pi_{N1\ldots N4})(A_{1g}) = (1/16)[(1)(4)(1)+(2)(0)(1)+(1)(0)(1)+(2)(-2)(1)(2)(0)(1)$

$$+(1)(0)(1)+(2)(0)(1)+(1)(-4)(1)+(2)(2)(1)+(2)(0)(1)]$$

$$=(1/16)[4+0+0-4+0+0+0-4+4+0]=0$$

- $\Gamma(\pi_{N1\ldots N4})(A_{2g}) = (1/16)[(1)(4)(1)+(2)(0)(1)+(1)(0)(1)+(2)(-2)(-1)+(2)(0)(-1)$
$\qquad +(1)(0)(1)+(2)(0)(1)+(1)(-4)(1)+(2)(2)(-1)+(2)(0)(-1)]$
$\qquad = (1/16)[4+0+0+4+0+0+0-4-4+0] = 0$

- $\Gamma(\pi_{N1\ldots N4})(E_g) = (1/16)[(1)(4)(2)+(2)(0)(0)+(1)(0)(-2)+(2)(-2)(0)+(2)(0)(0)$
$\qquad +(1)(0)(2)+(2)(0)(0)+(1)(-4)(-2)+(2)(2)(0)+(2)(0)(0)]$
$\qquad = (1/16)[8+0+0+0+0+0+0+8+0+0] = 1$

$\qquad \Gamma(\pi_{N1\ldots N4}) = A_{2u}+B_{2u}+E_g$

- Similarly, the reduction of $\Gamma(\pi_{C1}\ldots_{C4})$, $\Gamma(\pi_{C5}\ldots_{C12})$, $\Gamma(\pi_{C13}\ldots_{C20})$, and $\Gamma(\sigma_{N1}\ldots_{N4})$ into irreducible components of D_{4h} gives

$$\Gamma(\pi_{C1\ldots C4}) \; = A_{2u}+B_{1u}+E_g$$
$$\Gamma(\pi_{C5\ldots C12}) \; = A_{1u}+A_{2u}+B_{1u}+B_{2u}+2E_g$$
$$\Gamma(\pi_{C13\ldots C20}) = A_{1u}+A_{2u}+B_{1u}+B_{2u}+2E_g$$
$$\Gamma(\sigma_{N1\ldots N4}) \; = A_{1g}+B_{1g}+E_u$$

This gives a total of 24 molecular π orbitals ($2A_{1u}+4A_{2u}+3B_{1u}+3B_{2u}+6E_g$) and four molecular σ orbitals ($A_{1g}+B_{1g}+E_u$).

- Each molecular orbital wave function is written as a linear combination of atomic orbital wave functions, ϕ, by applying the projection operator:

$$\hat{P}^j_{\phi_i} = \frac{1}{h}\sum_R \chi(R)^j \hat{R}$$

- For A_{1u} of ($\pi_{C5}\ldots_{C12}$), to project out the arbitrary function ϕ_I,

$$\hat{P}^{A_{1u}}_{\phi_i} = \frac{1}{16}\sum_R \chi(R_{A_{iu}})\hat{R}$$

The numerical factor (1/16) may be ignored; it can be calculated later by the normalization procedure. Use the character table D_{4h}:

D_{4h}	E	$2C_4$	C_2	$2C_2'$	$2C_2''$	i	$2S_4$	σ_h	$2\sigma_v$	$2\sigma_d$
A_{1u}	1	1	1	1	1	−1	−1	−1	−1	−1

Or

D_{4h}	E	C_4	C_4^3	C_2	$(C_2')_x$	$(C_2')_y$	$(C_2'')_{x,y}$	$(C_2'')_{-x,y}$	i	S_4	S_4^3	σ_h	σ_v	σ_v'	σ_d	σ_d'
A_{1u}	1	1	1	1	1	1	1	1	−1	−1	−1	−1	−1	−1	−1	−1

$$\hat{P}^{A_{1u}}\phi_{C5} = (1)\hat{E}\phi_{C5} + (1)\hat{C}_4\phi_{C5} + (1)\hat{C}_4^3\phi_{C5} + (1)\hat{C}_2\phi_{C5} + [(1)\hat{C}_2'\phi_{C5}]_x$$
$$+ [(1)\hat{C}_2'\phi_{C5}]_y + [(1)\hat{C}_2''\phi_{C5}]_{x,y} + [(1)\hat{C}_2''\phi_{C5}]_{-x,y} + (-1)\hat{i}\phi_{C5} + (-1)\hat{S}_4\phi_{C5}$$
$$+ (-1)\hat{S}_4^3\phi_{C5} + (-1)\sigma_h\phi_{C5} + (-1)\sigma_v\phi_{C5} + (-1)\hat{\sigma}_v'\phi_{C5} + (-1)\hat{\sigma}_d\phi_{C5} + (-1)\hat{\sigma}_d'\phi_{C5}$$

$$\hat{P}^{A_{1u}}\phi_{C5} = \phi_{C5} + \phi_{C7} + \phi_{C11} + \phi_{C9} - \phi_{C10} - \phi_{C6} - \phi_{C12} - \phi_{C8} + \phi_{C9}$$
$$+ \phi_{C7} + \phi_{C11} + \phi_{C5} - \phi_{C10} - \phi_{C6} - \phi_{C8} - \phi_{C12}$$

$$\hat{P}^{A_{1u}}\phi_{C5} = 2\phi_{C5} - 2\phi_{C6} + 2\phi_{C7} - 2\phi_{C8} + 2\phi_{C9} - 2\phi_{C10} + 2\phi_{C11} - 2\phi_{C12}$$

$$\hat{P}^{A_{1u}}\phi_{C5} = 2[\phi_{C5} - \phi_{C6} + \phi_{C7} - \phi_{C8} + \phi_{C9} - \phi_{C10} + \phi_{C11} - \phi_{C12}]$$

The numerical factor can be established using the normalization process: If

$$\Psi_i = \sum_j a_{ij}\phi_j$$

requires that

$$\int \Psi_i \Psi_i \, d\tau = 1$$

then

$$\frac{1}{N^2} = \int \left(\sum_j a_{ij}\phi_j\right)^2 d\tau$$

$$\frac{1}{N^2} = \sum_j a_{ij}^2 \int \phi_j\phi_j \, d\tau + \sum_{\substack{j,k \\ (j \neq k)}} a_{ij}a_{ik} \int \phi_j\phi_k \, d\tau$$

The ϕ_j's are a basis set, and

$$\sum_{j,k} a_{ij}a_{jk} \int \phi_j\phi_k \, \partial\tau = 0$$

because overlap is assumed to be zero and

$$\int \phi_j\phi_j \, d\tau = 1$$

Then

$$\frac{1}{N^2} = \sum_j a_{ij}^2$$

$$N_i = \frac{1}{\sqrt{\sum_j a_{ij}^2}} \qquad a_{ij} = \pm 1$$

$$N = \frac{1}{\sqrt{n}}$$

In a cyclic molecule with rotational symmetry C_n, there will always be n π molecular orbitals:

$$\frac{1}{N^2}\int \phi_i\phi_i\,d\tau=1 \qquad \frac{1}{N^2}\int \phi_i\phi_j\,d\tau=0$$

Then

$$\frac{1}{N^2}\int \left(\phi_{C5}-\phi_{C6}+\phi_{C7}-\phi_{C8}+\phi_{C9}-\phi_{C10}+\phi_{C11}-\phi_{C12}\right)$$

$$\times\left(\phi_{C5}-\phi_{C6}+\phi_{C7}-\phi_{C8}+\phi_{C9}-\phi_{C10}+\phi_{C11}-\phi_{C12}\right)\partial\tau=1$$

$$\frac{1}{N^2}\int \phi_5\phi_5\,\partial\tau+\int \phi_6\phi_6\,\partial\tau+\int \phi_7\phi_7\,\partial\tau+\int \phi_8\phi_8\,\partial\tau+\int \phi_9\phi_9\,\partial\tau+\int \phi_{10}\phi_{10}\,\partial\tau$$

$$+\int \phi_{11}\phi_{11}\,\partial\tau+\int \phi_{12}\phi_{12}\,\partial\tau=1$$

$$\frac{1}{N^2}(8)=1 \qquad N=\sqrt{8}=2\sqrt{2}$$

$$\psi^{A_{1u}}_{C5-C12}=\frac{1}{2\sqrt{2}}\left[\phi_{C5}-\phi_{C6}+\phi_{C7}-\phi_{C8}+\phi_{C9}-\phi_{C10}+\phi_{C11}-\phi_{C12}\right.$$

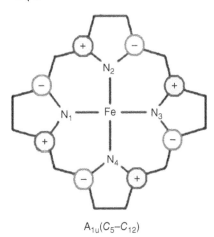

$$A_{1u}(C_5-C_{12})$$

- For A_{1u} of $(\pi_{C13}\cdots {}_{C20})$

$$\hat{P}^{A_{1u}}\phi_{C13}=(1)\hat{E}\phi_{C13}+(1)\hat{C}_4\phi_{C13}+(1)\hat{C}_4^3\phi_{C13}+(1)\hat{C}_2\phi_{C13}+[(1)\hat{C}_2'\phi_{C13}]_x$$

$$+[(1)\hat{C}_2'\phi_{C13}]_y+[(1)\hat{C}_2''\phi_{C13}]_{x,y}+[(1)\hat{C}_2''\phi_{C13}]_{-x,y}+(-1)\hat{i}\phi_{C13}+(-1)\hat{S}_4\phi_{C13}$$

$$+(-1)\hat{S}_4^3\phi_{C13}+(-1)\sigma_h\hat{\phi}_{C13}+(-1)\sigma_v\hat{\phi}_{C13}+(-1)\hat{\sigma}_v'\phi_{C13}+(-1)\hat{\sigma}_d\phi_{C13}$$

$$+(-1)\hat{\sigma}_d'\phi_{C13}$$

$$\hat{P}^{A_{1u}}\phi_{C13} = \phi_{C13} + \phi_{C13} + \phi_{C19} + \phi_{C17} - \phi_{C18} - \phi_{C14} - \phi_{C16} - \phi_{C20} + \phi_{C15}$$
$$+ \phi_{C19} + \phi_{C17} + \phi_{C13} - \phi_{C18} - \phi_{C14} - \phi_{C16} - \phi_{C20}$$
$$\hat{P}^{A_{1u}}\phi_{C13} = 2\phi_{C13} - 2\phi_{C14} + 2\phi_{C15} - 2\phi_{C16} + 2\phi_{C17} - 2\phi_{C18} + 2\phi_{C19} - 2\phi_{C20}$$
$$\hat{P}^{A_{1u}}\phi_{C13} = 2[\phi_{C13} - \phi_{C14} + \phi_{C15} - \phi_{C16} + \phi_{C17} - \phi_{C18} + \phi_{C19} - \phi_{C20}]$$

where $n = 8$. Then

$$N = \sqrt{8} = 2\sqrt{2}$$

$$\hat{P}^{A_{1u}}\phi_{C13} = \frac{1}{2\sqrt{2}}[\phi_{13} - \phi_{14} + \phi_{15} - \phi_{16} + \phi_{17} - \phi_{18} + \phi_{19} - \phi_{20}]$$

$$\psi^{A_{1u}}_{C13\cdots C20} = \frac{1}{2\sqrt{2}}[\phi_{13} - \phi_{14} + \phi_{15} - \phi_{16} + \phi_{17} - \phi_{18} + \phi_{19} - \phi_{20}]$$

$$\psi^{A_{1u}}_{C13-C20} = \frac{1}{2\sqrt{2}}[\phi_{C13} - \phi_{C14} + \phi_{C15} - \phi_{C16} + \phi_{C17} - \phi_{C18} + \phi_{C19} - \phi_{C20}]$$

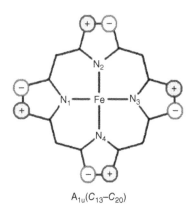

$A_{1u}(C_{13}-C_{20})$

- Using the same procedure, the following can be calculated:
  - A_{2u} of $(\pi_{N1}\cdots N4)$

$$\psi^{A_{2u}}_{(N_1-N_4)} = \frac{1}{2}(\phi_{N_1} + \phi_{N_2} + \phi_{N_3} + \phi_{N_4})$$

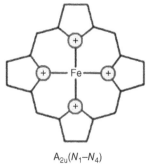

$A_{2u}(N_1-N_4)$

○ A_{2u} of $(\pi_{C1} \cdots C4)$

$$\psi^{A_{2u}}_{(C_1-C_4)} = \tfrac{1}{2}(\phi_{C_1} + \phi_{C_2} + \phi_{C_3} + \phi_{C_4})$$

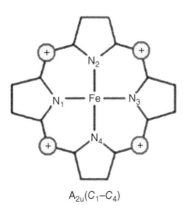

$A_{2u}(C_1-C_4)$

○ A_{2u} of $(\pi_{C5} \cdots C12)$

$$\psi^{A_{2u}}_{(C_5-C_{12})} = \frac{1}{2\sqrt{2}}(\phi_{C_5} + \phi_{C_6} + \phi_{C_7} + \phi_{C_8} + \phi_9 + \phi_{C_{10}} + \phi_{C_{11}} + \phi_{C_{12}})$$

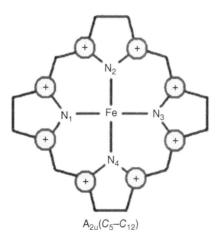

$A_{2u}(C_5-C_{12})$

○ A_{2u} of $(\pi_{C13} \cdots _{C20})$

$$\psi^{A_{2u}}_{(C_{13}-C_{20})} = \frac{1}{2\sqrt{2}} \left(\phi_{C_{13}} + \phi_{C_{14}} + \phi_{C_{15}} + \phi_{C_{16}} + \phi_{17} + \phi_{C_{18}} + \phi_{C_{19}} + \phi_{C_{20}} \right)$$

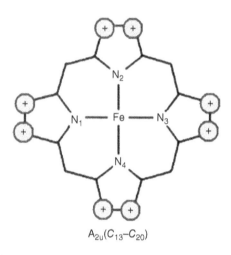

$A_{2u}(C_{13}-C_{20})$

○ B_{1u} of $(\pi_{C1} \cdots _{C4})$

$$\psi^{B_{1u}}_{(C_{1}-C_{4})} = \frac{1}{2} \left(\phi_{C_{1}} - \phi_{C_{2}} + \phi_{C_{3}} - \phi_{C_{4}} \right)$$

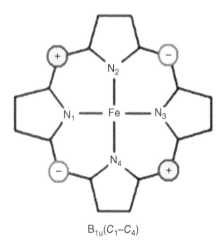

$B_{1u}(C_{1}-C_{4})$

○ B_{1u} of $(\pi_{C5} \ldots {}_{C12})$

$$\psi^{B_{1u}}_{(C_5 - C_{12})} = \frac{1}{2\sqrt{2}} \left(\phi_{C_5} - \phi_{C_6} - \phi_{C_7} + \phi_{C_8} + \phi_9 - \phi_{C_{10}} - \phi_{C_{11}} + \phi_{C_{12}} \right)$$

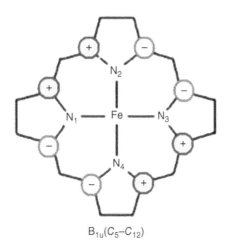

$B_{1u}(C_5 - C_{12})$

○ B_{1u} of $(\pi_{C13} \ldots {}_{C20})$

$$\psi^{B_{1u}}_{(C_{13} - C_{20})} = \frac{1}{2\sqrt{2}} \left(\phi_{C_{13}} - \phi_{C_{14}} - \phi_{C_{15}} + \phi_{C_{16}} + \phi_{17} - \phi_{C_{18}} - \phi_{C_{19}} + \phi_{C_{20}} \right)$$

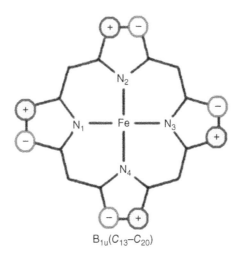

$B_{1u}(C_{13} - C_{20})$

○ B_{2u} of $(\pi_{N1} \cdots {}_{N4})$

$$\psi^{B_{2u}}_{(N_1-N_4)} = \frac{1}{2}(\phi_{N_1} - \phi_{N_2} + \phi_{N_3} - \phi_{N_4})$$

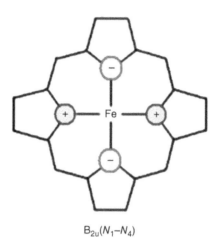

$$B_{2u}(N_1-N_4)$$

○ B_{2u} of $(\pi_{C5} \cdots {}_{C12})$

$$\psi^{B_{2u}}_{(C_5-C_{12})} = \frac{1}{2\sqrt{2}}(\phi_{C_5} + \phi_{C_6} - \phi_{C_7} - \phi_{C_8} + \phi_9 + \phi_{C_{10}} - \phi_{C_{11}} - \phi_{C_{12}})$$

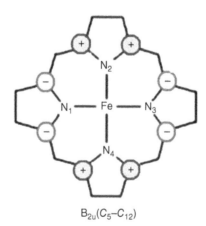

$$B_{2u}(C_5-C_{12})$$

- B_{2u} of $(\pi_{C13} \cdots C_{20})$

$$\psi^{B_{2u}}_{(C_{13}-C_{20})} = \frac{1}{2\sqrt{2}}(\phi_{C_{13}} + \phi_{C_{14}} - \phi_{C_{15}} - \phi_{C_{16}} + \phi_{17} + \phi_{C_{18}} - \phi_{C_{19}} - \phi_{C_{20}})$$

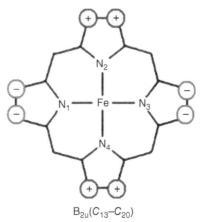

$B_{2u}(C_{13}-C_{20})$

- For E_g of $(\pi_{N1} \cdots N_4)$:
  - Assume each linear combination of atomic orbital (LCAO), ψ_i, must belong uniquely to a single one-dimensional representation.
  - The individual of $\psi^{E_g}_{(N1 \ldots N4)}$ easily converts to a pair of one-dimensional representations by passing to a pure rotational subgroup of D_{4h} (this is C_4 Point group):

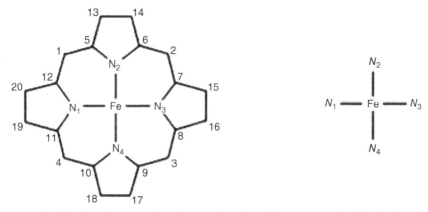

  - In C_4, the character of each operation in each representation is already split into two parts, x- and y-components in C_4 point group:

C_4	E	C_4	C_2	C_4^3
A	1	1	1	1
B	1	-1	1	-1
E	1	i	-1	$-i$
	1	$-i$	-1	i

o Then eliminate the imaginary coefficients by taking the linear combination of E_1 and E_2 (by adding and subtracting them and dividing out i):

$$\hat{P}^{E1}\phi_{N1} = \phi_{N1} + i\phi_{N2} - \phi_{N3} - i\phi_{N4}$$

$$\hat{P}^{E2}\phi_{N1} = \phi_{N1} - i\phi_{N2} - \phi_{N3} + i\phi_{N4}$$

Adding yields

$$2(\phi_{N1} - \phi_{N3}) \rightarrow n = 2 \rightarrow N = \sqrt{2} \rightarrow \frac{1}{\sqrt{2}}(\phi_{N1} - \phi_{N3})$$

Subtracting yields

$$2i(\phi_{N2} - \phi_{N4}) \rightarrow \frac{1}{\sqrt{2}}(\phi_{N2} - \phi_{N4})$$

$$\Psi^{E_g}_{(N1\ldots N4)} \rightarrow \left(\frac{1}{\sqrt{2}}(\phi_{N1} - \phi_{N3}), \frac{1}{\sqrt{2}}(\phi_{N2} - \phi_{N4})\right)$$

o E_g of $(\pi_{N1} \cdots {}_{N4})$

$$\psi^{E_g}_{(N_1 - N_4)} \rightarrow \begin{cases} \dfrac{1}{\sqrt{2}}\left[\phi_{N_2} - \phi_{N_4}\right] \\ \dfrac{1}{\sqrt{2}}\left[\phi_{N_1} - \phi_{N_3}\right] \end{cases}$$

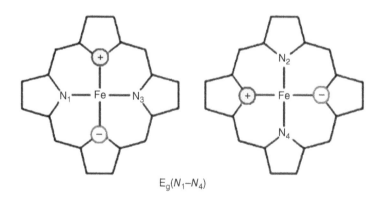

$E_g(N_1-N_4)$

- $\psi^{E_g}_{(\pi_{C1,C2,C3,C4})}$, $\{\psi^{E_g}_{(\pi_{C13,C14,C17,C18})}$ and $\psi^{E_g}_{(\pi_{C15,C16,C19,C20})}\}$, have the same symmetry, these molecular orbitals span in C_{2v} point group, as B_1 and B_2:

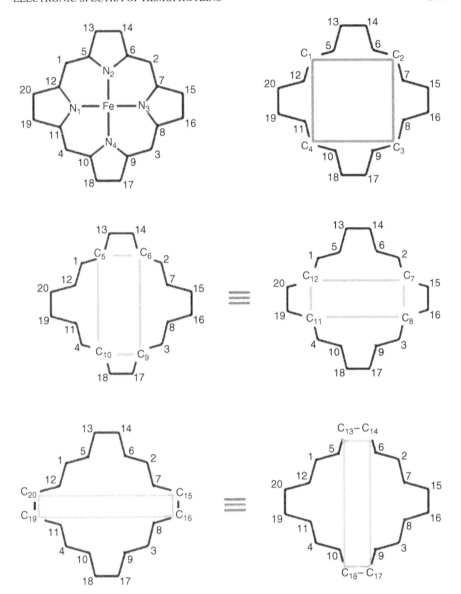

C_{2v}	E	C_2	σ_v^{xz}	σ_v^{yz}	
A_1	1	1	1	1	z
A_2	1	1	-1	-1	R_z
B_1	1	-1	1	-1	x, R_y
B_2	1	-1	-1	1	y, R_x

$$\hat{P}^{B1} = \phi_{C1} - \phi_{C3} + \phi_{C4} - \phi_{C2}$$

$$\hat{P}^{B1} = \phi_{C1} - \phi_{C2} - \phi_{C3} + \phi_{C4}$$

$$\Psi^{B1}_{C1\ldots C4} = \tfrac{1}{2}(\phi_{C1} - \phi_{C2} - \phi_{C3} + \phi_{C4})$$

$$\hat{P}^{B2} = \phi_{C1} - \phi_{C3} - \phi_{C4} + \phi_{C2}$$

$$\hat{P}^{B2} = \phi_{C1} + \phi_{C2} - \phi_{C3} - \phi_{C4}$$

$$\Psi^{B2}_{C1\ldots C4} = \tfrac{1}{2}(\phi_{C1} + \phi_{C2} - \phi_{C3} - \phi_{C4})$$

○ E_g of $(\pi_{C1} \ldots {}_{C4})$

$$\psi^{E_g}_{(C_1-C_4)} \rightarrow \left\{ \begin{array}{l} \psi^{B_1}_{(C_1-C_4)} = \tfrac{1}{2}\left[\phi_{C_1} - \phi_{C_2} - \phi_{C_3} + \phi_{C_4}\right] \\[6pt] \psi^{B_2}_{(C_1-C_4)} = \tfrac{1}{2}\left[\phi_{C_1} + \phi_{C_2} - \phi_{C_3} - \phi_{C_4}\right] \end{array} \right\}$$

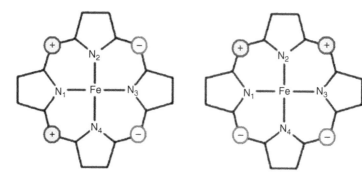

$$E_g(C_1-C_4)$$

○ E_g of $(\pi_{C5} \ldots {}_{C12})$

$$\psi^{E_g}_{(C_5-C_{12})} \rightarrow \left\{ \begin{array}{l} \left\{ \begin{array}{l} \psi^{B_1}_{(C_5-C_{12})} = \tfrac{1}{2}\left[\phi_{C_5} - \phi_{C_6} - \phi_{C_9} + \phi_{C_{10}}\right] \\[6pt] \psi^{B_2}_{(C_1-C_4)} = \tfrac{1}{2}\left[\phi_{C_7} - \phi_{C_8} - \phi_{C_{11}} + \phi_{C_{12}}\right] \end{array} \right\} \\[20pt] \left\{ \begin{array}{l} \psi^{B_1}_{(C_5-C_{12})} = \tfrac{1}{2}\left[\phi_{C_5} + \phi_{C_6} - \phi_{C_9} - \phi_{C_{10}}\right] \\[6pt] \psi^{B_2}_{(C_1-C_4)} = \tfrac{1}{2}\left[\phi_{C_7} + \phi_{C_8} - \phi_{C_{11}} - \phi_{C_{12}}\right] \end{array} \right\} \end{array} \right\}$$

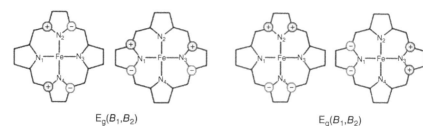

$$E_g(B_1,B_2) \qquad\qquad\qquad\qquad\qquad E_g(B_1,B_2)$$

○ E_g of $(\pi_{C13} \cdots {}_{C20})$

$$\psi^{E_g}_{(C_{13}-C_{20})} \rightarrow \begin{cases} \begin{cases} \psi^{B_1} = \frac{1}{2}\left[\phi_{C_{13}} - \phi_{C_{14}} - \phi_{C_{17}} + \phi_{C_{18}}\right] \\ \psi^{B_2} = \frac{1}{2}\left[\phi_{C_{15}} - \phi_{C_{16}} - \phi_{C_{19}} + \phi_{C_{20}}\right] \end{cases} \\ \begin{cases} \psi^{B_1} = \frac{1}{2}\left[\phi_{C_{13}} + \phi_{C_{14}} - \phi_{C_{17}} - \phi_{C_{18}}\right] \\ \psi^{B_2} = \frac{1}{2}\left[\phi_{C_{15}} + \phi_{C_{16}} - \phi_{C_{19}} - \phi_{C_{20}}\right] \end{cases} \end{cases}$$

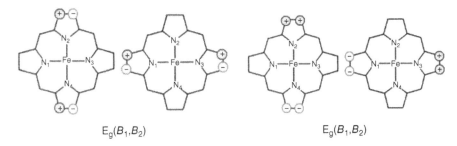

$E_g(B_1, B_2)$ $\qquad\qquad\qquad\qquad$ $E_g(B_1, B_2)$

○ In case of A_{1g} of $(\sigma_{N1} \cdots {}_{N4})$

$$\psi^{A_{1g}}_{(N_1-N_4)} = \frac{1}{2}(\phi_{N_1} + \phi_{N_2} + \phi_{N_3} + \phi_{N_4})$$

○ In case of B_{1g} of $(\sigma_{N1} \cdots {}_{N4})$

$$\psi^{B_{1g}}_{(N_1-N_4)} = \frac{1}{2}(\phi_{N_1} - + \phi_{N_3} - \phi_{N_4})$$

○ E_g of $(\sigma_{N1} \cdots {}_{N4})$

$$\psi^{E_g}_{(N_1-N_4)} \rightarrow \begin{cases} \dfrac{1}{\sqrt{2}}\left[\phi_{N_2} - \phi_{N_4}\right] \\ \dfrac{1}{\sqrt{2}}\left[\phi_{N_1} - \phi_{N_3}\right] \end{cases}$$

- When ligands are added, the symmetry of these molecular orbitals are no longer D_{4h}, and every representation can be given in C_{4v} or even in C_{2v}.
- The correspondence between labels are:

D_{4h}	C_{4v}	C_{2v}	D_{4h}	C_{4v}	C_{2v}
A_{1g}, A_{2g}	A_1	A_1	B_{1g}, B_{2g}	B_1	A_1
A_{2u}, A_{1u}	A_2	A_2	B_{2u}, B_{1u}	B_2	A_2
E_g, E_u	E	B_1, B_2			

- Consequently, the representation of the molecular orbital symmetry according to their D_{4h} classification can be given according to C_{4v} or C_{2v}.
- The next step is to combine these molecular orbitals linearly with the 3d orbitals of iron.
- These linear combinations are subject to the restriction that all orbital symmetries must transform according to the same irreducible representation of the symmetry group, where the correspondence between symmetry representations and d orbitals are:

In D_{4h}: $A_{1g} \rightarrow d_{z^2}$, $B_{1g} \rightarrow d_{x^2-y^2}$, $B_{2g} \rightarrow d_{xy}$, $A_{2u} \rightarrow p_z$, $E_g \rightarrow d_{xz}, d_{yz}$, $E_u \rightarrow p_x, p_y$
In C_{4v}: $A_1 \rightarrow d_{z^2}$, $B_1 \rightarrow d_{x^2-y^2}$, $B_2 \rightarrow d_{xy}$, $E \rightarrow p_x, p_y, d_{yz}, d_{xz}$
In C_{2v}: $A_1 \rightarrow p_z, d_{z^2}, d_{x^2-y^2}$, $A_2 \rightarrow d_{xy}$, $B_1 \rightarrow d_{xz}$, $B_2 \rightarrow d_{yz}$

- Description of the porphyrin transitions can be obtained from the Hückel LCAO.

Electronic Transitions

In the porphyrin molecule, the four top-filled orbitals are nearly degenerate and carry the symmetry labels a_{2u}, a_{1u}, a'_{2u}, and b_{2u}, while the lowest empty orbitals are degenerate and have e_g symmetry in D_{4h}. What are the expected electronic transitions? Characterize and cite the main factors that may affect these transitions.

- In the singly excited configuration, only one electron is promoted from the orbital it occupies in the ground configuration to an orbital of higher energy.

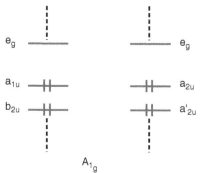

Six-orbital model (Makinen and Eaton, 1973.
Reproduced by permission of John Wiley & Sons.)

- The ground state of porphyrin molecule has A_{1g} symmetry: A_{1g}: . . . $a_{1u}^2 a_{2u}^2$.
- The lowest energy, singly excited configurations are:
  - $a_{1u}^2 a_{2u} e_g$
  - $a_{2u}^2 a_{1u} e_g$

  The symmetry of each excited configurations is found by forming the direct product $a_{2u} e_g$ and $a_{1u} e_g$ representations in the point group D_{4h} (Tables 5-9 and 5-10).

D_{4h}	E	$2C_4$	C_2	$2C_2'$	$2C_2''$	i	$2S_4$	σ_h	$2\sigma_v$	$2\sigma_d$		
A_{1g}	1	1	1	1	1	1	1	1	1	1		x^2+y^2, z^2
A_{2g}	1	1	1	-1	-1	1	1	1	-1	-1	R_z	
B_{1g}	1	-1	1	1	-1	1	-1	1	1	-1		x^2-y^2
B_{2g}	1	-1	1	-1	1	1	-1	1	-1	1		xy
E_g	2	0	-2	0	0	2	0	-2	0	0	(R_x, R_y)	(xz, yz)
A_{1u}	1	1	1	1	1	-1	-1	-1	-1	-1		
A_{2u}	1	1	1	-1	-1	-1	-1	1	1	1	z	
B_{1u}	1	-1	1	1	-1	-1	1	-1	-1	1		
B_{2u}	1	-1	1	-1	1	-1	1	-1	1	-1		
E_u	2	0	-2	0	0	-2	0	2	0	0	(x, y)	

$$E_u : \ldots a_{1u}^2 a_{2u} e_g \qquad E_u' : \ldots a_{2u}^2 a_{1u} e_g$$

- The transitions $A_{1g} \rightarrow E_u$ $(a_{1u} \rightarrow e_g)$ and $A_{1g} \rightarrow E_{u'}$ $(a_{2u} \rightarrow e_g)$ have almost similar ΔE.

- In these two excitations, the direct product for the ground and excited states is an irreducible representation, E_u.

- The E_u state has an electronic dipole moment in the direction of the x and y Cartesian coordinates.

TABLE 5-9   Direct Product of $a_{2u}e_g$ Representation

D_{4h}	E	$2C_4$	C_2	$2C_2'$	$2C_2''$	i	$2S_4$	σ_h	$2\sigma_v$	$2\sigma_d$
A_{2u}	1	1	1	-1	-1	-1	-1	-1	1	1
E_g	2	0	-2	0	0	2	0	-2	0	0
Direct product	2	0	-2	0	0	-2	0	2	0	0
E_u	2	0	-2	0	0	-2	0	2	0	0

TABLE 5-10   Direct Product of $a_{1u}e_g$ Representation

D_{4h}	E	$2C_4$	C_2	$2C_2'$	$2C_2''$	i	$2S_4$	σ_h	$2\sigma_v$	$2\sigma_d$
A_{1u}	1	1	1	1	1	-1	-1	-1	-1	-1
E_g	2	0	-2	0	0	2	0	-2	0	0
Direct product	2	0	-2	0	0	-2	0	2	0	0
E_u	2	0	-2	0	0	-2	0	2	0	0

- Therefore:
  - These electronic transitions are dipole allowed, with high transition moments.
  - The transition moments for these promotions are almost equal.
  - The produced excited configurations are of the same symmetry, E_u, and are strongly mixed by *configuration interaction*.

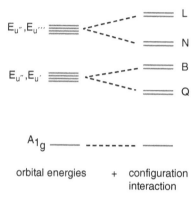

- This mixing is similar in a mathematic sense to the interaction of two hydrogen atoms in forming the H_2 molecule.
- We could write the secular determinant for the mixing of the two states as

$$\begin{vmatrix} E_1^0 - E & H_{12} \\ H_{21} & E_1^0 - E \end{vmatrix} = 0$$

- Solution of the secular determinant gives us two new energies after mixing:

$$E_1 = \tfrac{1}{2}\left\{ E_1^0 + E_2^0 + \left[\left(E_1^0\right)^2 + \left(E_2^0\right)^2 - 2E_1^0 E_2^0 + 4H_{12}^2 \right]^{1/2} \right\}$$

$$E_2 = \tfrac{1}{2}\left\{ E_1^0 + E_2^0 - \left[\left(E_1^0\right)^2 + \left(E_2^0\right)^2 - 2E_1^0 E_2^0 + 4H_{12}^2 \right]^{1/2} \right\}$$

where E_1 and E_2 are the energies of the excited states.

- Maximum absorption occurs when the electric vector of the light is parallel to the direction of the electric dipole moment transition moment in the molecule.
  - The *addition* of the transition moments for these excitations gives a large net transition moment for the higher frequency B band.
  - But the *subtraction* of these moments is for the low-frequency Q band.
- In a similar fashion, N and L bands can be explained. Note:
  - Light is polarized in any direction of the light beam.
  - In solution, molecules are randomly oriented, consequently, light that is polarized in any direction can be absorbed.

   o Polarized absorption and linear dichroism are techniques that are used to study the optical properties of oriented systems.

Vibronic Excitation

Give reasons for the vibronic excitation, β-band (Q_v), in electronic spectra of hemoprotein derivatives.

- Coupling of electronic and vibrational wave functions is called vibronic coupling.
- This coupling takes place when *both* electronic transitions and vibronic excitations are allowed.
- In order to find the allowed vibronic excitations, we first determine the symmetries of the normal modes of vibration:
  1. To generate the representation $\Gamma_{x,y,z}$ from a motional vector basis set, consider:
     (a) An atom whose nucleus (vector origin) is *shifted* from its original position by an operation of the group contributes *zero* to the character of that operation's matrix in the total representation, $\Gamma_{x,y,z} = 0$.
     (b) An atom that is *not shifted* by an operation and where none of its motional vectors are rotated about their local nuclear origin contributes +3 to the character of that operation's matrix in the total representation, $\Gamma_{x,y,z} = 3$.
     (c) If an atom is *unshifted* by an operation but is rotated about its own nucleus (vector origin), then it contributes to the character of that operation's matrix as follows:

Operation	Contribution to $\Gamma_{x,y,z}$	Operation	Contribution to $\Gamma_{x,y,z}$
C_n	$1 + 2\cos(2\pi/n)$	S_n	$-1 + 2\cos(2\pi/n)$
C_2	-1		$= (C_n \text{ value}) - 2$
C_3	0	$S_1 = \sigma$	1
C_4	1	$S_2 = i$	-3
C_6	2	S_3	-2
		S_4	-1
		S_6	0

    (d) Assume Fe-porphyrin belongs to the point group D_{4h}.
  2. Determine the number of atoms that do not change location during the symmetry operation.

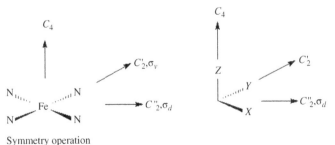

Symmetry operation

TABLE 5-11 **Total Representation of All Degrees of Freedom in D_{4h} Point Group**

D_{4h}	E	$2C_4$	C_2	$2C_2'$	$2C_2''$	i	$2S_4$	σ_h	$2\sigma_v$	$2\sigma_d$
$\Gamma_{x,y,z}$	3	1	-1	-1	-1	-3	-1	1	1	1
Unmoved atoms	5	1	1	3	1	1	1	5	3	1
Γ total	15	1	-1	-3	-1	-3	-1	5	3	1

3. For each operation, multiply the number of unmoved atoms by the character of $\Gamma_{x,y,z}$ at the bottom of the character table. This gives the character of Γ_{total}, the total representation of all degrees of freedom of the molecule, including transition, rotation, and vibration (Table 5-11).

4. Γ_{total}, a reducible representation, is the sum of irreducible representation, Γ_i, using

$$\Gamma_i = \frac{1}{h}\sum_R \chi(R)\chi_i(R)$$

where, $\chi(R)$ is the character of the reducible representation and $\chi_i(R)$ is the character of irreducible representation.

D_{4h}	E	$2C_4$	C_2	$2C_2'$	$2C_2''$	I	$2S_4$	σ_h	$2\sigma_v$	$2\sigma_d$		
A_{1g}	1	1	1	1	1	1	1	1	1	1		x^2+y^2, z^2
A_{2g}	1	1	1	-1	-1	1	1	1	-1	-1	R_z	
B_{1g}	1	-1	1	1	-1	1	-1	1	1	-1		x^2-y^2
B_{2g}	1	-1	1	-1	1	1	-1	1	-1	1		xy
E_g	2	0	-2	0	0	2	0	-2	0	0	(R_x, R_y)	(xz, yz)
A_{1u}	1	1	1	1	1	-1	-1	-1	-1	-1		
A_{2u}	1	1	1	-1	-1	-1	-1	-1	1	1	z	
B_{1u}	1	-1	1	1	-1	-1	1	-1	-1	1		
B_{2u}	1	-1	1	-1	1	-1	1	-1	1	-1		
E_u	2	0	-2	0	0	-2	0	2	0	0	(x, y)	

$$\Gamma_{total} = a_{1g} + a_{2g} + b_{1g} + b_{2g} + e_g + 2a_{2u} + b_{2u} + 3e_u$$

Using D_{4h} character table gives:

(a) $\Gamma_{transition} = a_{2u} + e_u$.

(b) $\Gamma_{rotation} = a_{2g} + e_g$,

(c) $\Gamma_{vibration} = a_{1g} + b_{1g} + b_{2g} + a_{2u} + b_{2u} + 2e_u$

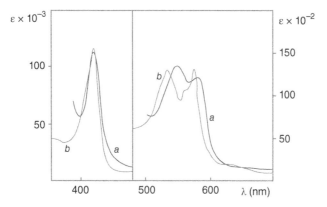

FIGURE 5-70 Spectral properties of Mb(II)NO (a), and Mb(III)No (b), pH 7, 25°C (Stephanos and Addison, 1989).

- Now we compare the symmetries of these modes of vibrations with the symmetries of the excited configurations to give Q and B bands:

$$E_u: \ldots a_{1u}^2 a_{2u} e_g \qquad E_{u'}: \ldots a_{2u}^2 a_{1u} e_g$$

- Coupling of the electronic and vibrational wave function is expected.
- Thus, the pure electronic transitions Q and B are allowed, as well as the vibronic excitations of symmetries E_u.

  (a) The atoms in molecules do not change position appreciably (*Franck–Condon principle*):

    (i) The transition from the ground state to the vibronic level occurs in a very short time (about 10^{-15} s).

    (ii) The momentum, which is close to zero in the ground state, is close to zero in the excited state.

    (iii) The electronic transition is rapid. The molecule will find itself with the same molecular configuration and vibrational kinetic energy in the excited state that it had in ground state at the moment of absorption of the photon.

  (b) In hemoprotein spectra, Q bands show vibrational structure, but the B and N bands do not (Fig. 5-70 and Table 5-12).

  (c) The Q_o and Q_v are often called α and β bands.

Symmetry and d-Orbitals

Explain how symmetry affects the splitting of 3d orbital energy levels within the heme?

- In free ion, the five 3d orbitals are degenerate.
- First, we imagine that all the ligands are identical and sit at the corners of an octahedron, O_h.

TABLE 5-12 Band Positions (Abs.) and Extinction Coefficient (ε) for Some Derivatives of Horse Heart Myoglobin at 25°C, pH 7

	Q	Q_v	B		Q	Q_v	B
Mb(III)H$_2$O				Mb(III)Im			
Abs. (nm)	573	503	408	Abs. (nm)		531	415
$10^{-3} \times \varepsilon$	9.02	19.6	166	$10^{-3} \times \varepsilon$		11.3	132
Mb(III)CN				MbO$_2$			
Abs. (nm)	570(sh)	538	421	Abs. (nm)	578	542	412
$10^{-3} \times \varepsilon$	9.02	11.3	117	$10^{-3} \times \varepsilon$	9.61	10.1	120
Mb(III)MTG				Mb(deoxy)			
Abs. (nm)	570(sh)	544	424	Abs. (nm)		553	430
$10^{-3} \times \varepsilon$	18.81	11.3	120	$10^{-3} \times \varepsilon$		13.8	158
Mb(III)N$_3$				Mb(II)CO			
Abs. (nm)		537	421	Abs. (nm)	576	539	421
$10^{-3} \times \varepsilon$		11.3	121	$10^{-3} \times \varepsilon$	13.4	15.4	191
Mb(III)NCO	580(sh)	522	414	Mb(II)MTG	578(sh)	558	419
Abs. (nm)	5.2	12.4	166	Abs. (nm)	9.90	12.8	124
$10^{-3} \times \varepsilon$				$10^{-3} \times \varepsilon$			
Mb(III)NCS	570(sh)	511	412	Mb(II)MEic	570	542	430
Abs. (nm)	9.24	10.2	132	Abs. (nm)	18.4	16.8	186
$10^{-3} \times \varepsilon$				$10^{-3} \times \varepsilon$			
Mb(III)NCSe	562	504	415	Mb(II)Buic	560	529	432
Abs. (nm)	8.62	9.60	110	Abs. (nm)	15.5	14.4	175
$10^{-3} \times \varepsilon$				$10^{-3} \times \varepsilon$			
Mb(III)Pyr				Mb(II)Bzic	564	532	411
Abs. (nm)		504	408	Abs. (nm)	15.7	14.9	129
$10^{-3} \times \varepsilon$		19.0	155	$10^{-3} \times \varepsilon$			

Source: Unpublished data, Addison and Stephanos. Pyr, pyridine; MEic, 2-morpholinoethylisocyanide; Buic, butylisocyanide; Bzic, benzylisocyanide; MTG, methylthioglycolate; Im, imidazole.

- The crystal field splits the 3d orbitals into a lower three degenerate set containing the d_{xy}, d_{yz}, and d_{xz} orbitals, which are labeled t_2 and upper two degenerate pair, d_{z^2} and $d_{x^2-y^2}$ labeled e_g.
- Both t_2 and e_g are split by a tetragonal crystal field complex with fourfold rotation axis, and the only degenerate orbitals are d_{xz} and d_{yz}.

- The destruction of the fourfold axis (rhombic perturbation) removes the degeneracy of d_{xz} and d_{yz} orbitals.

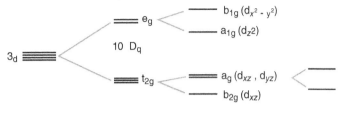

Free ion   Octahedral field   Tetragonal field   Rhombic field

Energies of iron 3d orbitals in octahedral, tetragonal, and rhombic fields

- In the weak ligand field, the spin of all five electrons of the ferric ion of d orbitals is parallel.
- In the strong ligand field, the five electrons are in the lower three d orbitals.
- The ground electronic configurations for the ferrous and ferric ions in weak and strong fields are shown below.

Fe^{3+} high spin Fe^{3+} low spin Fe^{2+} high spin Fe^{2+} low spin

Electronic Spectra

In the electronic spectrum of aquomethemoglobin, Hb(III)H$_2$O (Fig. 5-69), which bands are due to $\pi \to \pi^*$ and $d \to d$ transitions and charge transfer?

- Ferric high-spin spectra, Hb(III)H$_2$O, usually exhibit four bands in addition to the porphyrin $\pi \to \pi^*$ transitions (Fig. 5-71).
- The ground electronic state of ferric high-spin complexes are an orbitally nondegenerate, spin sextet, $^6A_1$, in O_h symmetry.
- This ground state is described by a single configuration $t_2^3 e^2$, in which all five electrons are parallel.
- All $d \to d$ transitions must result in spin pairing and are therefore forbidden by the selection rule, which requires no change in spin multiplicity.
- Consequently, the possibility that bands I–IV arise from $d \to d$ transitions can be eliminated (Fig. 5-71).

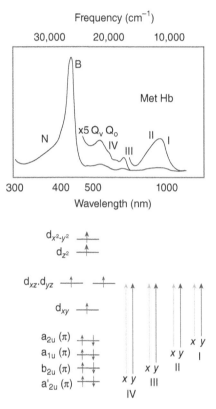

FIGURE 5-71 Absorption spectrum of methemoglobin and possible charge transfer bands. (Modified from Eaton and Hofichter, 1981. Reproduced by permission of Elsevier.)

- Bands I–IV must therefore be due to *charge transfer transitions*.
- Charge transfer is a transition in which an electron is transferred from one atomic or molecular orbital in the molecule to another and displays very intense bands, $\varepsilon_{max} \sim 10^4 \, mol^{-1} \, cm^{-1}$.
- All four bands are x, y polarized, indicating that the transitions are degenerate.
- Bands I and II are observed in the absorption spectra of Hb(III)F.
- The porphyrin $\pi \rightarrow d_{xy}$ transitions are forbidden in both D_{4h} and C_{4v} symmetries.
- Therefore, bands I–IV have been assigned as promotions from the top filled π orbitals to degenerate d_{xz}, d_{yz} iron orbitals.

In the electronic spectrum of Hb(III)CN (Fig. 5-69), which bands are due to $\pi \rightarrow \pi^*$ and $d \rightarrow d$ transitions and charge transfer?

- Hb(III)CN is a low-spin ferric complex (Fig. 5-72).
- In O_h symmetry, the ground electronic state is a spin doublet, $^2T_2$.

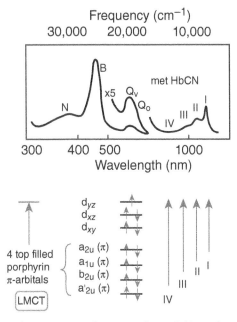

FIGURE 5-72 Absorption spectrum of cyanomethemoglobin and possible charge transfer bands. (Modified from Eaton and Hofichter, 1981. Reproduced by permission of Elsevier.)

- The electronic configuration described by the threefold orbitally degenerate, t_2^5, has only a single unpaired electron.
- There are now possibilities of spin-allowed $d \rightarrow d$ transitions.
- None of the bands I–IV exhibit measurable CD, indicating that they are magnetic dipole forbidden, (unlike the electronic spectra, $g \rightarrow g$ transitions are magnetically allowed in circular dichroism, CD).
- This excludes $d \rightarrow d$ assignments for bands I–IV.
- Bands I–IV have been assigned to transitions from the four top-filled porphyrin π orbitals into the iron d_{yz} orbital.
- Notice the splitting of d_{xz}, d_{yz} orbitals due to the tilt attachment of CN^-.

In the electronic spectrum of Hb(II), Fig. 5-69, which bands are due to $\pi \rightarrow \pi^*$ and $d \rightarrow d$ transition and charge transfer?

- Deoxyhemoglobin shows ferrous high-spin spectra.
- Ferrous high-spin hemoglobin contains four bands in addition to the porphyrin $\pi \rightarrow \pi^*$ transitions (Fig. 5-73).
- Band II, which is prominent in the CD spectrum ($g \rightarrow g$), has been assigned to the magnetically dipole allowed $d_{xz} \rightarrow d_{z^2}$.

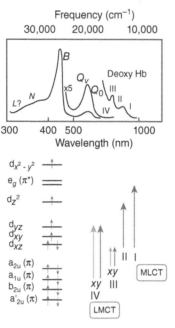

FIGURE 5-73 Absorption spectrum of deoxyhemoglobin. (Modified from Eaton and Hofichter, 1981. Reproduced by permission of Elsevier.)

- Band I is z polarized and has been assigned to $d_{xz} \rightarrow e_g$ (π^*), iron to porphyrin, charge transfer transition.
- Bands III and IV are x and y polarized and have been assigned as a_{2u} $(\pi) \rightarrow d_{xz}$ and a_{1u} $(\pi) \rightarrow d_{yz}$, respectively.

In the electronic spectrum of Hb(II)CO (Fig. 5-69), which bands are due to $\pi \rightarrow \pi^*$ and $d \rightarrow d$ transitions and charge transfer?

- Hb(II)CO displays a ferrous low-spin spectra (Fig. 5-74).
- Bands I and II at 625 and 560 nm are observed only in CD spectra.
- Therefore, these bands are magnetic dipole allowed: $d_{xy} \rightarrow d_{x^2-y^2}$ and $d_{xz}, d_{yz} \rightarrow d_{z^2}, d_{x^2-y^2}$ (Fig. 5-74).

In the electronic spectrum of Hb(II)O$_2$ (Fig. 5-69), which bands are due to $\pi \rightarrow \pi^*$ and $d \rightarrow d$ transitions and charge transfer?

- The absorption spectrum of Hb(II)O$_2$ exhibits a single additional very broad band, centered at 925 nm (Fig. 5-75).
- This band is composed of four separate electronic transitions in CD spectra, labeled I–IV.

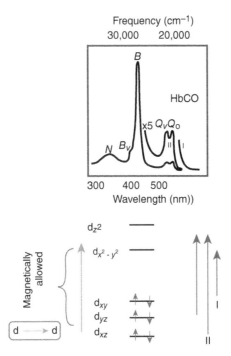

FIGURE 5-74 Absorption spectrum of carboxyhemoglobin. (Modified from Makinen and Eaton, 1973. Reproduced by permission of John Wiley & Sons.)

- Also, bands V and VI are observed in the CD spectrum.
- Bands VI and VII are observed in the crystal spectrum because of their z polarization.
- The assignment depends on an extended Hückel calculation.
- Unlike other hemoglobin derivatives, the d_{xy} orbital remains pure, but d_{xz} and d_{yz} orbitals are strongly mixed with the oxygen π_g.
- This combination produces four molecular orbitals (Fig. 5-75).
- Based on their frequency and characteristic appearance in the CD spectrum, bands I and II have been assigned to magnetic dipole allowed excitations: $d_{yz} + O_2(\pi_g) \rightarrow d_{xz} + O_2(\pi_g)$ and $d_{x^2-y^2} \rightarrow d_{xz} + O_2(\pi_g)$.
- Bands III and IV are more pronounced, indicating involvement of porphyrin orbitals. They are assigned to the x-polarized $a_{2u}(\pi) \rightarrow d_{xz} + O_2(\pi_g)$ and yz-polarized $a_{1u}(\pi) \rightarrow d_{xz} + O_2(\pi_g)$.
- Bands V and VI appear in the same frequency range as bands I and II of Hb(II)CO and are observable in the CD spectra, indicating that they are similar to d → d promotions: $d_{x^2-y^2} \rightarrow d_{xy}$ and $d_{xz} - O_2(\pi_g)$, $d_{yz} - O_2(\pi_g) \rightarrow d_{z^2}, d_{xy}$.
- Band VII is characterized by high z-polarized intensity and is assigned to the $O_2(\pi_u) \rightarrow d_{xz} + O_2(\pi_g)$ transition based on extended Hückel calculations.

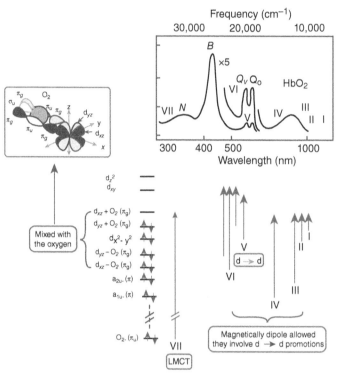

FIGURE 5-75 Absorption spectrum of oxyhemoglobin. (Modified from Eaton and Hofichter, 1981. Reproduced by permission of Elsevier.)

ESR SPECTRA OF HEMOPROTEINS

ESR Spectroscopy and Magnetic Moment

What is the relationship between the ESR spectroscopy and the magnetic moment?

ESR spectroscopy probes the magnetic moment (μ) associated with the unpaired electron(s) present in a paramagnetic heme:

$$\mu^2 = g^2 S(S + 1)$$

where $2S$ is the number of electrons present in paramagnet, the g factor specifies the strength of the interaction of the spin system with an applied magnetic field H_0, and $g^2 = \frac{1}{3}(g_x^2 + g_y^2 + g_z^2)$.

Symmetry and Splitting Parameters

Draw a diagram to show the splitting pattern of the d orbitals as the symmetries of low-spin ferric hemoproteins are reduced from cubic and indicate the relative splitting parameters.

- The electronic configuration of low-spin ferric hemoproteins is usually written as $(t_{2g})^5$ or d_{xz}^2, d_{yz}^2, and d_{xy}^1.

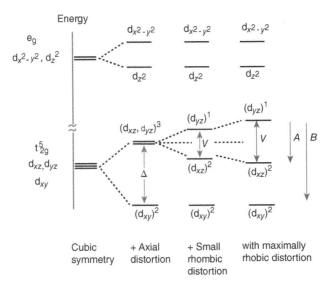

SCHEME 5-46   Non spherical splitting pattern of d orbital.

- The low-spin ferric hemoglobins are formed when:
  - H_2O of $Hb(III)H_2O$ becomes deprotonated to form hydroxide:
$$Hb(III)H_2O \rightarrow Hb(III)OH$$
  - Exogenous ligands such as N_3^- and CN^- are added:
$$Hb(III)H_2O + N_3^- \rightarrow Hb(III)N_3 + H_2O$$
$$Hb(III)H_2O + CN^- \rightarrow Hb(III)CN + H_2O$$
  - The protein structure is altered so that electron-rich endogenous ligands can bind the iron atom in the distal position.
- The nonspherical component of the ligand field (Scheme 5-46, can be decomposed as follows:
  - The large axial component is z directed and destabilizes d_{xz} and d_{yz} with respect to d_{xy} by an amount Δ.
  - The rhombic component produces an inequivalence in the x and y directions such that d_{xz} is more stable than d_{yz} by an amount V.
  - The relative stabilizations of d_{yz}, d_{xz}, and d_{xy} are 0, A, and B.
  - V/Δ is never larger than $\frac{2}{3}$.
  - The two energies splitting A and B are given in units of the spin–orbit coupling constant λ.

Splitting Parameters and g-Values

Write the expressions that correlate g values and different splitting parameters. If the following represents the ESR spectrum of $Hb_\alpha(III)OH$ at 77 K at which

the *g* values are **2.56, 2.18, and 1.88, find the value of** V/λ **and** Δ/λ**. How can these parameters be used as a structural property of low-spin ferric hemoproteins?**

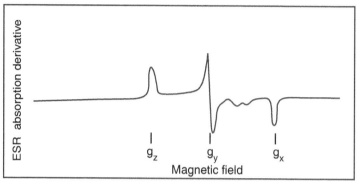

ESR spectrum of Hb$_\alpha$(III)OH at 77 K (From Peisach et al., 1973
Reproduced by permission of John Wiley & Sons.)

- Using the following useful relationships, the *relative values* of A, B, V, and Δ can be obtained. Assume a, b, and c are real numbers such that $a^2 + b^2 + c^2 = 1$ and b and c are approximately $\lambda/(2A)$ and $\lambda/(2B)$, respectively, and λ is the spin–orbital coupling, which has the value of 420 cm^{-1} for free ion (Fe). Then

  - $g_z = 2[(a+b)^2 - c^2]$
  - $g_y = 2[(a+c)^2 - b^2]$
  - $g_x = 2[a^2 - (b+c)^2]$
  - $g_z + g_y - g_x = 2(a+b+c)^2$
  - $a = \dfrac{g_z + g_y}{[8(g_z + g_y - g_x)]^{1/2}}$
  - $b = \dfrac{g_z - g_x}{[8(g_z + g_y - g_x)]^{1/2}}$
  - $c = \dfrac{g_y - g_x}{[8(g_z + g_y - g_x)]^{1/2}}$
  - $A = E_{yz} - E_{xz} = \dfrac{a+c}{2b} = \dfrac{g_x}{g_z + g_y} + \dfrac{g_y}{g_z - g_x} \rightarrow$ a measure of rhombicity

    $(E_{yz} - E_{xz})$
  - $B = \dfrac{a+b}{2c} = \dfrac{g_x}{g_z + g_y} + \dfrac{g_z}{g_y - g_x} \rightarrow$ a measure of $(E_{yz} - E_{xy})$
  - $\dfrac{V}{\lambda} = A$
  - $\dfrac{\Delta}{\lambda} = B - \dfrac{A}{2} = \dfrac{g_x}{2(g_z + g_y)} + \dfrac{g_z}{g_y - g_x} - \dfrac{g_y}{2(g_z - g_x)} \rightarrow$ a measure of the

  axial ligand field sensed by the iron orbitals, the *tetragonal field*.

- The following expressions allow one to compute any g value knowing the other two:

$$g_x^2 + g_y^2 + g_z^2 + g_y g_z - g_x g_z - g_y g_x - 4(g_z + g_y - g_x) = 0$$

- The calculated parameters of $Hb_\alpha(III)OH$ are:
  - $a = 0.991$
  - $b = 0.142$
  - $c = 0.063$
  - $a^2 + b^2 + c^2 = 1.006$
  - $g_z + g_y - g_x = 2(a + b + c)^2 = 2.86$
  - $g_z = 2[(a + b)^2 - c^2] = 2.56$
  - $g_y = 2[(a + c)^2 - b^2] = 2.18$
  - $g_x = 2[a^2 - (b + c)^2] = 1.88$
  - $A = V/\lambda = 3.6$
  - $B = 8.93$
  - $\Delta/\lambda = 7.13$
  - $V/\Delta = 0.5$
  - $g_x^2 + g_y^2 + g_z^2 + g_y g_z - g_x g_z - g_y g_x - 4(g_z + g_y - g_x) = 0$
- Absolute values for A, B, V, and Δ can only be obtained if the precise value of λ for the molecule under study is available. This is because λ decreases in magnitude as the covalency of the paramagnetic center increases.
- The values of V/Δ and Δ/λ were calculated for a large number of naturally occurring hemoproteins together with hemoprotein derivatives and appropriate model compounds. These values were then recorded on a plot of V/λ versus Δ/λ (Fig. 5-76).
- Derivatives that have similar ESR spectra are structurally similar.
- The results of a large amount of data fell into five major domains, labeled C, B, II, O, and P.
  - C-Domain: The proximal ligand is an imino nitrogen from imidazole and the distal ligand is α thio ether sulfur (methionine sulfur).
  - B-Domain: The proximal ligand is imidazole and the distal ligand is N-imidazole.
  - H-Domain: The proximal ligand is imidazole and the distal ligand is HN-imidazole.
  - O-Domain: The proximal ligand is imidazole and the distal ligand is OH.
  - P-Domain: The proximal ligand is imidazole and the distal ligand is thiolate.

High-Spin Ferric Hemoproteins

Give examples of high-spin ferric hemoproteins. Find the crystal field splitting in cubic symmetry, the effect of the axial/rhombic distortion, the spin–orbital

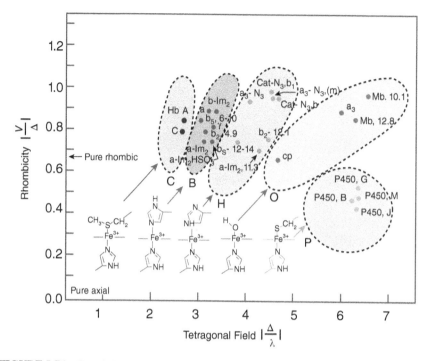

FIGURE 5-76 Correlation diagram for low-spin hemoproteins. (Modified from Peisach et al., 1973. Reproduced by permission of John Wiley & Sons.)

coupling, and the impact of the magnetic field on spin–orbital coupling and how to determine the rhombicity.

- Almost without exception, the ESR of high-spin ferrihemoproteins is characterize by $g_\perp \approx 6$, and $g_\parallel \approx 2$ (Fig. 5-77). The spectrum A is the integrated presentation of spectrum B. Both A and B exhibit signals at $g = 6$ and $g = 2$, but B is more prominent.
- The electronic configuration of the iron is $t_{2g}^3 e_g^2$ with one electron in each of the five 3d wave functions (^6A, $S = \frac{5}{2}$).
- In octahedral symmetry, the alternative configurations:

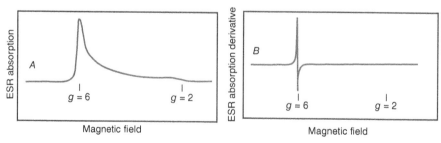

FIGURE 5-77 ESR spectrum of hemin chloride at 1.5 K, A = absorption derivative, B = absorption. (From Peisach et al., 1973. Reproduced by permission of John Wiley & Sons.)

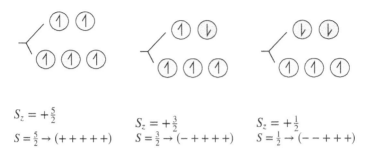

$$S_z = +\tfrac{5}{2}$$
$$S = \tfrac{5}{2} \to (+ + + + +)$$

$$S_z = +\tfrac{3}{2}$$
$$S = \tfrac{3}{2} \to (- + + + +)$$

$$S_z = +\tfrac{1}{2}$$
$$S = \tfrac{1}{2} \to (- - + + +)$$

are *equienergetic* in the absence of magnetic field.

- In cubic symmetry, the excited state is $^4T_1 \left[(d_{xy}, d_{xz}, d_{yz})^4, (d_{z^2})^1 \right]$, where the electron pairing is less than the ligand field splitting, $\pi < \Delta$.
- In real compounds, the symmetry is lowered, with the possibility of both axial and rhombic contributions to the ligand field.
- In the axial ligand field, ^4T is split into $^4A \left[(d_{xy})^2, (d_{xz})^1, (d_{yz})^1, (d_{z^2})^1 \right]$ and $^4E \left[(d_{xz}, d_{yz})^3, (d_{xy})^1, (d_{z^2})^1 \right]$ (Fig. 5-78).
- The effect of the spin–orbit interaction is to stabilize the $M_s = \pm\tfrac{1}{2}, \pm\tfrac{3}{2}$ states with respect to the $\pm\tfrac{5}{2}$ states. The relative changes are usually written parametrically in the form:

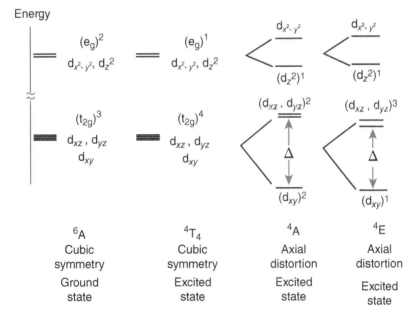

FIGURE 5-78 The crystal-field splitting pattern by spin–orbital interaction for high-spin Fe^{3+}-complex of Δ/λ large and negative. (After Maltempo, 1974.)

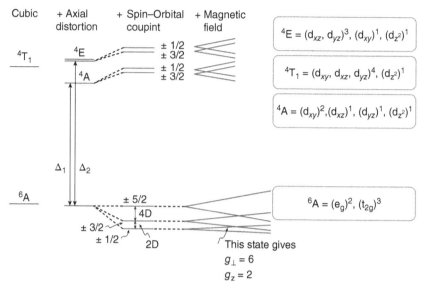

FIGURE 5-78 (*Continued*)

$$E = D[S_z^2 - \tfrac{1}{3} S (S + 1)]$$

where

$$D = \frac{\lambda}{5} \left(\frac{1}{\Delta_1} - \frac{1}{\Delta_2} \right)$$

Δ_1 and Δ_2 are the excitation energies for $^6A \xrightarrow{\Delta_1} {}^4A_2$, and $^6A \xrightarrow{\Delta_2} {}^4E$, respectively, for $\Delta_2 > \Delta_1$, D is positive.

D is the zero-field splitting for methemoglobin; $D \sim 10\,\mathrm{cm}^{-1}$.

$D \sim kT$ at $T = 4\,\mathrm{K}$ will yield 100% of $M_s = \pm\tfrac{1}{2}$.

$S_s = M_z$, $S = S_{\text{high spin}} \rightarrow$ before spin–orbital interaction.

- For $M_s = S_z = \pm\tfrac{1}{2}$ in O_h, $S_{\text{high spin}} = \tfrac{5}{2}$:

$$S_z^2 - \tfrac{1}{3}S(S + 1) = \tfrac{1}{4} - \tfrac{1}{3}(\tfrac{5}{2})(\tfrac{5}{2} + 1)$$

$$= \tfrac{1}{4} - \tfrac{35}{12} = -\tfrac{8}{12} = -\tfrac{8}{3}$$

$$\therefore E_{\pm 1/2} = -\tfrac{8}{3} D$$

- For $S_z = \pm\tfrac{3}{2}$ in O_h, $S_{\text{high spin}} = \tfrac{5}{2}$:

$$S_z^2 - \tfrac{1}{3}S(S + 1) = \tfrac{9}{4} - \tfrac{1}{3}(\tfrac{5}{2})(\tfrac{5}{2} + 1)$$

$$= \tfrac{9}{4} - \tfrac{35}{12} = -\tfrac{8}{12} = -\tfrac{2}{3}$$

$$\therefore E_{\pm 3/2} = -\tfrac{2}{3} D$$

- For $S_z = \pm\frac{5}{2}$ in O_h, $S_{high\ spin} = \frac{5}{2}$:

$$S_z^2 - \tfrac{1}{3}S(S+1) = \tfrac{25}{4} - \tfrac{1}{3}(\tfrac{5}{2})(\tfrac{5}{2}+1)$$
$$= \tfrac{25}{4} - \tfrac{35}{12} = \tfrac{40}{12} = +\tfrac{10}{3}$$
$$\therefore E_{\pm5/2} = +\tfrac{10}{3}D$$

- Now the spin–orbit interaction of ^4A and ^4E with the ground state is different for various components of ^6A such that the $M_s = \pm\frac{1}{2}$ states become more stable by an amount $6D$, the $M_s = \pm\frac{3}{2}$ states become more stable by $4D$, and $M_s = \pm\frac{5}{2}$ states are unshifted.
  - $E_{\pm5/2} - E_{\pm3/2} = +\tfrac{10}{3}D - (-\tfrac{2}{3}D) = 4D$
  - $E_{\pm3/2} - E_{\pm1/2} = -\tfrac{2}{3}D - (-\tfrac{8}{3}D) = 2D$
  - ESR is only observed from the lowest $M_s = \pm\frac{1}{2}$ substates. With H_0, $g_{||} = 2$ and $g_\perp = 6$ (in general, when $D > g_e\beta H_0$, $g_\perp = 2S+1$).
- If the local symmetry is reduced from tetragonal to rhombic, the resultant distortion in the heme plane may cause the energies of d_{xz} and d_{yz} to be different and ^4E split into E_x and E_y.

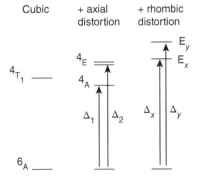

- The consequences of the resultant spin–orbit interaction are then more complicated:

$$D[S_z^2 - \tfrac{1}{3}S(S+1)] + E(S_x^2 - S_y^2)$$

and

$$D = (\lambda/5)[1/\Delta_1 - 1/\Delta_2)]$$

the same as before because

$$\Delta_2 = \tfrac{1}{2}(\Delta_x + \Delta_y)$$

where Δ_x and Δ_y are the excitation energies from $^6A_1$ to E_x and E_y, respectively,

$$E = \frac{\lambda^2}{10}\left(\frac{1}{\Delta_x} - \frac{1}{\Delta_y}\right)$$

and λ is the spin coupling constant.

- The expressions for the g values are given as

$$g_{x,y} = 3g_e \pm 24\frac{E}{D} - \frac{168\,E^2}{9\,D^2}$$

$$g_z = g_e - \frac{304\,E^2}{9\,D^2}$$

- The geometry of these low-symmetry centers can be defined in terms of their rhombicity, $R = E/D$.
- From the equations of g values

$$R = \frac{E}{D} = \frac{g_x - g_y}{48} \times 100$$

or, in proper coordinate system, $E/D < \frac{1}{3}$, or $g_x - g_y < 16$, and

$$R = \frac{g_x - g_y}{16} \times 100$$

Examples are given in Table 5-13.

- The hydroxylating enzyme cytochrome P-450 is most unusual in that $\Delta g = 4.4$, equivalent to 30% rhombicity.

Mixed-Spin States in Hemoproteins

What are the possible implications of mixed–spin states in hemoproteins?

1. The spectrum of the sample under study is the weighted *sum* of the two component spin systems.
   (a) Mixed spins are induced by exogenous ligand, but the amount of ligand present in solution is *insufficient* to saturate the heme.

TABLE 5-13   Rhombicity of High-Spin Ferrihemoproteins

Compound	Rhombicity (%)
Hemin	0.0
Ferrihemoglobin A	0.8
Cytochrome oxidase	1.4
Horseradish peroxidase	4.3
Cytochrome *c* peroxidase	4.9
Bacterial catalase	6.5
Cytochrome P-450cam	26.0

- In such cases, the concentration of ligand needed to obtain full occupancy of heme may be very high.
- Such high concentrations may not be attainable with less soluble compounds.

(b) Mixed spins are induced when the *ligand field strength* of the axial ligands is almost adequate to induce a *high spin to low spin transition*.

- The relative contributions of the two components can be modulated by temperature.
- For example:

$$Mb(III)H_2O \rightleftarrows Mb(III)OH + H^+$$

$$\text{high-spin} \quad \rightleftarrows \quad \text{low spin}$$

$$(1-a) \qquad\qquad (a) \qquad \text{at high pH}$$

The proportions of the two states are temperature dependent:

$$K = \frac{a}{1-a} = \frac{\varepsilon_l - \varepsilon}{\varepsilon - \varepsilon_h}$$

where, ε_l and ε_h are the molar absorptivity associated with the high- and low-spin conformations, respectively. Using the observed ε and T at a single given wavelength, the best combination of values ε_l, ε_h, $\Delta H°$, and $\Delta S°$ can be obtained by fitting the data to the van't Hoff equation:

$$\ln K = \ln \frac{\varepsilon_l - \varepsilon}{\varepsilon - \varepsilon_h} = -\frac{\Delta H°}{RT} + \Delta S°T$$

2. The quantum mechanically mixed-spin system gives a spectrum that is the *average* of the participating species.

(a) The combination of a relatively *weak* axially directed field with a *strong* in-plain ligand field *raises* the energy of the $d_{x^2-y^2}$ orbital. Consequently, the ^4A *intermediate spin-configuration* $(d_{xy})^2$, $(d_{xz})^1$, $(d_{yz})^1$ $(d_{z^2})^1$ may become more stable to the usual ^6A high–spin state (Fig. 5-79).

- This intermediate-spin, ^4A, state is also split by spin–orbital interaction into $\pm1/2$ and $\pm3/2$ levels.
- The g values of the lowest doublet, $\pm1/2$, are $g_{||} = 2$ and $g_\perp = 4$.

(b) When the axial field is only *adequate* to make the ^4A and ^6A states of comparable stability, the ^4A and ^6A configurations are *mixed* by spin–orbit coupling (Fig. 5-80).

- The g values exhibited by this *quantum mechanical* mixed state are $g_{||} = 2$ and $g_\perp = 6a^2 + 4b^2$, where $a^2 + b^2 = 1$.
- In cytochrome c' from *chromatium* protein, $g_\perp$ was found to be 4.75, implying that the ground state is composed of approximately 30% ^6A and 70% ^4A state.

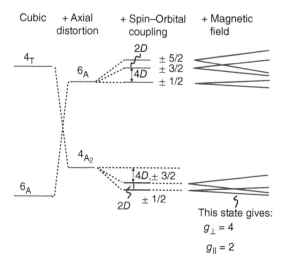

FIGURE 5-79 The crystal-field splitting pattern by spin–orbital interaction for intermediate-spin Fe^{3+} -complex of Δ/λ large and positive. (After Maltempo, 1974.)

- The magnetic moment is predicted from the relationship

$$\mu^2 = a^2 g^2 S_1(S_1 + 1) + b^2 g^2 S_2(S_2 + 1)$$

where, $S_1 = \frac{1}{2}$ and $S_2 = \frac{3}{2}$. Then

$$\mu^2 = 35a^2 + 15b^2, \quad \text{if } a^2 = 30\%, \quad b^2 = 70\%$$

Here $\mu^2 = 24$, a value comparable with the 26.5 measured experimentally at room temperature.

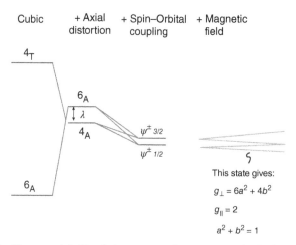

FIGURE 5-80 The crystal-field splitting pattern of quantum mechanically mixed of ^4A and ^6A states by spin–orbital interaction for intermediate-spin Fe^{3+} of $\Delta/\lambda = 1$. (After Maltempo, 1974.)

Reduction of Cytochrome c

Use the following data to explain the distinct reduction behavior of cytochrome *c* in alkaline pH. At neutral pH, the interconversion between reduced and oxidized forms of cytochrome *c* does not involve any alterations in the coordination sphere of the iron ion or in the heme moiety. Assume the pH is raised to the values of 9–10. The ESR signal at *g* = 3.05 characteristic of the native protein has been decreased. A new prominent signal at *g* = 3.4 was observed. This resonance is similar to that of Mb(III)–NH$_3$ complexes. From NMR spectroscopy, the contact shifted resonance assigned to the methionine CH$_3$ protons has disappeared. The weak optical absorption at 698 nm disappears at pH 9.3. The ascorbate is unable to reduce cytochrome *c*.

- Possible explanations to these data are:
  - ○ The electronic transition at 698 nm, which disappears with a p*K* of 9.3, is supposed to be a consequence of the iron–methionine sulfur interaction, the C-form.
  - ○ Therefore, at alkaline pH the methionine, Met^{80}, no longer coordinates as the sixth ligand to the iron atom in this protein.
  - ○ In alkaline pH, probably Lys^{79} loses a proton from the amino group, the B-form (Scheme 5-47).
  - ○ Then Lys^{79} displaces Met^{80} as a ligand to the metal (the C-form), which cannot be reduced by ascorbate. However, it can be reduced by other reagents such as dithionite.
- A biphasic kinetics observed above pH 8 is proposed:
  - ○ In the first phase, the proportion of Fe^{3+} Cyt. *c* in the A-form is rapidly reduced.
  - ○ This is followed by a slow second phase consisting of the sequence C → B → A → B in which the conversion of C to A is rate limiting.

Cytochrome P-450 Spectra

ESR spectra of Cyt P-450 indicates signals with *g* values of 1.92, 2.27, and 2.47. At higher sensitivity, small signals also were observed at 8.46 and 4.3. Similar ESR spectra can be produced by addition of the thiol compounds to myoglobins.

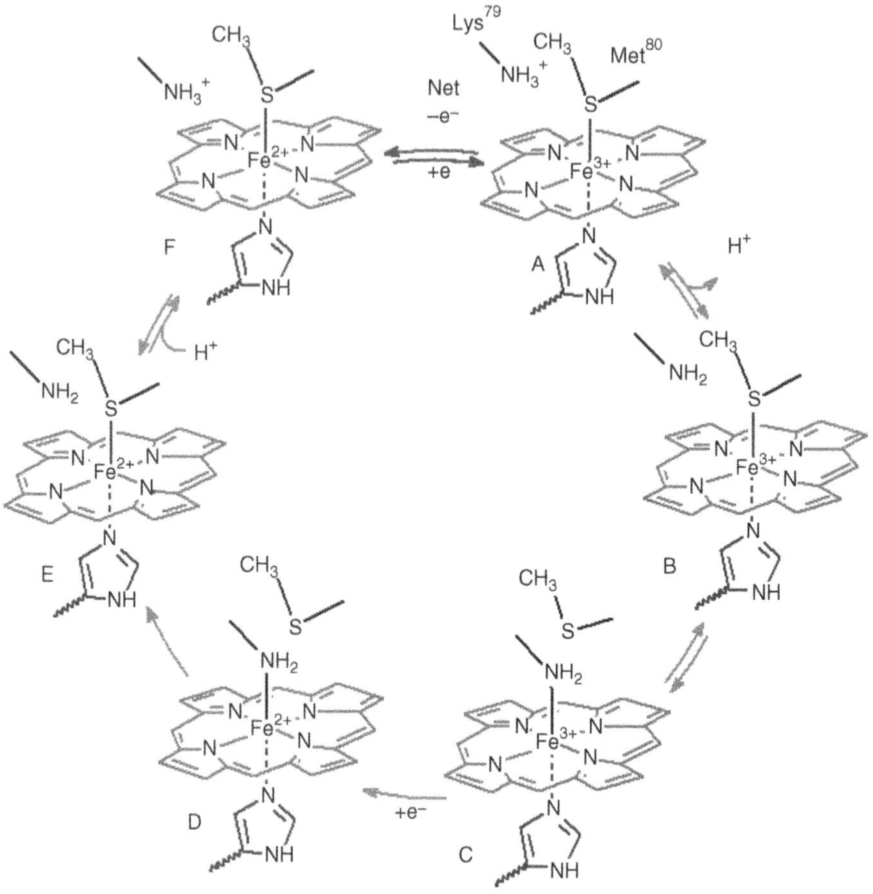

SCHEME 5-47 The redox cycle of cytochrome *c*. (Modified from Lambeth et al., 1973.)

After addition of camphor, Cyt. P-450 shows signals at 7.8, 3.9, and 1.8. These signals disappear when mercurials are added. How can ESR spectra be employed to probe the structural description of Cyt. P-450?

- The ESR spectrum of isolated Cyt. P-450 is characteristic of low-spin ferri-hemoproteins, with typical *g* values of 2.47, and 2.27, and 1.92.
- The high-spin resonance is most unusual with $g_x = 8.4$, $g_y = 4.3$, with 26% rhombicity.
- Addition of camphor converts Cyt. P-450 to the high-spin form of *g* values: 7.8, 3.9, and 1.8.
- These *g* values exhibit the following:
  - The smallest spread found in the P group indicates the thiolate ligation.
  - Addition of mercurials destroys these ESR spectra.

- ○ Similar ESR spectra can be produced by addition of the thiol compounds to myoglobins. Consequently, it has been concluded that cysteine functions as an axial ligand to the iron, thus causing the small g shifts in ESR.
- The ESR properties suggest that P-450 contains histidine and mercaptide as axial ligands.

Hyperfine Splitting

How can you explain the possible hyperfine splitting that might associate the binding of:

(i) **^{15}NO, and ^{14}NO with HRP and HRP contains either ^{57}Fe or proximal ^{14}N-imidazole.**

(ii) **NO in Fe(II)(NO)–tetraphenylporphine and in mixed-ligands hemoglobin derivatives, $\alpha NO^{II}\beta X^{II}$, $(Hb_\alpha{}^{II} (NO) Hb_\beta{}^{II}(X))$.**

- Nitric oxide is a 15-electron molecule: $\sigma_1^2(1S), \sigma_1^{*2}(1S), \sigma_2^2(2S), \sigma_2^{*2}(2S), \sigma_2^2(2p_x), \pi_2^2(2p_y), \pi_2^2(2p_z), \pi_2^{*1}(2p_y)$. Thus, NO is paramagnetic with net spin $\frac{1}{2}$.
- When NO is added to *ferrihemoproteins*, the product protein is *diamagnetic*:

$$Fe^{3+} + NO \quad \rightleftarrows \quad Fe^{2+}NO^+$$

- Nitric oxide also reacts with *ferrohemoproteins* and the product of this reaction is *paramagnetic* with $S = \frac{1}{2}$:

$$Fe^{2+}(\text{high spin}) + NO(S = \tfrac{1}{2}) \rightleftarrows Fe^{2+}NO(\text{low spin}, S = \tfrac{1}{2})$$

- The ESR of the nitrosyl product exhibits hyperfine structure at g_z, which arises from both the nitrogen of NO and the nitrogen of the proximal imidazole.
- The spectrum is centered at $g = 2$ and is anisotropic (Fig. 5-81). The extreme values of g_x and g_y being 2.08 and 1.95, respectively, g_z is relatively constant at 2.005 (see Yonetani and Yamamoto, 1973; Palmer, 1979).
  - ○ When ^{15}NO ($I = \frac{1}{2}$, $2I + 1 = 2$) is used as the external ligand and the proximal ^{14}N-imidazole ($I = 1$, $2I + 1 = 3$), (A in Fig. 5-81) HRP–^{15}NO, the hyperfine is clearly *a doublet of a triplet* (Fig. 5-81).
  - ○ When ^{15}NO ($I = \frac{1}{2}$, $2I + 1 = 2$) is used as the external ligand and the heme containing ^{57}Fe ($I = \frac{3}{2}$, $2I + 1 = 4$), (B) ^{57}HRP – ^{15}NO, the hyperfine is clearly *a doublet of a quartet.*
  - ○ When ^{14}NO ($I = 1$, $2I + 1 = 3$) is used as the external ligand and the proximal ^{14}N-imidazole ($I = 1$, $2I + 1 = 3$), (C) HRP – ^{14}NO, the hyperfine is clearly *a triplet of a triplet*, a nine-line spectra.

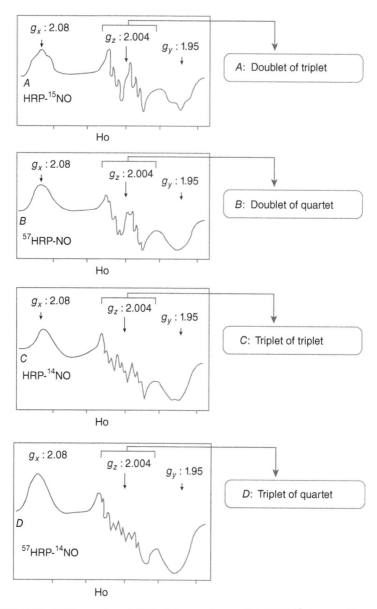

FIGURE 5-81 ESR spectra of HRP–NO complexes: (*A*) HRP–^{15}NO; (*B*) ^{57}HRP–NO; (*C*) HRP–^{14}NO; (*D*) ^{57}HRP–^{14}NO. (Data from Yonetani and Yamamoto, 1973, and G. Palmer, 1979. Reproduced by permission of Elsevier.)

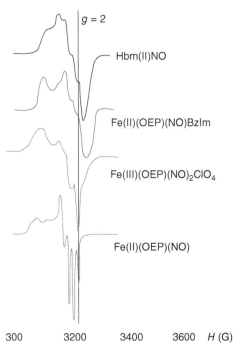

FIGURE 5-82   ESR spectra of Fe(II)(OEP)NO, Fe(III)(OEP)(NO)$_2$CLO$_4$, Fe(II)(OEP)NOB-zIm, and Hb$_m$(II)NO, OEP = octaethylporphyrine, BzIm = benezimidazole, Hb$_m$ = glycera dibranchiate, a monomeric hemoglobin (Stephanos, 1989).

- ○ When ^{14}NO ($I = 1$, $2I + 1 = 3$) is used as the external ligand and the heme containing ^{57}Fe ($I = \frac{3}{2}$, $2I + 1 = 4$), (D) ^{57}HRP–^{15}NO, the hyperfine is clearly *a triplet of a quartet.*
- Model compounds have been prepared from heme, NO, and a variety ρ-substituted pyridines. Nitrosyl derivatives in which the pyridine was ρ-substituted with an electron-releasing group exhibit the nine-line spectra (C, a triplet of a triplet).
- However, in derivatives with a ρ-substituent that is electron withdrawing, a three-line hyperfine is observed.
- The unpaired electron in nitroso ferroproteins is located in the d$_{z^2}$ of the iron atom. Thus the electron is able to interact with both axial ligands. Three very pronounced hyperfines have been attributed to five-coordinate species in which the proximal protein ligand is no longer attached to the metal ion. Such a spectrum is similar to that of octaethylporphine–Fe(II)NO (Fig. 5-82).
- Such variation might be achieved in the protein by varying the protonation of the N$_1$ status of the proximal histidine.
- A deprotonated histidine is electron releasing and vice versa.
- The ESR spectra of α_{NO}^{II} and β_{NO}^{II} chains of hemoglobin are different and the spectrum of HbNO is the sum of the contributions from isolated α and β subunits

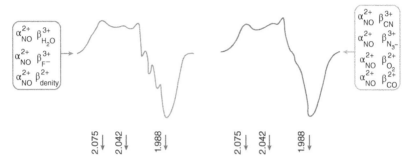

FIGURE 5-83   ESR spectra of α_{NO} hybrids. (Modified from G. Palmer, 1979. Reproduced by permission of Elsevier.)

- For the ESR spectra of mixed chains, $\alpha_{NO}^{II}\beta_X^{II}$, when the β chain is in the low-spin form (O_2, CO, CN^-, N_3^-), the ESR of α_{NO}^{II} was comparable to that of the free α chain (Fig. 5-83).
- When the β chain is in the high-spin state (F^-, H_2O, β^{II}-deoxy), hyperfine splitting is observed (Fig. 5-83).
- Therefore, the α-heme depends in some way on the spin state of the β-heme, and high-spin β-heme imposes an irrelevant confirmation on the α chain.

REFERENCES

A. W. Addison and J. J. Stephanos, *Biochemistry*, 25, 4104 (1986).

A. D. Allen and C. V. Senoff, *Chem. Commun.*, p. 621 (1965).

R. J. Angelici, "Stability of coordination compound," in *Inorganic Biochemistry*, G. L. Eichhorn Ed., Vol. 1, p. 106, Elsevier, New York (1973).

L. Banci, I. Bertini, A. Dikiy, D. H. Kastrau, C. Luchinat, and P. Sompornpisut, *Biochemistry*, 34, 206–219 (1995).

W. S. Beck, Vitamins and Mormons, 20. 413 (1968).

G. I. Berglund, G. H. Carlsson, A. T. Smith, H. Szoke, A. Henriksen, and J. Hajdu, *Nature*, 417, 463 (2002).

Yu. G. Borodko, , and A. E. Shilov, *Uspekhi Khim.*, 38, 761 (1969).

R. Breslow and L. N. Lukens, *Am. Chem. Soc.*, 235, 292 (1960).

H. Brintzinger, *Biochemistry*, 5, 3947 (1966).

J. H. Brock, "Transferrins," in *Metalloproteins, Part 2*, P. M. Harrison, Ed., Macmillan, Houndmills (1985).

M. Brunori, M. Coletta, and B. Giardina, in *Metalloproteins*, P. M. Harrison, Ed., Part 2, p. 271, Macmillan, Hong Kong (1985).

G. W. Bushnell, G. V. Louie, and G. D. Brayer, *J. Mol. Biol.* 214, 585–595 (1990).

C. E. Castro, "Routes of electron transfer," in *The Porphyrin*, D. Dolphin, Ed., Vol. 5, Academic, New York (1978).

C. E. Castro, E. W. Bartnicki, C. Robertson, R. Havlin, H. Davis, and T. Osborn, *Ann. N.Y. Acad. Sci.*, 244, 132 (1975).

C. E. Castro, C. Robertson, and H. F. Davis, *Bioorg. Chem.*, 3, 343 (1974).

T. Chatake, K. Kurihara, I. Tanaka, I. Tsyba, R. Bau, F. E. Jenney, M. W. Adams, and N. Niimura, *Acta Crystallogr., Sect. D*, 60, 1364–1373 (2004).

I. A. Cohen, C. Jung, and T. Governo, *J. Am. Chem. Soc.*, 94, 3003 (1972).

M. J. Coon and R. E. White, "Cytochrome P-450, a versatile catalyst in monooxygenation reaections," in *Metal Ions in Biology, Metal Ion Activation of Dioxygen*, T. G. Spiro, Ed., Vol. 2, p. 73, Wiley, New York (1980).

F. A. Cotton and J. Wilkinson in *Advanced Inorganic Chemistry a Comprehensive Text*, 4th Eds, John Wiley & Sons Inc., New York (1980).

P. A. Cox, in *The Elements*, Oxford University Press, Oxford (1989).

D. Dolphin, Z. Muljiani, K. Rousseau, D. C. Borg, J. Fajer, and R. H. Felton, *Ann. N. Y. Acad. Sci*, 206, 177 (1973).

J. Dupuy, A. Volbeda, P. Carpentier, C. Darnault, J. M. Moulis, and J. C. Fontecilla-Camps, *Structure*, 14, 129–139 (2006).

W. A. Eaton and J. Hofrichter, "Polarized absorption and linear dichroism spectroscopy of hemoglobin," in *Hemoglobins, Methods in Enzymology*, E. Antonini, L. Rossi-Bernardi, and E. Chiancone, Eds., Vol. 76, Academic, New York (1981).

B. E. Eckenroth, A. N. Steere, N. D. Chasteen, S. J. Everse, and A. B. Mason, *Proc. Natl. Acad. Sci. USA*, 108, 13089–13094 (2011).

R. W. Estabrook, J. Baron, J. Peterson, and Y. Ishimura, in *Biological Hydroxylation Mechanisms*, G. S. Boyd and R. M. S. Smelie, Eds., p. 159, *Biochem. Soc. Symp.* No. 34, Academic, New York (1972).

G. Fermi, M. F. Perutz, B. Shaanan, and R. Fourme, *J. Mol. Biol.*, 175, 159–174 (1984).

J. J. Fraústo da Silva and R. J. P. Williams, in *The Biological Chemistry of the Elements, the Inorganic Chemistry of Life*, 2nd ed., Oxford University Press, New York (2004).

P. Giastas, N. Pinotsis, G. Efthymiou, M. Wilmanns, P. Kyritsis, J.-M. Moulis, and I. M. Mavridis, *J. Biol. Inorg. Chem.*, 11, 445–458 (2006).

B. Giardnia and G. Amiconi, "Hemoglobin," in *Methods in Enzymology*, E. Antonini and L. Rossi Bernardi, Eds., Vol. 76, Academic, London (1981).

C. Greenwood, "Cytochrome c, and cytochrome c containing enzymes," in *Metalloproteins, Part 1*, P. M. Harrison, Ed., Macmillan, Houndmills (1985).

C. Greenwood, in *Metalloproteins*, P. M. Harrison, Ed., Part 1, p. 43, Macmillan, Houndmills (1985).

R. R. Grinstead, *J. Am. Chem. Soc.*, 82, 3472 (1960).

L. P. Hager, D. L. Doubek, R. M. Silverstein, J. H. Hargis, and J. C. Martin, *J. Am. Chem. Soc.*, 94, 4364 (1972).

G. A. Hamilton and J. P. Friedman, *J. Am. Chem. Soc.*, 85, 1008 (1963).

G. A. Hamilton, J. P. Friedman, and P. M. Campbell, *J. Am. Chem. Soc.*, 88, 5266 (1966).

G. A. Hamilton, J. W. Hanifin, Jr., and J. P. Friedman, *J. Am. Chem. Soc.*, 88, 5269 (1966).

G. A. Hamilton, *Adv. Enzymol.*, 33, 55 (1969).

H. A. Harbury and R. H. L. Marks, "Cytochromes b and c," in *Inorganic Biochemistry*, G. L. Eichhorn, Ed., Vol. 2, p. 908, Elsevier, Amsterdam (1973).

M. Haridas, B. F. Anderson, and E. N. Baker, *Acta Crystallogr.,* Sect. D, 51, 629–646 (1995).

R. W. Hardy, R. D. Holsten, E. K. Jackson, and R. C. Burns, *Plant Physiol.,*43, 1185 (1968).

R. W. F. Hardy, R. C. Burns, and G. W. Parshall, *Adv. Chem.,* 100, 219 (1971).

R. W. F. Hardy, R. C. Burns, and G. W. Parshall, "Bioinorganic chemistry of dinitrogen fixation," in *Inorganic Biochemistry,* G. L. Eichhorn, Ed., Vol. 2, Chapter 23, Elsevier, New York (1973).

O. Hayashi, M. Katagiri, and S. Rothberg, *J. Am. Chem. Soc.,* 77, 5450 (1955).

S. M. Heald, E. A. Stern, B. A. Bunker, E. M. Holt, and S. L. Holt, *J. Am. Chem. Soc.,* 101, 67 (1979).

H. A. O. Hill, D. R. Turner, and G. Pellizer, *Biochem. Biophys. Res. Commun.* 56, 739 (1974).

M. A. Holmes, I. Le Trong, S. Turley, L. C. Sieker, and R. E. Stenkamp, *J. Mol. Biol.,* 218, 583–593 (1991).

M. N. Hughes, in *Inorganic chemistry of Biological Processes,* 2nd ed., p. 252, Wiley, New York (1981).

G. B. Jameson, B. F. Anderson, G. E. Norris, D. H. Thomas, and E. N. Baker, *Acta Crystallogr.,* Sect. D, 54, 1319–1335 (1998).

H. Kameda, K. Hirabayashi, K. Wada, and K. Fukuyama, *PLOS ONE,* 6, e21947–e21947 (2011).

W. Kerler, *Z. Phys.,* 167, 194 (1962).

W. Kerler and W. Neuwirth, *Z. Phys.,* 167, 176 (1962).

T. P. Ko, M. K. Safo, F. N. Musayev, M. L. Di Salvo, C. Wang, S. H. Wu, and D. J. Abraham, *Acta Crystallogr.,* Sect. D, 56, 241–245 (2000).

N. Kunishima, K. Fukuyama, H. Matsubara, H. Hatanaka, Y. Shibano, and T. Amachi, *J. Mol. Biol.,* 235, 331–344 (1994).

D. M. Kurtz, Jr., D. F. Shriver, and I. M. Koltz, *Coord. Chem. Rev.,* 24, 145 (1977).

D. Lambeth, K. Cambell, R. Zand, and G. Palmer, *J. Biol. Chem.,* 248, 8130 (1973).

R. W. Lane, J. A. Ibers, R. B. Frankel, and R. H. Holm, *Proc. Natl. Acad. Sci. USA,* 72, 2868 (1975).

D. M. Lawson, P. J. Artymiuk, S. J. Yewdall, J. M. Smith, J. C. Livingstone, A. Treffry, A. Luzzago, S. Levi, P. Arosio, G. Cesareni, C. D. Thomas, W. V. Shaw, and P. M. Harrison, *Nature,* 349, 541–544 (1991).

S. J. Lloyd, H. Lauble, G. S. Prasad, and C. D. Stout, *Protein Sci.,* 8, 2655–2662 (1999).

D. J. Lowe, R. N. F. Thorneley, and B. E. Smith, in *Metalloproteins,* P. M. Harrison, Ed., Part 1, p. 207, Macmillan, Hong Kong (1985).

M. W. Makinen and W. A. Eaton, *Ann. N.Y. Acad. Sci.,* 206, 214 (1973).

M. M. Malley, *Ann. N.Y. Acad. Sci.,* 206, 268 (1973).

M. Maltempo, *J. Chem. Phys.,* 61, 2540 (1974).

A. E. Martell and M. M. Taqui Khan, "Metal ion catalysis of reactions of molecular oxygen," in *Inorganic Biochemistry,* G. L. Eichhorn, Ed., Vol. 2, Chapter 20, Elsevier (1973).

P. A. Mayes, "The citric acid cycle," in *Review of Physiological Chemistry,* H. A. Harper, V. W. Rodwell, and P. A. Mayes Eds., 16th ed., Lange Medical Publications, Los Altos, California (1977).

J. J. Mayerle, S. E. Dermark, B. V. DePamphilis, J. A. Ibers, and R. H. Holm, *J. Am. Chem. Soc.,* 97, 1032 (1975).

J. J. Mayerle, R. B. Frankel, R. H. Holm, J. A. Ibers, W. D. Phillips, and J. F. Weiher, *Proc. Nat. Acad. Sci.,* 70, 2429 (1973).

M. Mohammadi, J. Viger, P. Kumar, D. Barriault, J. T. Bolin, and M. Sylvestre, *J. Biol. Chem.*, 286, 27612 (2011).

R. I. Murray, M. T. Fisher, P. G. Debrunner and S. G. Sligar, "Structrue and chemistry of cytochrome P-450," in *Metalloproteins, Part 1*, P. M. Harrison, Ed., Chapter 5, p. 185, Macmillan, Houndmills (1985).

E. Ochiai, in *Bioinorganic Chemistry: Asurvey*, Elsevier, Amsterdam (2008).

D. H. Ohlendorf, A. M. Orville, and J. D. Lipscomb, *J. Mol. Biol.*, 244, 586–608 (1994).

M. Y. Okamura and I. M. Klotz, "Hemerythrin," in *Inorganic Biochemistry*, G. L. Eichhorn, Ed., Vol. 1, Chapter 11, p. 322, Elsevier (1973).

W. H. Orme-Johnson, in *Inorganic Biochemistry*, G. L. Eichhorn, Ed., Vol. 2, Chapter 22, Elsevier, New York (1973).

A. M. Orville, J. D. Lipscomb, and D. H. Ohlendorf, *Biochemistry*, 36, 10052–10066 (1997).

D. C. Owsley and G. K. Helmkamp, *J. Am. Chem. Soc.*, 89, 4558 (1967).

G. Palmer, in *The Porphyrins*, D. Dolphin, Ed., Vol. IV, pp. 313–353, Academic, New York (1979).

J. Peisach, W. E. Blumberg, and A. Adler, *Ann. N.Y. Acad. Sci.*, 206, 311 (1973).

S. E. Phillips, *J. Mol. Biol.*, 142, 531 (1980).

T. L. Poulos, B. C. Finzel, and A. J. Howard, *Biochemistry*, 25, 5314–5322 (1986).

L. Que, Jr., J. R. Anglin, M. A. Bobrik, A. Davison, and R. H. Holm, *J. Am. Chem. Soc.*, 96 (19), 6042 (1974).

K. N. Raymond, in *Bioinorganic Chemistry*, Vol. II, Advances in Chemistry Series 162, American Chemical Society, Washington, DC (1977).

V. Richard, G.G. Dodson, and Y. Mauguen, *J. Mol. Biol.*, 233, 270–274 (1993).

A. C. Rosenzweig, H. Brandstetter, D. A. Whittington, P. Nordlund, S. J. Lippard, and C. A. Frederick, *Proteins*, 29, 141–152 (1997).

B. Schmid, O. Einsle, H. J. Chiu, A. Willing, M. Yoshida, J. B. Howard, and D. C. Rees, *Biochemistry*, 41, 15557–15565 (2002).

B. Shaanan, *J. Mol. Biol.*, 171, 31 (1983).

A. E. Shilov and A. K. Shilova, *Kinet. Catalysis*, 10, 1163 (1970).

T. G. Spiro, in *Inorganic Biochemistry*, G. L. Eichhorn, Ed., Vol. 1, Chapter 17, Elsevier, New York (1973).

J. J. Stephanos, *"Hemo proteins: Reactivity and spectroscopy,"* Ph.D. Thesis, Drexel University, Philadelphia, PA (1989).

J. J. Stephanos and A. W. Addison, "Thermochromism of monomeric heme proteins," *J. Inorg. Biochem.*, 39, 351–369 (unpublished spectra) (1990).

K. R. Strand, S. Karlsen, M. Kolberg, A. K. Rohr, C. H. Gorbitz, and K. K. Andersson, *J. Biol. Chem.*, 279, 46794–46801 (2004).

K. Sugimoto, T. Senda, H. Aoshima, E. Masai, M. Fukuda, and Y. Mitsui, *Structure Fold. Des.*, 7, 953–965 (1999).

G. R. Schonbaum and F. Jajczay, "Selective Modification of Catalases", in *Hemes and Hemoproteins*, B. Chance, R. W. Estabrook, and T. Yonetani, Academic Press, New York (1966).

E. E. Tamelen, G. Boche, and R. Greeley, *J. Am. Chem. Soc.*, 90, 1677 (1968).

T. Tanabe and S. Sugano, *J. Phys. Soc. Japan*, 9, 753 (1954).

R. K. Thauer and P. Schönheit, in *Iron-Sulfur Proteins*, T. G. Spiro, Ed., Chapter 8, Wiley, New York (1982).

D. Turner, M. Whitfield, and A. G. Dickson, *Geochim. Acta* 45, 855 (1981).

J. Uppenberg, F. Lindqvist, C. Svensson, B. Ek-Rylander, and G. Andersson, *J. Mol. Biol.*, 290, 201–211 (1999).

J. S. Valentine, "Dioxygen reactions," in *Bioinorganic Chemistry*, I. Bertini, H. B. Gray, S. J. Lippard, and J. S. Valentine, Eds., p. 283, University Science Books, Mill Vally, CA (1994).

P. Venkateswara Rao and R. H. Holm, *Chem. Rev.*, 104, 527 (2004).

J. Vojtechovsky, K. Chu, J. Berendzen, R. M. Sweet, and I. Schlichting, *Biophys. J.*, 77, 2153 (1999).

A. Volbeda, E. Garcin, C. Piras, A. L. De Lacey, V. M. Fernandez, E. C. Hatchikian, M. Frey, and J. C. Fontecilla-Camps, *J. Am. Chem. Soc.* 118, 12989 (1996).

R. S. Wade and C. E. Castro, *J. Am. Chem. Soc.*, 95, 226 (1973).

J. Wally P. J. Halbrooks, C. Vonrhein, M. A. Rould, S. J. Everse, A. B. Mason, and S. K. Buchanan, *J. Biol. Chem.*, 281, 24934–24944 (2006).

M. E. Winfield, *Rev. Pure Appl. Chem.*, 5, 217 (1955).

J. M. Wood, in *Metal Ions in Biology*, T. G. Spiro, Ed., Vol. 2, Wiley, New York (1980).

N. Yang, H. Zhang, M. Wang, Q. Hao, and H. Sun, *Sci. Rep.* 2: 999–999.

T. Yonetani, H. Schleyer, and B. Chance, "The chemical nature of complex ES of cytochrome c oxidase," in *Hemes and Hemoproteins*, B. Chance and R. W. Estabrook Eds., p. 293, Academic, New York (1966).

T. Yonetani and H. Yamamoto, in *Oxidases and Related Enzymes Systems*, T. E. King, H. S. Mason, and M. Morrison, Eds., p. 279, University Park Press, Baltimore, MD (1973).

SUGGESTIONS FOR FURTHER READING

Electronic Spectra

1. A. B. P. Lever, in *Inorganic Electronic Spectroscopy*, American Elsevier, New York (1984).

2. C. N. R. Rao and J. R. Ferraro, in *Spectroscopy in Inorganic Chemistry*, Academic, New York, Vol. I (1970), Vol. II (1971).

3. D. C. Harris and M. D. Bertolucci, in *Symmetry and Spectroscopy, An Introduction to Vibrational and Electronic Spectroscopy*, Oxford University Press, New York (1978).

4. F. A. Cotton, in *Chemical Applications of Group Theory*, 2nd Ed., Wiley Eastern Limited, New Delhi (1970).

5. R. D. Gillard, *Chem. in Brit.*, 1(1967).

6. J. E. Crooks, in *The Spectrum in Chemistry*, p. 155, Academic, London, 1978.

7. P. Crabbe, in *ORD and CD in Chemistry*, Academic, New York (1972).

Mössbauer Spectra

8. R. F. Gould, Ed., *Advances in Chemistry*, No. 68, "The Mössbauer effect and its application to chemistry," American Chemical Society, Washington, DC (1967).

9. R. S. Drago, *Physical Methods of Chemistry*, Saunders, Philadelphia (1977).

ESR Spectroscopy

10. A. Abragam, and B. Bleany, in *Electron Paramagnetic Resonance of Transition Ions*, Oxford University Press, New York (1970).

11. N. M. Atherton, in *Electron Spin Resonance*, Ellis Horwood, Chichester (1973).

12. R. S. Alger, in *Electron Spin Resonance*, Wiley Interscience, New York (1968).

Iron Bioavailability

13. B. E. Douglas, D. H. McDaniel, and J. J. Alexander, in *Concept and Models of Inorganic Chemistry*, 3rd ed., p. 890, Wiley, New York (1994).

Siderophores

14. B. F. Matzanke, G. Muller, and K. N. Raymond, in *Iron Carriers Iron Proteins*, T. M. Loehr, Ed., pp. 1–121, VCH Publishers, New York (1989).

15. V. Braun and M. Braun, Active transport of iron and siderophores antibiotics, *Curr. Opin. Microbiol.* 5, 194–201 (2002).

16. D. J. Kosman, *Mol. Microbiol.*, 47, 1185–1197 (2003).

17. P. Aisen, C. Enns, and M. Wessling-Resnik, *Int. J. Biochem. Cell Biol.*, 33, 940–959 (2001).

18. J. B. Neilands, Ed., *Microbial Iron Metabolism*, Academic, New York (1974).

19. K. Burger, I. T. Millar, and D. W. Allen, in *Coordination Chemistry: Experimental Methods*, CRC Press, Boca Raton, FL (1973).

20. T. Emery, *Adv. Enzymol.*, 33, 135 (1971).

21. J. B. Neilands, *Struct. Bonding*, 1, 59 (1966).

Ferritin

22. P. Aisen, *Int. J. Biochem. Cell Biol.*, 36, 2137–2143 (2001).

23. K. Zeth, S. Offermann, L. O. Essen, and D. Oesterhelt, *Proc. Natl. Acad. Sci. USA*, 101, 13780–13785 (2004).

24. E. C. Theil, *Adv. Inorg. Biochem.*, 5, 1 (1983).

25. E. C. Theil and P. Aisen, in *Iron Transport in Microbes, Plants, and Animals*, D. van der Helm, J. Neilands, and G. Winkelmann, Eds., p. 421, VCH Publishers, Wien-heim, Federal Republic of Germany (1987).

26. E. C. Theil, *Ann. Rev. Biochem.*, 56, 289 (1987).

27. E. C. Theil, *Adv. Enzymol.*, 63, 421 (1990).

28. G. C. Ford, P. M. Harrison, D. W. Rice, J. M. Smith, A Treffry, J. L. White, and J. Yariv, *Philos. Trans. Roy. Soc. Lond. B.*, 304, 551–565 (1984).

29. D. C. Harris and P. Aisen, in *Iron Carriers and Iron Proteins*, T. M. Loehr, Ed., pp. 239–352, VCH Publishers, New York (1989).

30. D. C. Harris and T. M. Lilley, in *Iron Carriers and Iron Proteins*, T. M. Loehr, Ed., pp. 123–238, VCH Publishers, New York (1989).

31. P. M. Harrison, and T. G. Hoy, in *Inorganic Biochemistry*, G. L. Eichhorn, Ed., Vol. 1, Chapter 8, Elsevier, New York (1973).

32. R. R. Crichton et al., in *Proteins of Iron Metabolism*, E. B. Brown, P. Aisen, J. Feilding, and R. R. Crichton, Eds., Grune and Statton, New York (1977).

33. P. M. Harrison, *Semin. Haematol.*, 14, 55 (1977).

34. R. R. Crichton, "Structure and function of ferritin," *Angew. Chem. Int. Ed.*, 12, 57 (1973).

Transferrin

35. P. G. Thakurta, D. Choudhury, R. Dasgupta, and J. K. Dattagupta, *Biochem. Biophys. Res. Commun.*, 316, 1124–1131 (2004).

36. P. G. Thakurta, D. Choudhury, R. Dasgupta, and J. K. Dattagupta, *Acta. Crystallogr., Sect. D*, 59, 1773–1781 (2003).

37. P. Aisen, *Int. J. Biochem. Cell. Biol.*, 36, 2137–2143 (2004).

38. P. Aisen, C. Enns, and M. Wessling-Resnick, *Int. J. Biochem. Cell. Biol.* 33, 940–959 (2001).

39. N. D. Chasteen, *Adv. Inorg. Biochem.*, 5, 201 (1983).

40. P. Aisen and I. Listowsky, *Annu. Rev. Biochem.*, 49, 357 (1980).

41. E. N. Baker, B. F. Anderson, and H. M. Baker, *Int. J. Biol. Macromol.*, 13, 122 (1991).

42. S. Bailey, R. W. Evans, R. C. Garratt, B. Gorinsky, S. Hasnain, C. Horsburgh, H. Jhoti, P. F. Lindley, and A. Mydin, *Biochemistry*, 27, 5804–5812 (1988).

43. P. Aisen and I. Listowsky, "Iron transport and storage proteins," *Ann. Rev. Biochem.*, 49, 357 (1980).

44. J. H. Brock, in *Metalloproteins*, P. M. Harrison, Ed., Part 2, p. 183, Macmillan, Hong Kong (1985).

45. P. Aisen, in *Inorganic Biochemistry*, G. L. Eichhorn, Ed., Vol. 1, Chapter 9, Elsevier, New York (1973).

Dioxygenase

46. G. A. Hamilton, "Chemical models and mechanism for oxygenases (R)," in *Molecular Mechanisms of Oxygen Activation (B)*, O. Hayaishi, Ed., pp. 405–451, Academic, New York (1974).

47. H. Taube, *J. Gen. Physiol.*, 49, Part 2, 29–52 (1965).

48. L. Que, Jr., "The catechol dioxygenases (R)," in *Iron Carriers and Ion Proteins*, T. M. Loehr, Ed., pp. 467–524, VCH Publishers, New York (1989).

49. L. Que, Jr., *J. Chem. Ed.*, 62, 938–943 (1985).

50. J. W. Whittaker J. D. Lipscomb, T. A. Kent, and E. Miinck, *J. Biol. Chem.* 259, 4466–4475 (1984).

51. Y. Tomimatsu, S. Kint, and J. R. Scherer, *Biochemistry*, 15, 4918–4924 (1976).

52. D. H. Ohlendorf, J. D. Lipscomb, and P. C. Weber, *Nature*, 336, 403–405 (1988).

53. D. D. Cox and L. Que, Jr., *J. Am. Chem. Soc.*, 110, 8085–8092 (1988).

54. Y. Sawaki and C. S. Foote, *J. Am. Chem. Soc.*, 105, 5035–5040 (1983).

55. A. M. Orville, N. Elango, J. D. Lipscomb, and D. H. Ohlendorf, *Biochemistry*, 36, 10039–10051 (1997).

56. L. Que, Jr., and R. Y. N. Ho, *Chem. Rev.*, 96, 2607–2624 (1996).

57. T. D. Bugg, *Curr. Opin. Chem. Biol.*, 5, 550–555 (2001).

58. E. G. Kovaleva and J. D. Lipscomb, *Science*, 316, 453–457 (2007).

59. R. Yoshida, K. Hori, M. Fujiwara, Y. Saeki, H. kagamiyama, and M. Nozaki, *Biochemistry*, 15, 4048 (1976).

60. W. E. Blumberg and Peisach, *Ann. N.Y. Acad. Sci.*, 222, 539 (1973).

61. L. Que. J. D. Lipscomb, R. Zimmerman, E. Munck, N. R. Orme–Johnson, and W. H. Orme–Johnson, *Biochim. Biophys. Acta.*, 452, 320 (1976).

62. R. T. Ruettinger, G. R. Griffith, and M. J. Coon, *Arch. Biochem. Biophys.*, 183, 528 (1977).

Iron Sulfur Proteins

63. A. J. Thomson, in *Metalloproteins*, P. M. Harrison, Ed., Part 1, p. 79, Macmillan, Hong Kong (1985).

64. N. M. Atherton, K. Garbett, R. D. Gillard, R. Mason, S. J. Mayhew, J. L. Peel, and J. E. Estangroom, *Nature*, 212, 590 (1966).

65. B. Schmid, O. Einsle, H. J. Chiu, A. Willing, M. Yoshida, J. B. Howard, and D. C. Rees, *Biochemistry*, 41, 15557–15565 (2002).

66. R. H. Holm, *Chem. Rev.*, 10, 455 (1981).

67. R. H. Holm, *Endeavour*, 34, 38 (1975).

68. R. H. Holm, *Accts. Chem. Res.*, 10, 427 (1977).

69. R. Mason and J. A. Zubieta, "Iron–sulfur proteins: Structural chemistry of their chromophores and related systems," *Angew. Chem. Int. Ed.* 12, 390 (1973).

70. R. H. Holm, "Identification of active sites in iron–sulfur proteins," in *Biological Aspects of Inorganic Chemistry*, A. W. Addison, W. R. Cullen, D. Dolphin and B. R. James, Eds., Wiley, New York (1976).

71. G. Palmer, in *The Enzymes*, P. D. Boyer, Ed., Vol. 12, p. 1, Academic, New York (1975).

72. W. Lovenberg, Ed., in *Iron–Sulphur Proteins*, Vols. I, II, and III, Academic, New York (1973, 1977).

73. W. V. Sweeney and J. C. Rabinowitz, "Proteins containing 4Fe–4S clusters: An overview," *Ann. Rev. Biochem.* 49, 139 (1980).

74. E. T. Adman, J. C. Sieker, L. H. Jensen, M. Bruschi, and J. Le Gall, *J. Mol. Biol.*, 112, 113 (1977).

75. W. Lovenberg and W. M. Williams, *Biochemistry*, 8, 141 (1968).

76. W. D. Phillips, M. Poe, J. F. Weiker, C. C. McDonald, and W. Lovenberg, *Nature*, 227, 574 (1970).

77. W. A. Eaton and W. Lovenberg, *J. Am. Chem. Soc.*, 92, 7195 (1970).

78. T. Kimura, *Structure and Bonding*, Vol. 5, Springer–Verlag, Berlin (1999).

79. H. Beinert, R. H. Holm, and E. Münck, *Science*, 277, 653–659 (1997).

80. H. Beinert, M. C. Kennedy, and D. D. Stout, *Chem. Rev.*, 96, 2335–2373 (1996).

81. L. Noodleman, C. Y. Peng, D. A. Case, and J.-M. Mousesca, *Coord. Cem. Rev.*, 144, 199–244 (1995).

82. C. Lieutayd, W. Nitschke, A. Vermeglio, P. Parot, and B. Schoepp-Cothenet, *Biochim. Biophys. Acta*, 1557, 83–90 (2003).

83. A. Dey, T. Glaser, M. M.-J. Couture, I. D. Eltis, R. H. Holm, B. Hedman, K. O. Hodgson, and E. I. Solomon, *J. Am. Chem. Soc.*, 126, 8320–8324 (2004).

84. D. H. Flint, and R. M. Allen, *Chem. Rev.*, 96, 2315–2334 (1996).

Nitrogenase

85. R. R. Eady, *Chem. Rev.*, 96, 3013–3030 (1996).

86. M. K. Chan, J. Kim, and D. C. Rees, *Science*, 260, 792 (1993).

87. J. R. Postgate, Ed., in *The Chemistry and Biochemistry of Nitrogen Fixation*, Plenum, New York (1971).

88. J. Chat, J. R. Dilworth and R. L. Richards, "Recent advances in the chemistry of nitrogen fixation," *Chem. Rev.*, 78, 589 (1978).

89. L. E. Mortenson and R. N. F. Thorneley, "Structure and function of nitrogenase," *Ann. Rev. Biochem.*, 48, 387 (1979).

90. R. W. F. Hardy, F. Bottomley, and R. C. Burns, Eds., in *A Treatise on Dinitrogen Fixation, Section I and II: Inorganic and Physical Chemistry and Biochemistry*, Wiley, New York (1977).

91. W. E. Newton and W. H. Orme–Johnson, Eds., in *Nitrogen Fixation*, Vols. I and II, University Park Press, Baltimore, MD (1980).

92. A. H. Gibson and W. E. Newton, Eds., *Current Perspectives in Nitrogen Fixation*, Australian Academy of Science, Canberra (1981).

93. A. D. Allen and F. Bottomeley, Inorganic nitrogen fixation, nitrogen compounds of the transition metals, *Acc. Chem. Res.*, 1, 360 (1968).

94. A. D. Allen and F. Bottomely, *Acc. Chem. Res.*, 1, 360 (1968).

95. L. E. Mortenson, in *The Bacteria*, I. C. Gunsalus and R. Y. Stanier, Eds., Vol. 3, p. 718, Academic, New York (1962).

Binuclear Fe Proteins

96. L. Que, Jr., in *Bioinorganic Catalysis*, J. Reedijk, Ed., pp. 347–393, Marcel Dekker, New York (1993).

97. J. Sanders-Loehr, in *Iron Carriers and Iron Proteins*, T. M. Loehr, Ed., Vol. 5, pp. 373–466, VCH Publishers, New York (1989).

98. M. N. Hughes, in *Inorganic Chemistry of Biological Processes*, 2nd ed., Wiley, New York (1980).

99. W. Haase, M. Dietrich, H. Witzel, R. Löcke, and B. Krebs, *Inorg. Chim. Acta*, 252, 13 (1996).

100. R. E. Stenkamp, *Chem. Rev.*, 94, 715 (1994).

101. A. C. Rozenzweig, C. A. Frederick, S. J. Lippard, and P. Nordlund, *Nature*, 366, 537 (1993).

102. B. Krebs and N. Sträter, *Angew. Chem. Int. Ed. Engl.*, 33, 841 (1994).

103. A. L. Feig and S. J. Lippard, *Chem. Rev.*, 94, 759 (1994).

104. B. J. Waller and J. D. Lipscomb, *Chem. Rev.*, 90, 1447 (1996).

105. D. M. Kurtz, Jr., *Chem. Rev.*, 90, 585 (1990).

106. J. B. Vincent, G. L. Oliver-Lilley, and B. A. Averill, *Chem. Rev.*, 90, 1447 (1990).

107. L. Que, Jr., and Y. Dong, *Acc. Chem. Res.*, 29, 190 (1996).

108. A. Uehara and M. Suzuki, *J. Am. Chem. Soc.*, 118, 701 (1996).

109. H. B. Gray and H. J. Schurar, "Electronic structure of iron complexes," in *Inorganic Biochemistry*, G. L. Eichhorn, Ed., Vol. 1, Elsevier, New York (1973).

110. M. Brunori, M. Coletta, and B. Giardina, in *Metalloproteins*, P. M. Harrison, Ed., Part. 2, p. 263, Macmillan, Hong Kong (1985).

111. M. Y. Okamura and I. M. klotz, in *Inorganic Biochemistry*, G. L. Eichhorn, Ed., Vol. 1, Chapter 11, Elsevier, New York (1973).

Myoglobin and Hemoglobin

112. M. F. Perutz, *Q. Rev. Biophys.*, 22, 139 (1989).

113. M. F. Perutz, *Annu. Rev. Physiol.*, 52, 1 (1990).

114. B.A. Springer, S. G. Sligar, J. S. Olson and G. N. Phillips, Jr., *Chem. Rev.*, 94, 699–714 (1994).

115. M. N. Hughes, in *Inorganic Chemistry of Biological Processes*, 2nd ed., Wiley, New York (1980).

116. J. M. Rifkind, in *Inorganic Biochemistry*, G. L. Eichhorn, Ed., Vol. 2, Chapter 25, Elsevier, New York (1973).

117. E. Antonini and M. Brunori, in *Hemoglobin and Myoglobin in Their Reactions with Ligands*, North Holland Publishing, Amsterdam (1971).

118. A. Kleinzeller and G. F. Springer, in *Molecular Biology Biochemistry and Biophysics*, Vol. 15, Springer-Verlag, New York (1974).

119. M. F. Perutz,"A review of the structure of hemoglobin and Perutz theory of cooperativity", *Brit. Med. Bull.*, 32, 193 (1976).

120. R. E. Dickerson and I. Geiss, in *The Structure and Action of Protein*, Harper & Row, New York (1969).

121. K. M. Smith, Ed., in *Porphyrins and Metallophyrins*, Elsevier, Amesterdam (1975).

122. M. Brunori, M. Coletta, and B. Giardina, in *Metalloproteins*, P. M. Harrison, Ed., Part 2, p. 263, Macmillan, Hong Kong (1985).

123. K. Imai, in *Allosteric Effect in Haemoglobin*, University Press, Cambridg (1982).

124. A. W. Addison and J. J. Stephanos, "Nitrosyliron(III) hemoglobin: Autoreduction and spectroscopy," *Biochemistry*, 25, 4104–4113 (1986).

125. A. W. Addison and J. J. Stephanos, "Nitrosyliron(III) hemoglobin: Autoreduction and spectroscopy," *Rev. Port. Quim.*, 27, 201–202 (1985).

126. A. W. Addison and J. J. Stephanos, "Spin and other equilibria of monomeric heme proteins derivatives," *Recueil des Travaux Chimiques des Pay-Bas*, 106, 324 (1987).

127. J. J. Stephanos, and A. W. Addison, "Spectroscopic and kinetic aspects of *Elephas Maximus* hemoglobin," *Eur. J. Biochem.*, 189, 185–191 (1990).

128. T. J. DiFeo, A. W. Addison, and J. J. Stephanos, "Kinetic and spectroscopic studies of hemoglobin and myoglobin from *Urechis-Caupo*-distal residue," *Biochem. J.*, 269, 739–747 (1990).

129. J. J. Stephanos, T. J. DiFeo, and A. W. Addison, "Cooperative binding of xanthine drugs and s-donors to monomeric *Glycera* hemoglobin," *Proc. of IEEE*, 12, 1677–1678 (1990).

130. J. J. Stephanos, "Drug–protein interaction: Two-site binding of heterocyclic and anionic ligands to monomeric hemoglobin," *Inorg. Biochem.*, 62, 155–169 (1996).

131. J. J. Stephanos, S. A. Farina, and A. W. Addison, "Iron ligand recognition by monomeric hemoglobin," *Biochem. Biophys. Acta*, 1295 (2), 209–221 (1998).

132. J. J. Stephanos, S. A. Farina, and A. W. Addison, "Triangular kinetic schemes applied to the stability of a heme-globin complex," *J. Inorg. Biochem.*, 66, 83–98 (1996).

133. J. J. Stephanos, L. M. Jackson, and A. W. Addison, "Copper (II) Schiff-base complexes and apoglobin stability," *J. Inorg. Biochem.*, 73, 137–144 (1999).

Synthetic Models

134. M. Momenteau and C. A. Reed, *Chem. Rev.*, 94, 659 (1994).

135. E. Tsuchida, T. Komatsu, E. Hasegawa, and H. Nishide, *J. Chem. Soc. Dalton Trans.*, 9, 2713 (1990).

136. E. Oldfield, H. C. Lee, C. Coretsopoulos, F. Adebodum, K. D. Park, S. Yang, J. Chung, and B. Phillips, *J. Am. Chem. Soc.*, 113, 8680 (1991).

137. J. P. Collman, X. Zhang, K. Wong, and J. I. Brauman, *J. Am. Chem. Soc.*, 116, 6245 (1994).

138. F. Basolo, B. M. Hoffman, and J. A. Ibers, "Synthetic oxygen carriers of biological interest," *Acc. Chem. Res.*, 8, 384 (1975).

139. J. P. Collman, "Synthetic models for oxygen binding hemoproteins," *Acc. Chem. Res.*, 10, 265 (1977).

140. G. McLendon and A. E. Martell, *Coord. Chem. Rev.*, 19, 1 (1977).

141. J. P. Collman, R. R. Gagne, C. A. Reed, T. R. Hulbert, G. Lang, and W. T. Rorbinson, *J. Am. Chem. Soc.*, 97, 1427 (1975).

142. J. P. Collman and K. S. Suslick, *Pure Appl. Chem.*, 50, 951 (1978).

143. J. Almog. J. E. Baldwin, R. L. Dyer, and M. Peters, *J. Am. Chem. Soc.*, 97, 226 (1975); J. Almog, J. E. Baldwin, and J. Huff, *J. Am. Chem. Soc.*, 97, 227 (1975).

144. J. C. Stevens, P. J. Jackson, W. P. Schammel, G. C. Christoph and D. H. Bush, *J. Am. Chem. Soc.*, 102, 3283 (1980).

145. J. H. Wang, *J. Am. Chem. Soc.*, 80, 3168 (1958).

146. A. G. Sykes, "Function properties of the biological oxygen carriers," *Adv. Inorg. Bioinorg. Mech.*, 2, 121 (1982).

Cytochrome *c*

147. H. A. Harbury and R. H. L. Marks, in *Inorganic Biochemistry*, G. L. Eichhorn, Ed., Vol. 2, Chapter 26, Elsevier, New York (1973).

148. E. Margolaish and A. Schejter, *Adv. Protein Chem.*, 21, 113 (1966).

149. R. E. Dickerson and R. Timkovich, in *The Enzymes*, P. D. Boyer, Ed., Vol. XI, p. 397, Academic Press, Inc., New York (1975).

150. F. R. Salemme, *Ann. Rev. Biochem.*, 46, 299 (1977).

151. E. Margoliash, *Adv. Chem. Phys.*, 29, 191 (1977).

152. D. Dolphin and R. H. Felton, *Acc. Chem. Res.*, 7, 26 (1974).

153. D. Dolphin, Z. Muljiani, K. Rousseau, D. C. Borg, J. Fajer, and R. H. Felton, *Ann. N.Y. Acad. Sci.*, 206, 177 (1973).

154. G. M. Brown, F. R. Hopf, J. A. Ferguson, T. J. Meyer, and D. G. Whtten, *J. Am. Chem. Soc.*, 95, 5930 (1973).

155. J. Fajer, D. C. Borg, A. Forman, R. H. Felton, L. Vegh, and D. Dolphin, *Ann. N. Y. Acad. Sci.*, 206, 177 (1973).

156. J. H. Fuhrhop and D. Maurzerall, *J. Am. Chem. Soc.*, 91, 4174 (1969).

157. G. L. Closs and I. E Closs, *J. Am. Chem. Soc.*, 85, 818 (1963).

158. D. Lexa, M. Momenteal, and J. Mispelter, *Biochim. Biophys. Acta*, 338, 151 (1974).

Peroxidases and Catalases

159. A. Decker and E. I. Solomon, *Curr. Opin. Chem.*, 9, 152–163 (2005).

160. D. Dolphin and R. H. Felton, *Acc. Chem. Res.*, 7, 26–32 (1974).

161. M. Sivaraja, D. B. Goodwin, M. Smith, and B. M. Hoffman, *Science*, 245, 738–740 (1989).

162. J. Everse, K. E. Everse, and M. B. Grisham, Eds., in *Peroxidases in Chemistry and Biology*, Vols. 1 and 2, CRC Press, Boca Raton, FL (1991).

163. B. Meunier, *Chem. Rev.*, 92, 1411 (1992).

164. D. Ostovic and T. Bruice, *Acc. Chem. Res.*, 25, 314 (1992).

165. M. Momenteau and C. A. Reed, *Chem. Rev.*, 94, 659 (1994).

166. M. J. Coon, R. E. White, "Cytochrome P-450, a versatile catalyst in monooxygenation reaections," in *Metal Ions in Biology, Metal Ion Activation of Dioxygen*, T. G. Spiro, Ed., Vol. 2, p. 73, Wiley, New York (1980).

167. B. C. Saunders, in *Inorganic Biochemistry*, G. L. Eichhorn, Ed., Vol. 2, Chapter 28, Elsevier, New York (1973).

168. H. B. Dunford and J. S. Stillman, *Coord. Chem Revs*, 19, 187 (1976).

169. M.-Y. R. Wang, B. M. Hoffman, and P. F. Hollenberg, *J. Biol. Chem.*, 252, 6268 (1977).

170. I. Morishima and S. Ogawa, *Biochemistry*, 17, 4384 (1978).

171. J. H. Wang, *J. Am. Chem. Soc.*, 77, 822, 4715 (1955).

172. E. Steiner and H. B. Dunford, *Eur. J. Biochem.*, 82, 543 (1978).

173. G. R. Schonbaum and F. Jajczay, "Selective modification of catalase" in *Hemes and Hemoproteins*, B. Chance and R. W. Estabrook, Eds., p. 328, Academic, New York (1966).

Cytochrome P-450

174. M. Sono, M. P. Roach E. D. Coulter, and J. H. Dawson, *Chem. Rev.* 96, 2841–2887 (1996).

175. M. J. Gunter and P. Turner, *Coord. Chem. Rev.*, 108, 115 (1991).

176. N. Ueyama, N. Nishikawa, Y. Yamada, T.-aki Okamura, and A. Nakamura, *J. Am. Chem. Soc.*, 118, 12826 (1996).

177. R. Sato and T. Omura, Eds., in *Cytochrome p-450*, Academic, New York (1978).

178. R. E. White and M. J. Coon, *Ann. Rev. Biochem.*, 49, 315 (1980).

179. L. S. Alexander and H. M. Goff, "Chemicals, cancer and cytochrome p-450," *J. Chem. Ed.* 59, 179 (1980).

180. J. P. Collman and S. E. Groh, " 'Mercaptan-tail' porphyrins: Synthetic analogues for the active site of cytochrome p-450," *J. Am. Chem. Soc.*, 104, 1391 (1982).

181. M. Sono, M. P. Roach, E. D. Coulter, and J. H. Dawson, *Chem. Rev.*, 96(7), 2841 (1996).

182. H. Kon, *Biochem. Biophys. Acta*, 379, 105 (1975).

183. Y. Henry and R. Banerjee, *J. Mol. Biol.*, 73, 469 (1973).

Electronic Spectra of Hemoproteins

184. A. Kleinzeller, G. F. Springer, and H. G. Wittmann, in *Molecular Biology Biochemistry and Biophysics*, p. 58, Springer-Verlag, Berlin Heidelberg New York (1974).

185. J. D. Mahoney, S. C. Ross, and J. Wyman, *J. Biol. Chem.*, 187, 393 (1977).

ESR of Hemoproteins

186. G. Palmer, in *Iron Porphyrin, Part II*, A. B. P. Lever and H. B. Gray, Eds., Addison-Wesley, Reading, MA (1983).

187. M. Kotani, *Prog. Theor. Phys. Suppl.*, 17, 4–13 (1961).

188. M. Weissbluth, *Hemoglobin: Coopertivity and Electronic Processes*, Springer, Berlin and New York (1973).

189. C. P. S. Tylor, *Biochem. Biophys. Acta*, 491, 137–149 (1977).

190. T. H. Bohan, *J. Magen. Res.*, 26, 109–118 (1977).

191. W. E. Blumberg and J. Peisach, in *Probes of Structure and Function of Macromolecules and Membranes*, B. Chance, T. Yonetani, and A. S. Mildvan, Eds., Vol. 2, pp. 215–229, Academic, New York (1971).

192. M. Maltempo, T. H. Moss, and M. Cuanovich, *Quart. Rev. Biophys.*, 9, 181–215 (1970).

193. G. Feher, in *Electron Paramagnetic Resonance with Applications to Selected Problems in Biology*, Gordon & Breach, New York (1970).

194. J. Peisach, W. E. Blumberg, E. T. Lode, and M. Coon, *J. Biol. Chem.*, 246, 5877 (1971).

195. T. Iizuka and T. Yonetani, *Adv. Biophys.*, 1, 155–178 (1970).

196. P. R. Weber, R. G. Bartsch, M. A. Cuasanovich, R. C. Hamlin, A. Howard, S. R. Jordon, M. D. Kamen, T. E. Meyer, D. W. Weatherford, N. Xuong, and F. R. Salemme, *Nature*, 286, 302–304 (1980).

197. C. A. Reed, T. Mashiko, S. P. Bentley, M. E. Kastner, W. R. Scheidt, K. Spartalian, and G. Lang, *J. Am. Chem. Soc.*, 101, 2948–2958 (1979).

198. Q. H. Gibson, *Prog. Biophys. Biophys. Chem.*, 9, 1–53 (1960).

199. Q. H. Gibson, C. Greenwood, D. C. Wharton, and G. J. Palmer, *J. Biol. Chem.*, 240, 888–894 (1965).

200. R. H. Morse and S. I. Chan, *J. Biol. Chem.*, 255, 7876–7882 (1980).

201. P. Reisberg, J. S. Olson, and G. J. Palmer, *J. Biol. Chem.*, 251, 4379–4383 (1976).

202. R. Hille, G. Palmer, and J. S. Olson, *J. Biol. Chem.*, 252, 403–405 (1977).

203. C. D. Coryell, L. Pauling, and R. W. Dodson, *J. Phys. Chem.*, 43, 825–829 (1939).

204. J. Kon and N. Nataoka, *Biochemistry*, 8, 4757–4762 (1969).

205. J. W. Chien, *J. Chem. Phys.*, 51, 4220–4227 (1969).

206. T. Shiga, K.-J. Hwang, and I. Tyuman, *Biochemistry*, 8, 378–382 (1969).

207. P. Reisberg, J. S. Olson, and G. Palmer, *J. Biol. Chem.*, 251, 4379 (1976).

6

VITAMIN B$_{12}$

Vitamin B$_{12}$ has a large number of well-diversified and well-established biological roles. The significance and importance of the biological functions of B$_{12}$ have always been well emphasized. It is essential for purine, pyrimidine, and methionine metabolism and in transmethylation. B$_{12}$ has been strongly appreciated in the physiological process associated with natural red cell formation and synthesis of nucleic acid. It is required for enzymatic activity of methylmalonyl–CoA mutase, methyltetrahydrofolate oxidoreductase, homocysteine methyl transferase, and ribonucleotide reductase.

Coronary failure, macrocytic anemia, and polycythemia (abnormally large number of red bloods) are associated with excess vitamin B$_{12}$. Pernicious anemia and characteristic lesions (damage) in the nervous system (combined system disease) arise from vitamin B$_{12}$ deficiency.

Vitamin B$_{12}$ is absorbed from ileum (the lowest of the three portions of the small intestine) but requires intrinsic factor and hydrochloric acid contributed by the stomach.

Important sources of B$_{12}$ are food of animal origin: liver, kidney, muscle meat, eggs, milk, and cheese. No significant amounts are found in higher plants. B$_{12}$ also is synthesized within the intestine by the activity of microorganisms. B$_{12}$ is stored mainly in the liver, for long periods, and excreted in feces. B$_{12}$ is labile to heat, acids, alkali, and light.

Vitamin B$_{12}$ contains cobalt, which is the only known function for this element. The total content of cobalt averages about 1.1 mg, and it is widely distributed throughout the body.

Chemistry of Metalloproteins: Problems and Solutions in Bioinorganic Chemistry, First Edition.
Joseph J. Stephanos and Anthony W. Addison.
© 2014 John Wiley & Sons, Inc. Published 2014 by John Wiley & Sons, Inc.

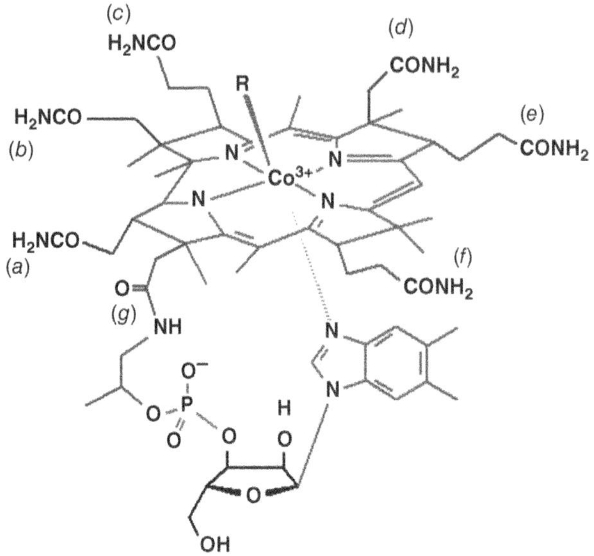

FIGURE 6-1   Vitamin B$_{12}$.

The purpose of this chapter is to focus attention on the chemistry of the different forms of vitamin B$_{12}$. Considerable interest is expressed in their enzymatic reactions and the possible mechanisms. Further, the main chemical properties and model compounds of this vitamin are highlighted.

What are the main structural differences among vitamin B$_{12}$, B$_{12r}$, B$_{12s}$, B$_{12a}$, and B$_{12}$ coenzyme?

- Co^{3+} binds to a 15-membered *corrin ring* (Fig. 6-1), one of the methylidyne carbons is missing, and there is *only one NH group* at the center.
- The carbons at the periphery of the corrin ring are saturated and bear seven amide groups as substituents.
- Three of these (*a*, *b*, *d*) are acetamides, and three (*c*, *e*, *f*) are propionamides.
- The last amide (*g*) is N-substituted propionamide.
  - This amide substituent is a propyl group connected through the 2–carbon to an unusual *ribonucleotide*.
  - What is unusual in this ribonucleotide is the base unit, *5,6-dimethylbenzimidazole*.
- The vitamin B$_{12}$ family includes:
  - Cobalamin, in which benzimidazole nucleotide is present.
  - Cobinamide, in which benzimidazole nucleotide is absent.
  - B$_{12}$, R $=$ CN^{-}, Co^{3+}
  - B$_{12r}$, Co^{2+}

- B_{12s}, Co^+
- B_{12a}, aquacobalamin, $R = H_2O$
- B_{12} coenzyme, adenosyl coenzyme:

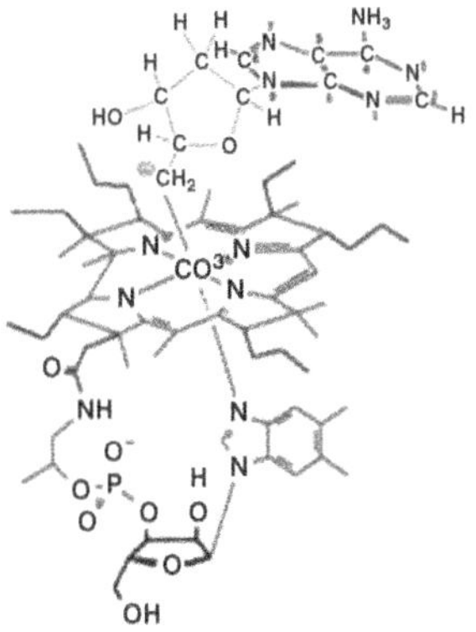

B_{12} coenzyme

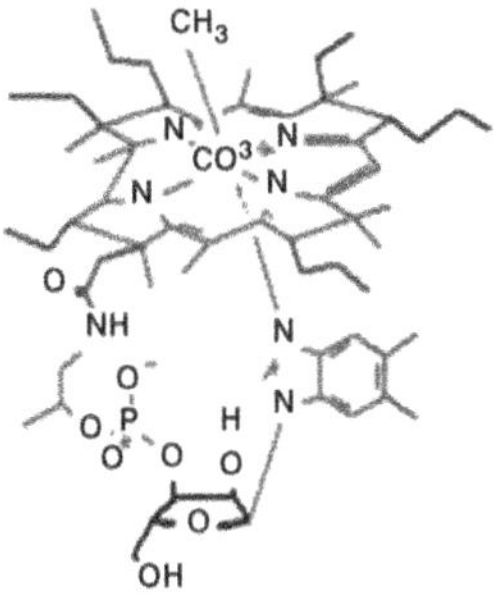

Methyl-B_{12}

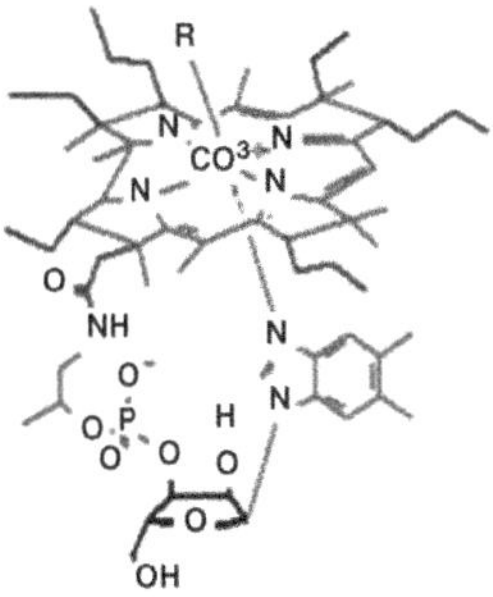

Alkyl derivative B_{12}

What are the enzymatic reactions of cobalt corrinoid compounds?

1. One-carbon transfer
   (a) The formation of *methyl mercury* compound (Scheme 6-1)

$$\underset{(Co^{3+})}{CH_3\text{–}B_{12}} + HgCl_2 \longrightarrow CH_3HgCl + \underset{(Co^{+})}{B_{12s}} + Cl^-$$

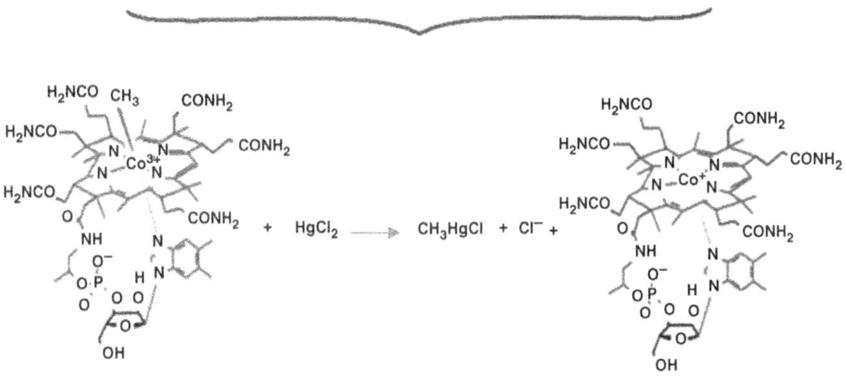

CH_3–B_{12}
(Co^{3+})

B_{12s}

SCHEME 6-1 Formation of methyl mercury.

(b) The formation of *methionine* from homocysteine (Scheme 6-2)

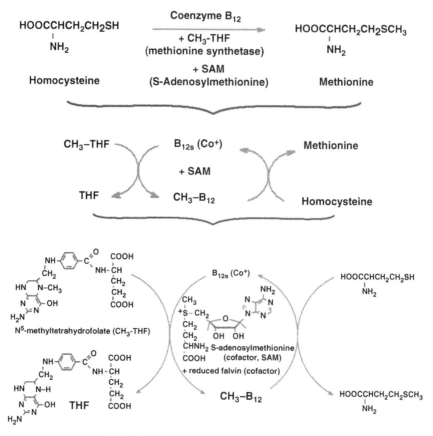

SCHEME 6-2   Action of methionine synthetase.

- Methionine synthetase (Fig. 6-2) catalyzes the transfer of a methyl group from N^5-methyl-tetrahydrofolate (Me-THF) to homocysteine to form methionine and tetrahydrofolate.
- The cobalamin cofactor, which serves as both acceptor and donor of the methyl group, is oxidized and reactivated by reduced flavodoxin and a methyl group from S-adenosyl-L-methionine (SAM).
- The role of the enzyme is to lower the reduction potential of the Co^{2+}/Co^+ couple by elongating the bond between the cobalt and its upper axial water ligand, effectively producing the cobalt 4-coordinate as a transient state.
- This transient state may be methylated to form methylcobalamin, which returns to the 6-coordinate state, triggering the rearrangement to a catalytic conformation.

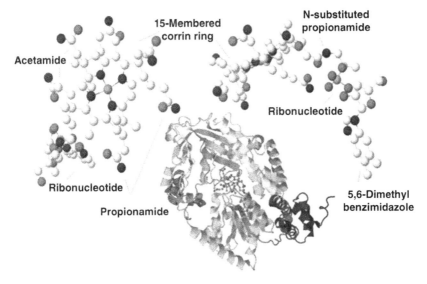

FIGURE 6-2 Structure of B$_{12}$-dependent methionine synthetase, PDB 3IV9 (Koutmos et al., 2009).

(c) The formation of *serine* from glycine:

$$\underset{\text{Glycine}}{\underset{\overset{|}{H}}{NH_2-CH-COOH}} \xrightarrow[\text{+ enzyme}]{\text{coenzyme B}_{12}} \underset{\text{Serine}}{\underset{\overset{|}{CH_2OH}}{NH_2-CH-COOH}}$$

(d) The conversion of CO_2 to acetate:

$$\underset{\overset{|}{CoL}}{CH_3} \xrightarrow{CO_2} \underset{\overset{|}{CoL}}{CH_2-COOH} \xrightarrow{[H]} CoL + CH_3COOH$$

2. Isomerase reactions: These involve 1,2–shift of a C, N, or O atom:

$$R_1-\overset{\textcircled{R_2}\ \textcircled{H}}{\underset{\overset{|}{H}}{C}}-\overset{}{\underset{\overset{|}{H}}{C}}-R_3 \rightleftharpoons R_1-\overset{\textcircled{H}\ \textcircled{R_2}}{\underset{\overset{|}{H}}{C}}-\overset{}{\underset{\overset{|}{H}}{C}}-R_3$$

$$\overset{X\quad H}{C_1-C_2} \rightleftharpoons \overset{H\quad X}{C_1-C_2}$$

(a) The interconversion of glutamic acid and methyl aspartic acid:

$$
\begin{array}{c}
\text{COOH} \\
\text{HC} - \text{NH}_2 \\
\text{H}_2\text{C} - \text{CH}_2 - \text{COOH}
\end{array}
\quad
\xrightarrow[\text{+ enzyme}]{\text{coenzyme B}_{12}}
\quad
\begin{array}{c}
\text{COOH} \\
\text{HC} - \text{NH}_2 \\
\text{H}_3\text{C} - \text{CH} - \text{COOH}
\end{array}
$$

Glutamic acid   Methyl aspartic acid

(b) The interconversion of propionaldehyde and propylene glycol:

$$
\begin{array}{c}
\text{HCHOH} \\
\text{HC} - \text{OH} \\
\text{CH}_3
\end{array}
\quad
\xrightarrow[\text{+ enzyme}]{\text{coenzyme B}_{12}}
\quad
\begin{array}{c}
\text{CH(OH)}_2 \\
\text{HC} - \text{H} \\
\text{CH}_3
\end{array}
\quad \longrightarrow \quad
\begin{array}{c}
\text{HCO} \\
\text{CH}_2 \\
\text{CH}_3
\end{array}
$$

Propylene glycol   Propionaldehyde

(c) The interconversion of succinyl–coenzyme A and L–methylmalonyl coenzyme A:

Succinyl-CoA   Methylmalonyl CoA

(d) Those reactions that require a co-5′-deoxyadenosyl cobamide and are involved in *hydrogen transfer* are summarized in the following:

Glutamate mutase

Methylmalonyl-CoA-
mutase

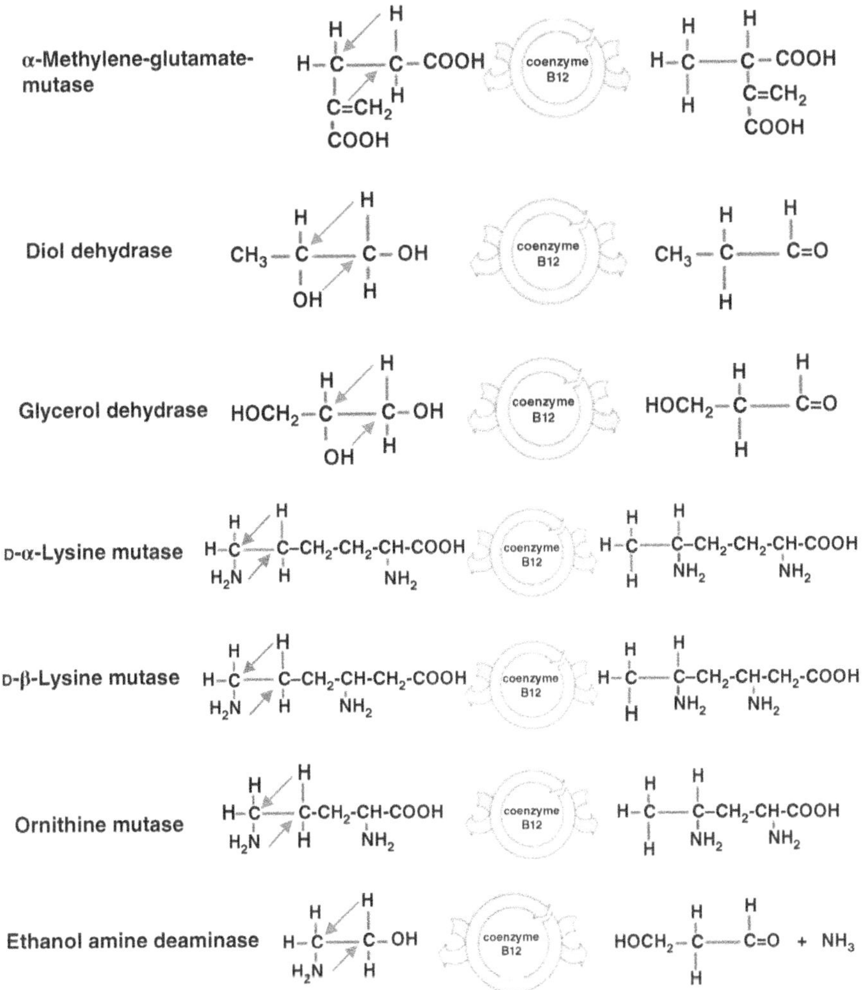

α-Methylene-glutamate-mutase

$$H-\overset{H}{\underset{C=CH_2}{C}}-\overset{H}{\underset{COOH}{C}}-COOH \quad \text{coenzyme B12} \quad H-\overset{H}{\underset{H}{C}}-\overset{H}{\underset{COOH}{C}}-COOH,\ C=CH_2$$

Diol dehydrase

$$CH_3-\overset{H}{\underset{OH}{C}}-\overset{H}{\underset{H}{C}}-OH \quad \text{coenzyme B12} \quad CH_3-\overset{H}{\underset{H}{C}}-\overset{H}{C}=O$$

Glycerol dehydrase

$$HOCH_2-\overset{H}{\underset{OH}{C}}-\overset{H}{\underset{H}{C}}-OH \quad \text{coenzyme B12} \quad HOCH_2-\overset{H}{\underset{H}{C}}-\overset{H}{C}=O$$

D-α-Lysine mutase

$$H-\overset{H}{\underset{H_2N}{C}}-\overset{H}{\underset{H}{C}}-CH_2-CH_2-\overset{}{\underset{NH_2}{CH}}-COOH \quad \text{coenzyme B12} \quad H-\overset{H}{\underset{H}{C}}-\overset{H}{\underset{NH_2}{C}}-CH_2-CH_2-\overset{}{\underset{NH_2}{CH}}-COOH$$

D-β-Lysine mutase

$$H-\overset{H}{\underset{H_2N}{C}}-\overset{H}{\underset{H}{C}}-CH_2-\overset{}{\underset{NH_2}{CH}}-CH_2-COOH \quad \text{coenzyme B12} \quad H-\overset{H}{\underset{H}{C}}-\overset{H}{\underset{NH_2}{C}}-CH_2-\overset{}{\underset{NH_2}{CH}}-CH_2-COOH$$

Ornithine mutase

$$H-\overset{H}{\underset{H_2N}{C}}-\overset{H}{\underset{H}{C}}-CH_2-\overset{}{\underset{NH_2}{CH}}-COOH \quad \text{coenzyme B12} \quad H-\overset{H}{\underset{H}{C}}-\overset{H}{\underset{NH_2}{C}}-CH_2-\overset{}{\underset{NH_2}{CH}}-COOH$$

Ethanol amine deaminase

$$H-\overset{H}{\underset{H_2N}{C}}-\overset{H}{\underset{H}{C}}-OH \quad \text{coenzyme B12} \quad HOCH_2-\overset{H}{\underset{H}{C}}-\overset{H}{C}=O\ +\ NH_3$$

Reproduced from Hill, (1973) by permission of Elsevier.

3. The reduction of the CHOH group of ribonucleotide triphosphates to CH$_2$ (Scheme 6-3)

Suggest general possible mechanisms for enzymatic reactions of coenzyme B$_{12}$.

Three mechanisms for these unusual reactions can be represented:

1. Through the dissociation of the substrate and *olefin complex formation* (Scheme 6-4):

    Step 1: Dissociation of 5-deoxyadenosyl group (as a radical or carbanion, A) and consequent hydrogen or proton abstraction from the substrate.

    Step 2: The substrate forms the transalkylated B$_{12}$ derivative.

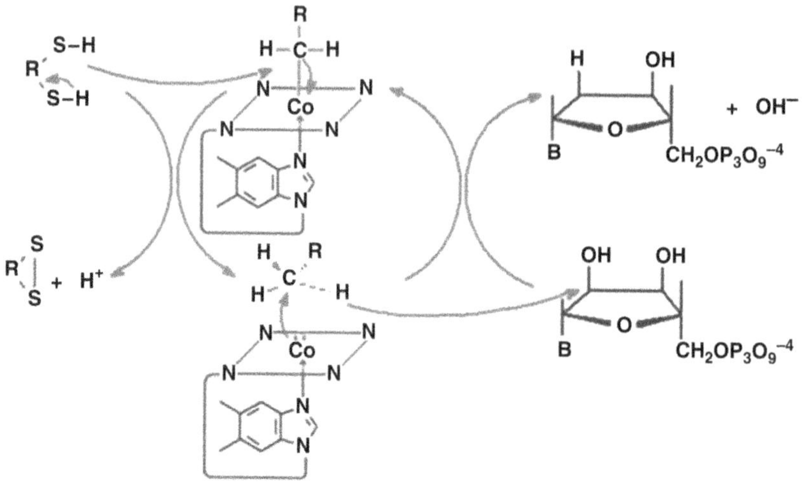

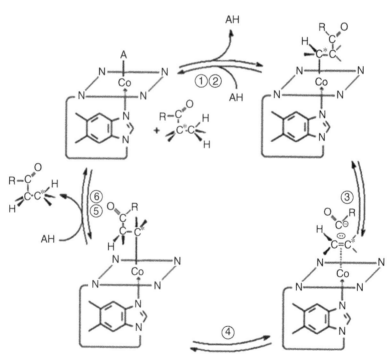

SCHEME 6-3 Reduction of the CHOH group of ribonucleotide triphosphates to CH$_2$. (Modified from Beck, 1968.)

SCHEME 6-4 Dissociation of substrate and olefin complex formation. (Based on Abeles and Dolphin, 1976, and Silverman and Dolphin, 1976.)

Step 3: May involve dissociation of the β-acyl group and formation of olefin complex.

Step 4: Migration of β-acyl group and attack of olefin complex.

Step 5: The rearranged substrate dissociates.

Step 6: Proton abstracts from 5-deoxyadenosine and regenerates B$_{12}$ coenzyme.

2. Through the dissociation of the substrate into a *radical or carbanion.*

- The dissociation of step 1 is followed by the formation of radical or carbanion in step 2:

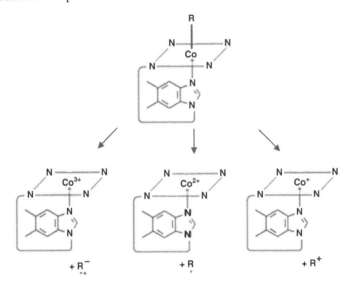

- The three ways of Co–C bond cleavage in alkylcobalamines explain the reactions of the enzyme.
- The β-acyl migration would occur in uncoordinated substrate radicals or carbanions in place of steps 2–6 (Scheme 6-5).

3. A general possible mechanism for enzymatic reactions of coenzyme B$_{12}$ is suggested in Scheme 6-6.

Step 1: Cleavage.

Step 2: Involves hydrogen transfer as a hydride (*a*), hydrogen atom (*b*), or proton (*c*) to give 5′-deoxyadenosine (A–CH$_3$) in (*d*), (*e*), and (*f*).

Step 3: Subsequent 1, 2-shift of R.

Step 4: Co(I) corrinoid acts as a nucleophile, displacing hydride from the C$_{5'}$ of deoxyadenosine to re-form the coenzyme and give the "product."

What are the main chemical properties of B$_{12}$?

There are two one-step electron reductions of B$_{12}$.

- From B$_{12a} \rightarrow$ B$_{12r}$ (Co^{2+}) $\rightarrow$ B$_{12s}$ (Co$^+$)

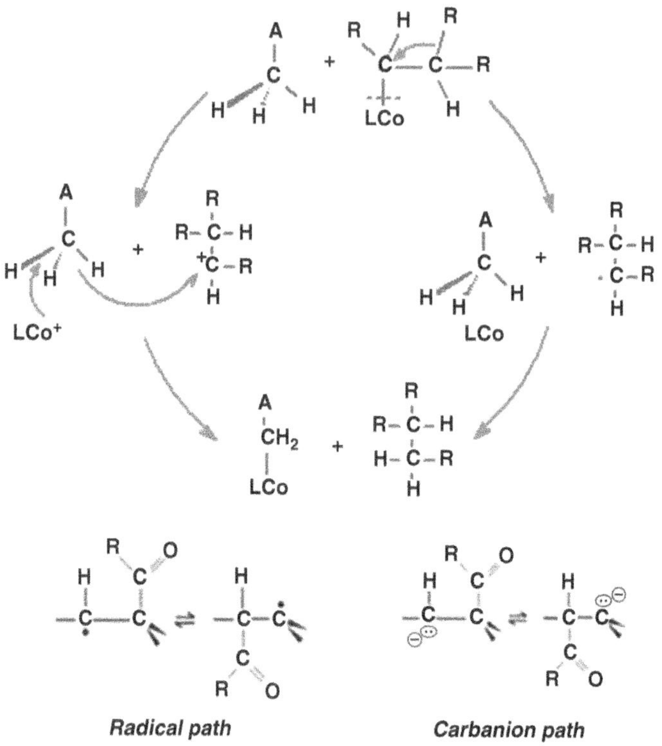

SCHEME 6-5 Dissociation of substrate into cation, radical, or carbanion.

- B_{12a}, aquacobalamin, $R = H_2O$, is reduced to B_{12r} by:

  - $SnCl_2$
  - Zn in CH_3COOH
  - $Cr(CH_3COO)_2$, pH 5
  - Thiols

  - Ascorbic acid
  - An electrochemical process
  - Catalytic hydrogenation

- B_{12a} takes part in many reactions:
  - Loss of H^+ to give hydroxocobalamin, $pK_a = 7.8$
  - Exchange of H_2O for a wide variety of ligands to give the vitamin: N_3^-, SO_3^{2-}, I^-, and $SeCN^-$
  - Displacement of 5, 6-dimethylbenzimidazole in strong acid by CN^- to give dicyanocobalamin
  - Reaction of sulfitocobalamin with CH_3I and β-naphthylamine under anaerobic conditions to give methylcobalamin:

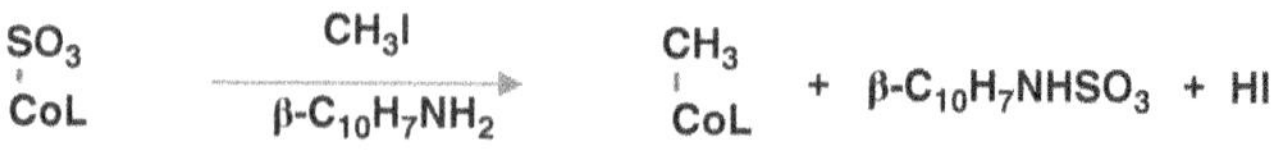

$$\begin{array}{c} SO_3 \\ | \\ CoL \end{array} \xrightarrow[\beta\text{-}C_{10}H_7NH_2]{CH_3I} \begin{array}{c} CH_3 \\ | \\ CoL \end{array} + \beta\text{-}C_{10}H_7NHSO_3 + HI$$

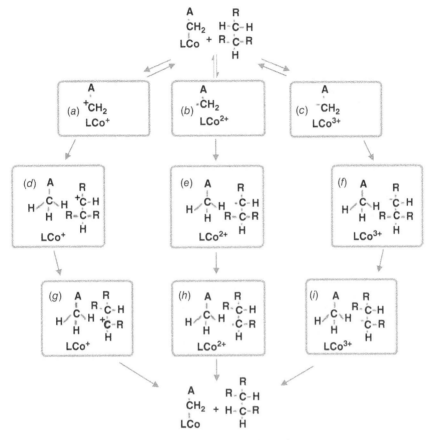

SCHEME 6-6 Hydrogen cleavage as hydride, hydrogen atom, or proton.

○ Reaction of aquocobalamin with dithiol:

- B$_{12r}$ is reduced to B$_{12s}$ by NaBH$_4$, Zn in (NH$_4$)Cl, Cr(CH$_3$COO)$_2$, pH 7, or electrochemically:
  ○ B$_{12r}$ is a low-spin Co^{2+} complex, d^7, and has the electronic configuration $(d_{yz})^2(d_{zx})^2(d_{x^2-y^2})^2(d_{z^2})^1(d_{yz})^0$.
  ○ B$_{12r}$ behaves like a free radical due to the odd electron in the d$_{z^2}$ orbital and combines reversibly with dioxygen.
- B$_{12s}$ is a Co$^+$ model with $(d_{yz})^2(d_{zx})^2(d_{x^2-y^2})^2(d_{z^2})^2(d_{yz})^0$.
  ○ The d$_{z^2}$ electrons behave as strongly lone pairs.
  ○ The nucleophilicity is represented in Scheme 6-7.

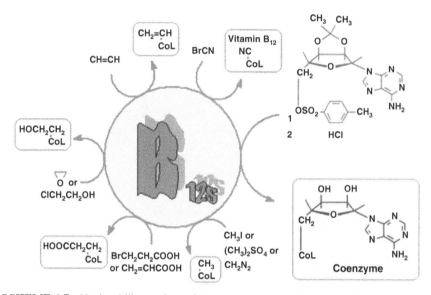

SCHEME 6-7 Nucleophilic reactions of B$_{12s}$. (Data from Smith et al., 1962; Johnson et al., 1963; and Müller and Müller, 1962, 1963.) (See the color version of this scheme in Color Plates section.)

Give an example of a naturally occurring organometallic compound. Explain the stability, and exemplify model compounds.

- B$_{12}$ coenzyme is orange-yellow, diamagnetic, and the only known naturally occurring organometallic compound.
- In the organometallic compound, the bond between the Co atom and the alkyl group is mainly of σ type.
- These compounds are kinetically unstable. The possible β-elimination pathway explains their decomposition:

$$H\!-\!\!-CH_2 \qquad\qquad H \qquad\ CH_2$$
$$M\!-\!\!-CH_2 \quad\longrightarrow\quad M \ \ + \ \ CH_2$$

- These compounds are stable if:
  - the transition metals are in low oxidation states and
  - a π-acid ligand such as CO is present.
- A large variety of σ-bonded alkyl derivatives of Co^{3+} have been prepared (Scheme 6-8).
- All of these complexes have a fair degree of π-conjugation to accommodate the extra charge placed on the cobalt.
- The high formation constant with carbon ligands implicates that the cobalt in corrinoids can be described as a soft ion.

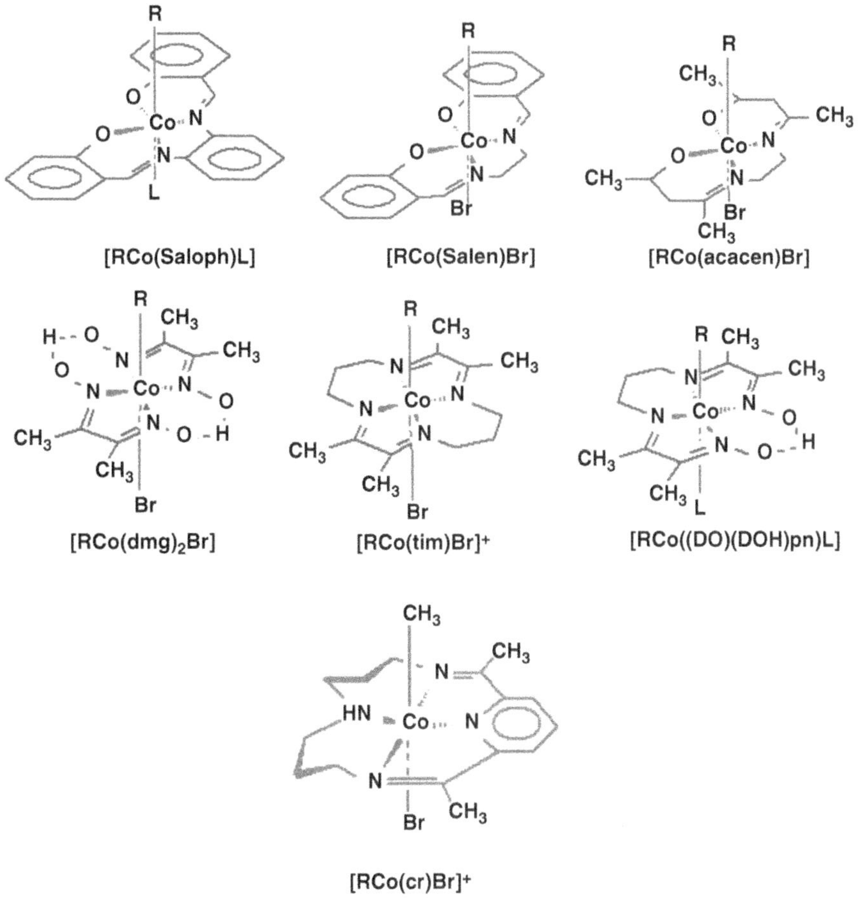

[RCo(Saloph)L]   [RCo(Salen)Br]   [RCo(acacen)Br]

[RCo(dmg)₂Br]   [RCo(tim)Br]⁺   [RCo((DO)(DOH)pn)L]

[RCo(cr)Br]⁺

SCHEME 6-8 Compounds of B_{12}.

REFERENCES

R. H. Abeles and D. Dolphin, *Accts. Chem. Res.*, 9, 114 (1976).

W. S. Beck, *Vitamins and Hormones*, 20, 413 (1968).

H. A. Hill, *Inorganic Biochemistry*, G. L. Eichhorn, Ed., Vol. 2, Chapter 30, Elsevier, New York (1973).

A. W. Johnson, L. Mervyn, N. Shaw, and E. Lester Smith, *J. Chem. Soc.*, 4146 (1963).

M. Koutmos, S. Datta, K. A. Pattridge, J. L. Smith, and R. G. Mathews, *Proc. Natl. Acad. Sci. U.S.A.*, 106, 18527 (2009).

O. Müller and G. Müller, *Biochem. Z.*, 336, 299 (1962).

O. Müller and G. Müller, *Biochem. Z.*, 337, 179 (1963).

R. B. Silverman and D. Dolphin, *J. Am. Chem. Soc.*, 98, 4626 (1976).

E. Lester Smith, L. Mervyn, A. W. Johnson, and N. Shaw, *Nature*, 194, 1175 (1962).

SUGGESTIONS FOR FURTHER READING

1. J. M. Pratt, *Inorganic Chemistry of Vitamin B_{12}*, Academic, London (1972).

2. H. A. O. Hill, J. M. Pratt, and R. J. P. Williams, *Chem. Brit.*, 4, 156 (1968).

3. B. T. Golding, in *Comprehensive Organic Chemistry* vol. 5, E. Haslam, Ed., Pergamon, New York (1979).

4. R. Breslow, P. J. Duggan, and J. P. Light, *J. Am. Chem. Soc.*, 114, 3982 (1992).

5. H. Flohr, W. Pannhorst, and J. Rétey, *Angew. Chem. Int. Ed. Engl.*, 15, 561 (1976).

6. E-I. Ochiai, "Adenosylcobalamin (vitamin B_{12} coenzyme)-dependent enzymes," in *Metal Ions in Biological System*, H. Sigel and A. Sigel, Eds., Vol. 30, pp. 255–278, Marcel Dekker, New York (1994).

7. J. Stubbe, *Ann. Rev. Biochem.*, 58, 257–285 (1989).

8. J. Stubbe, *Adv. Enzymol.*, 63, 349–419 (1990).

9. R. G. Finke, D. A. Schiraldi, and B. J. Mayer, *Coord. Chem. Rev.*, 54, 1–22 (1984).

10. G. Choi, S-C. Choi, A. Galan, B. Wilk, and P. Dowd, *Proc. Natl. Acad. Sci.*, 87, 3174–3176 (1990).

11. G. H. Reed, *Curr. Opin. Chem. Biol.*, 8, 448–477 (2004).

12. E. N. Marsh and C. L. Drennan, *Curr. Opin. Chem. Biol.*, 5, 499–501 (2001).

7

CHLOROPHYLL

Progress has been made in the clarification of the chemical aspects of the photo-synthetic reduction of carbon dioxide to carbohydrate. In this process, the light energy is converted to chemical energy. The details of the processes responsible for the conversion of the energy of light quanta to oxidation and reduction power become the focal point of modern photosynthesis studies. Several proposals have achieved universal acceptance, and there is general agreement that chlorophyll is closely involved in the light conversion act.

The chlorophylls are considered to be the primary photoreceptors in plants and represent a small group of plant pigments with closely related chemical structures.

In plants, chlorophylls are structured into photosynthetic units, each of which is composed of perhaps 300 chlorophyll molecules. The bulk of the chlorophyll is photochemically inert and is often referred to as antenna chlorophyll because its function primarily is to collect light quanta. A small number ($\sim$1%) of chlorophyll molecules appear to act as reaction centers, where the primary light conversion step occurs.

Photosynthetic bacteria, which perform photosynthesis without evolution of oxygen, contain bacterochlorophylls (tetrahydroporphyrins) closely related to but not identical with chlorophyll (dihydroporphyrins).

In this chapter, our concern will be directed exclusively to the structure of the bulk antenna, the nature of the reaction center of the chlorophyll, and the major processes of photosynthesis. How the adapted structure of the chlorophyll serves its function as well as models for the photosynthetic process and synthetic leaf are also the subject of this chapter.

Chemistry of Metalloproteins: Problems and Solutions in Bioinorganic Chemistry, First Edition.
Joseph J. Stephanos and Anthony W. Addison.
© 2014 John Wiley & Sons, Inc. Published 2014 by John Wiley & Sons, Inc.

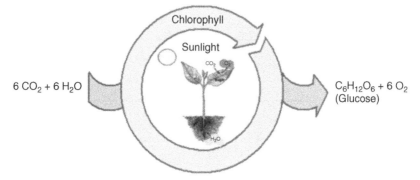

FIGURE 7-1 Typical plant showing inputs and outputs of photosynthetic process.

Write the general equation for the photosynthesis process and indicate the role of the reactants and the products.

- The *photosynthetic process* in green plants consists of splitting H_2O followed by reduction of CO_2:

$$2H_2O \rightarrow [2H_2] + O_2$$
$$CO_2 + [H_2] \rightarrow 1/x[CH_2O]_x + H_2O$$

- ○ $[H_2]$ does not imply free hydrogen, but the reducing capability formed by the oxidation-reduction of H_2O.
- ○ The reaction between carbon dioxide and water is catalyzed by sunlight to produce *glucose* and *oxygen*, the reverse of the respiration process.
- ○ The raw materials of photosynthesis, water and carbon dioxide, enter the cells of the leaf, and the products of photosynthesis, sugar and oxygen, leave the leaf (Fig. 7-1).
- In photosynthetic bacteria, the species oxidized is not H_2O, but an organic molecule, or H_2S. In the latter case, the element S is formed instead of O_2.
- In plants, NO_3^- can be used as an electron acceptor instead of CO_2, and NH_3 is formed instead of sugar. Certain organisms use N_2 to form NH_3 or H^+ to form H_2.

What are the molecular structures of the chlorophyll molecules and the role that chlorophylls play to initiate the light-dependent reactions.

- *Chlorophyll* is the molecule that traps the energy from the sun and is called a *photoreceptor*.
- There are two types of chlorophyll named *a* and *b*. They differ only slightly, in the composition of a side chain (in *a* it is CH_3, in *b* it is CHO) (Fig. 7-2).
- Chlorophyll is found in the chloroplasts of green plants.

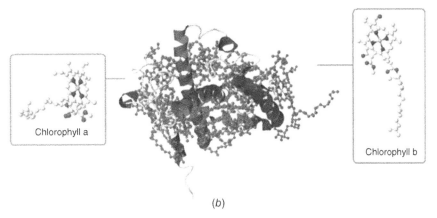

Chlorophyll a: R = –CH₃

Chlorophyll b: R = –CHO

(a)

Chlorophyll a

Chlorophyll b

(b)

FIGURE 7-2 (*a*) Molecular structure of chlorophylls. (*b*) Crystal structure of spinach major light-harvesting complex at 2.72 Å resolution, PDB 1RWT (Liu et al., 2004).

- The basic structure of a chlorophyll molecule is a porphyrin ring, coordinated to a central atom.
- Chlorophyll is very similar in structure to the heme group found in hemoglobin, except that in heme the central atom is iron, whereas in chlorophyll it is magnesium.
- All photosynthetic organisms (plants, prochlorobacteria, and cyanobacteria) have chlorophyll *a*.
- Chlorophyll *a* absorbs its energy from the violet-blue and reddish orange-red wavelengths and little from the intermediate (green-yellow-orange) wavelengths (Fig. 7-3).
- Antenna chlorophyll absorbs its energy in the red region of the spectrum, near 680 nm; it is red shifted relative to chlorophyll solutions in polar solvents, which generally have their red absorption maxima at 662–665 nm. The chlorophyll molecules of the reaction center absorb near 700 nm.

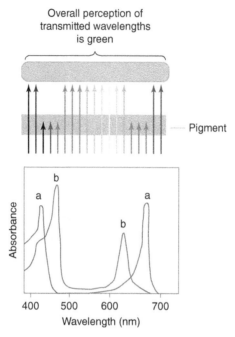

FIGURE 7-3   Absorption spectra of chlorophyll *a* and *b*. (Modified from Treibs, 1973. Reproduced by permission of John wiley & Sons.)

- The different side groups in the two chlorophylls "tune" the absorption spectrum to slightly different wavelengths, so that light that is not significantly absorbed by chlorophyll *a* will instead be captured by chlorophyll *b*.
- Thus, these two kinds of chlorophyll *complement* each other in absorbing sunlight.
- If a pigment absorbs light energy, one of three things will occur:
  - Energy is dissipated as *heat*.
  - The energy may be emitted immediately as a *longer wavelength*, a phenomenon known as *fluorescence*.
  - Energy may trigger a chemical reaction known as *phosphorescence*, as in photosynthesis.
- In order for phosphorescence to occur, there must be an *excited state with a finite lifetime*.
- The presence of the metal atom is necessary for phosphorescence to take place. Free porphyrins show only fluorescent emission.
- Spin–orbital coupling by the metal ion allows mixing of the excited singlet and triplet states and promotes the formation of the relatively stable triplet state, which is the source of the phosphorescence (and of the energy for photosynthesis).

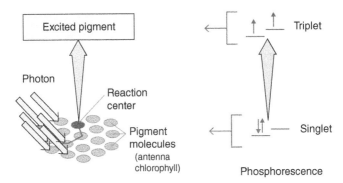

FIGURE 7-4   In phosphorescence, the excited triplet state must have a finite lifetime.

- *Excited pigment* molecule (such as chlorophyll) not only is a better *reducing agent* than the ground-state molecule (because the excited electron is more readily removed) but also is a better *oxidizing agent* because the of positive hole resulting from the excitation of the electron) (Fig. 7-4).

List the two major processes of photosynthesis and state what occurs in these sets of reactions.

- Photosynthesis is a two-stage process (Fig. 7-5):
  - The first process, the light-dependent process (*light reaction*), requires the direct energy of light to make energy carrier molecules that are used in the second process.

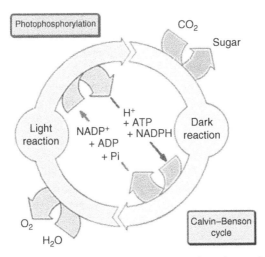

FIGURE 7-5   Overview of steps in photosynthesis. (See the color version of this figure in Color Plates section.)

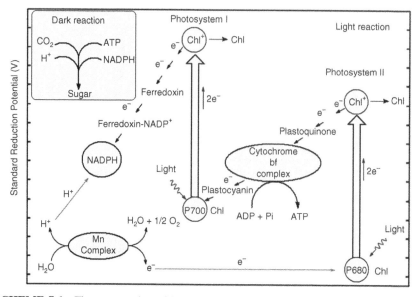

SCHEME 7-1   The regeneration of bioenergetic ATP and NADPH molecules in the light dependent process. (Modified from Lehninger et al., 1993.)

- ○ The light-independent process (or *dark reaction*) occurs when the products of the light reaction are used to form C–C covalent bonds of carbohydrates.
- In the first process, there are two light reaction paths in O_2-producing plants (Scheme 7-1), which can be summarized as in Scheme 7-2.
- These reactions are called photosystems I and II.
- Photosystem I uses chlorophyll *a*, in the form referred to as P700. Photosystem II uses a form of chlorophyll *a*, known as P680, and associates the formation of O_2, where 680 and 700 designate excitation photon wavelengths.
- A photon of light hitting a molecule of chlorophyll in either photosynthesis system provides the energy for a series of redox reactions.
- Photosystem I produces a *strong reducing species* (RED_I, e^-) and a *weak oxidizing agent* (OX_I, Chl^+). Photosystem I is the most efficient photoelectric apparatus in nature, exhibiting a quantum efficiency of almost 100% (Fig. 7-6).
- However, photosysytem II provides a *strong oxidizing agent* (OX_{II}, Chl^+) and a *weaker reducing agent* (RED_{II}, e^-). Photosystem II is exceptional in its power to extract electrons from water.
- RED_I transfers its electron through a series of carriers such as the cytochromes (Fe^{+2}/Fe^{+3}), ferredoxin (Fe^{+2}/Fe^{+3}), and plastocyanin (Cu^+/Cu^{+2}), eventually forming NADPH (Scheme 7-1).

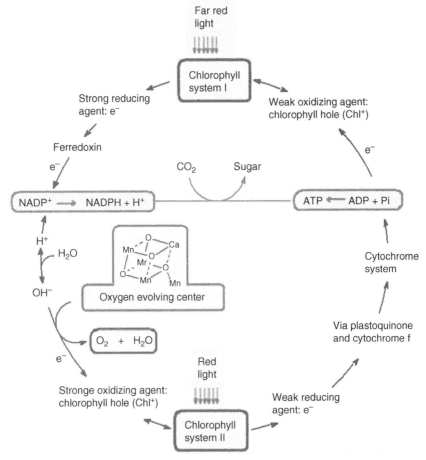

SCHEME 7-2 Photosynthesis process. (Modified from Johnson et al., 1969.) (See the color version of this scheme in Color Plates section.)

NADPH

- OX_I (Chl^+) and RED_{II} react with each other to complete the regeneration of Chl. Again a series of electron carriers are involved and energy-rich ATP is synthesized:

$$ADP^{3-} + HPO_4^{2-} \rightarrow ATP^{4-} + H_2O \qquad \Delta G = 9.7 \, \text{kcal/mol}$$

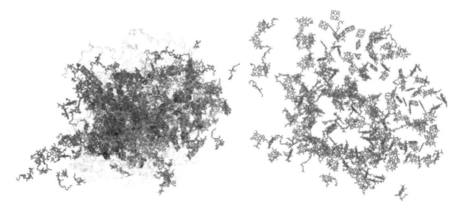

FIGURE 7-6 Structure of plant photosystem I complex together with its light-harvesting complex, PDB 2WSC (Amunts et al., 2010).

ATP

- OX$_{II}$ is responsible for the production of molecular oxygen. A manganese complex is involved in this process (Fig. 7-7). Scheme 7-3 presents a possible mechanism.

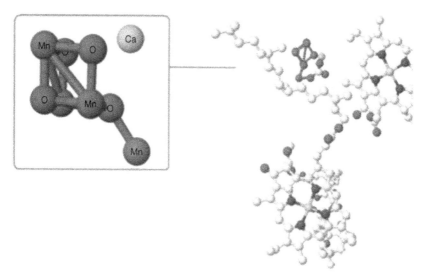

FIGURE 7-7 Structure of photosynthetic oxygen-evolving center, PDB 1S5L (Ferreira et al., 2004).

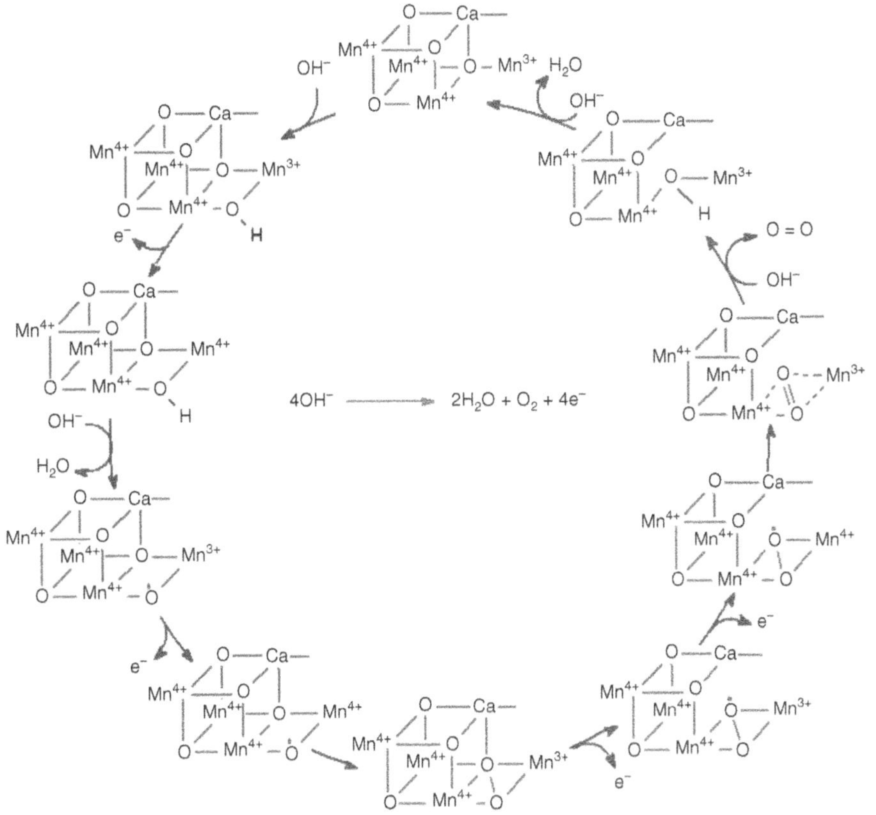

SCHEME 7-3 Manganese complex and oxygen production. (Modified from Ochiai, 2008.)

- In the dark reaction, or light-independent process, NADPH and ATP serve as stable sources of reducing capacity to convert CO_2 to carbohydrate.

How does the adapted structure of the chlorophyll serve its function?

- The structure of the chlorophyll molecule is implicit in its function:
  1. The two kinds of chlorophyll complement each other in absorbing sunlight.
  2. Both of these two chlorophylls are very effective photoreceptors because they contain a network of alternating single and double bonds and the orbitals can delocalize stabilizing the structure.
  3. This *extensive conjugation* of the porphyrin ring *lowers* the energy of the electronic transition and shifts the absorption maximum into the region of *visible* light.
     - The maximum intensity of light reaching Earth is in the visible region; UV is absorbed by O_2 and O_3, and IR is absorbed by CO_2 and H_2O. Out of the

visible region (760–380 nm) the intensity drops to less than one-half of the maximum intensity.

- Shorter wavelengths (with more energy) do not penetrate much below 5 m in seawater. The photosynthetic organisms in the sea, algae, are able to absorb some energy from the longer (hence more penetrating) wavelengths.

4. The alkyl groups in the structure of the chlorophyll shield the macrocycle from irrelevant chemistry at unsubstituted positions.

5. The chelated magnesium makes the excited state a powerful reductant and stabilizes the resulting cation.

6. Esterification of the two carboxyl groups increases the solubility in lipid-rich areas such as membranes. This is emphasized by the presence of phytol as one of the esters.

7. The reduction of the porphyrin ring to chlorin was a simple way to obtain an allowed transition, and thus a large optical cross section, in the red region of the spectrum.

8. Reduction to the chlorin ring creates a stereochemical environment favorable to the formation of the cyclopentanone ring. The cyclopentanone ring may be the basis of oxygen production.

Design a model for:

(a) photosynthetic process
(b) synthetic leaf

(a) Photosynthetic process (Fig. 7-8)

- The photocapturing pigment is zinc tetraphenylporphyrin deposit on a clean aluminum surface, Zn TPP.
- The electron carriers in the solution are $K_4[Fe(CN)_6]$ and $K_3[Fe(CN)_6]$.
- TPP-Zn is activated by orange light and captured energy is used to reduce NADP and oxidized H_2O to O_2.
- When the circuit shown in Fig. 7-8 (1) is closed, the two light-harvesting photoelectrodes will be connected electrically in series, and a potential of 2.2–2.6 V can be obtained.
- If the circuit is closed to an electrolytic cell containing NADP and NADP-reductase, the NADP is reduced and oxygen gas evolves.

(b) Synthetic leaf

- The construction of the synthetic leaf (Fig. 7-9) is as follows:
  ○ A chlorophyll–water adduct is either impregnated in a polymer membrane or deposited on one side of the metal foil.
  ○ The membrane or foil separates the two compartments of the photoelectric cell.

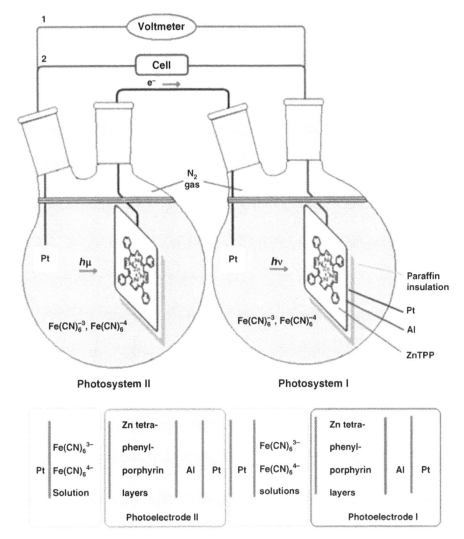

FIGURE 7-8 Model representation for the photosynthetic process. The model has two light-harvesting photoelectrodes connected electrically in series. Upon illumination by amber light, charge transfer take place with a photoelectromotive force of 1.1 to 1.3 V per subunit. (Based on Wang, 1969.)

    ◦ One compartment contains an oxidizing agent (such as tetramethylphe-nylene diamine), and the other contains a reducing agent (sodium ascorbate).

    ◦ Each compartment is connected to an external circuit by electrodes.

    ◦ Irradiation with light produces a potential.

    ◦ The cell responds to red light by generating a potential difference of 422 mV.

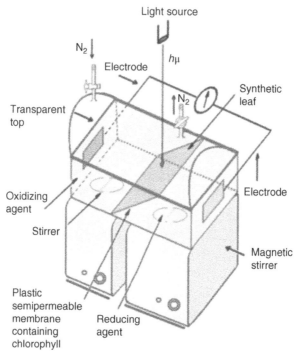

Light source

N₂

*h*μ

Electrode

Synthetic leaf

N₂

Transparent top

Oxidizing agent

Stirrer

Electrode

Magnetic stirrer

Plastic semipermeable membrane containing chlorophyll

Reducing agent

FIGURE 7-9   Experimental set up of semipermeable membrane containing chlorophyll separates two compartments. One compartment contains oxidizing agent and the other contains reducing agent. The compartments are connected to an external circuit by electrodes. Upon illumination by red light, a potential of 422 mV can be obtained. (Modified from Katz et al., 1976.)

What is the effect of the solvent (environments) on the association of the chlorophyll molecules?

- The chlorophyll molecule is *dimerized* in solvent that poorly coordinates Mg^{2+} (poor Lewis basicity) and poorly solvates the porphyrin moiety:

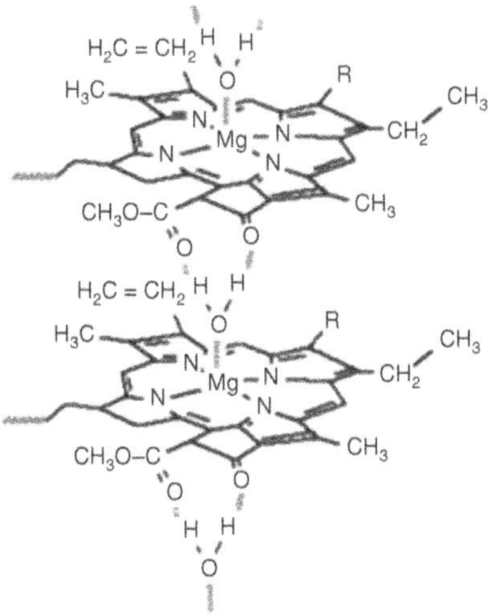

FIGURE 7-10 Chlorophyll oligomer.

- The keto group at carbon 9 of ring V coordinates the Mg^{2+} of a second Chl.
- In aqueous solution these dimers dissociate to form $(Chl \cdot H_2O)_n$ oligomer. Since H_2O can act as a Lewis base and acid, it may bridge Chl monomers, as in Fig. 7-10.
- The significant aspect of such H_2O–bridged Chl molecules is that they (and not monomers or dimers) are the only Chl form known to become *paramagnetic* (singlet $\rightarrow$ triplet) on photolysis.
- The origin of this paramagnetism seems to arise from photoinduced H atoms that transfer from a "sandwiched" H_2O to the carbonyl of Chl.
- The hydrogen transfer leaves an unpaired electron in the porphyrin ring (actually, the odd electron appears to "hop" rapidly from one ring to the next).

Describe the current hypothesis for the photo–oxidation of chlorophyll.

- It is proposed that in vivo photosynthesis involves hundreds of "antenna" molecules that capture photons and pass the energy along to the "active center."
- The antenna Chl (light harvesting complex) consists of $(Chl)_n$ chains terminating in a $(Chl \cdot H_2O \cdot Chl)$ unit, the active center (Fig. 7-11).
- In the antenna chain, it is suggested that the carbonyl group in the ring V is coordinated to the magnesium atom in an adjacent molecule.
- The antenna Chl harvests light quanta by either direct photo excitation or energy transfer from carotene that was previously photoexcited.
- This excitation energy is passed on to the $(Chl \cdot H_2O \cdot Chl)$ unit at the active site containing the ferredoxins, cytochromes, and so forth.

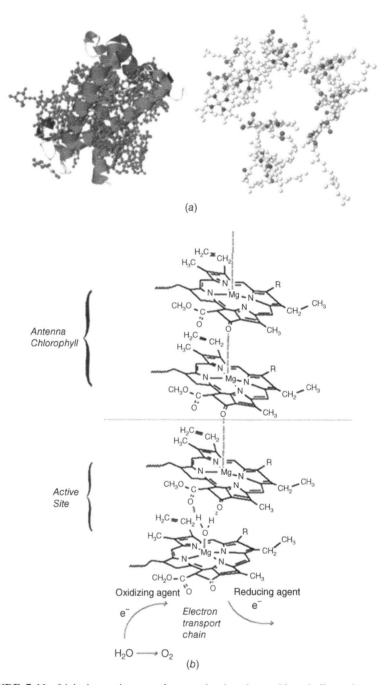

FIGURE 7-11 Light harvesting complex proteins is a large chlorophyll-protein complex, comprising of several subunits. Light energy captured by these complexes is transferred to the dimeric active center. (*a*) Structural insights of light-harvesting complex from spinach, PDB 3PL9 (Pan et al., 2011). (*b*) Antenna, light-harvesting (Chl)$_n$ chains and active center (Chl·H$_2$O·Chl). (Modified from Purcell and Kotz, 1977.)

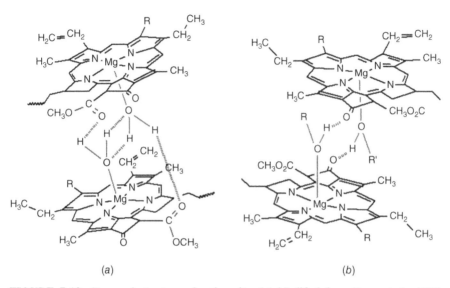

FIGURE 7-12 Proposed structure of active site: (*a*) Modified from Fong et al., 1976; (*b*) Modified from Shipman et al., 1976.

- In the active center, one water molecule is coordinated to the magnesium atom and forms a hydrogen bond with the keto group of the adjacent molecule.
- Now the excited (Chl·H_2O·Chl) unit is easily oxidized or reduced by quinines or ferredoxins, which diffused away to begin the conversion of CO_2 to glucose or to form O_2 molecules.
- The nature of the active center is a matter of some disagreement, but it is generally agreed that two chlorophyll molecules and two water molecules are involved.
- Two possible structures are shown in Fig. 7-12.

REFERENCES

A. Amunts, H. Toporik, A. Borovikova, and N. Nelson, *J. Biol. Chem.*, 285, 3478 (2010).

K. N. Ferreira, T. M. Iverson, K. Maghlaoui, J. Barber, and S. Iwata, *Science*, 303, 1831–1838 (2004).

F. K. Fong, V. J. Koester, and J. S. Polles, *J. Am. Soc.*, 98, 6406 (1976).

W. H. Johnson, L. E. Delanney, and T. A. Cole, in *Essentials of Biology*, Holt, Rinehart and Winston, New York (1969).

J. J. Katz et al., *Chem. Eng. News.*, 54(7), 32 (1976).

A. L. Lehninger, D. L. Nelson, and M. M. Cox, in *Principles of Biochemistry*, 2nd ed., p. 582, Worth, New York (1993).

Z. Liu, H. Yan, K. Wang, T. Kuang, J. Zhang, L. Gui, X. An, and W. Chang, *Nature*, 428, 287–292 (2004).

E. Ochiai, in *Bioinorganic Chemistry: A Survey*, Academic, Elsevier, Amsterdam (2008).

X. W. Pan, M. Li, T. Wan, L. F. Wang, C. J. Jia, Z. Q. Hou, X. L. Zhao, J. P. Zhang, and W. R. Chang, *Nat. Struct. Mol. Biol.*, 18, 309 (2011).

K. F. Purcell and J. C. Kotz, in *Inorganic chemistry*, Saunders, Philadelphia (1977).

L. L. Shipman, T. M. Cotton, J. R. Norris, and J. J. Katz, *Proc. Natl. Acad. Sci. USA*, 73, 1791 (1976).

A. Treibs, *Ann. N. Y. Acad. Sci.*, 206, 97 (1973).

J. H. Wang, *Proc. Natl. Acad. Sci. U. S.*, 62, 653 (1969).

SUGGESTIONS FOR FURTHER READING

1. J. Deisenhofer, H. Michel, and R. Huber, *Trends Biochem. Sci.*, 10, 243–248 (1985).

2. H. Zuber, R. Brunisholz, and W. Sidler, in *New Comprehensive Biochemistry: Photosynthesis*, J. Amesz, Ed., pp. 233–271, Elsevier, Amsterdam (1987).

3. P. Mathis and A. W. Rutherford, in *New Comprehensive Biochemistry: Photosynthesis*, J. Amesz, Ed., pp. 63–96, Elsevier, Amsterdam (1987).

4. G. C. Dismukes, *Photochem. Photobiol.*, 43, 99 (1986).

5. J. T. Babcock, in *New Comprehensive Biochemistry: Photosynthesis*, J. Amesz, Ed., pp. 125–158, Elsevier, Amsterdam (1987).

6. G. W. Brudvig, *J. Bioeng. Biomembr.*, 19, 91 (1987).

7. G. W. Brudvig and R. H. Crabtree, *Proc. Natl. Acad. Sci. U.S.A.*, 83, 4586 (1987).

8. G. W. Brudvig and R. H. Crabtree, *Prog. Inorg. Chem.*, 37, 99–142 (1989).

9. L. Que, Jr., and A. E. True, *Prog. Inorg. Chem.*, 38, 97–200 (1990).

10. G. M. Maggiora and L. L. Ingrham, *Struct. Bonding. (Berlin)*, 2, 126 (1967).

11. J. C. Hindman, R. Kugel, A. Svirmickas, and J. J. Katz, *Proc. Natl. Acad. Sci. U.S.A.*, 74, 5–9 (1977).

12. M. Calvin, *Am. Sci.*, 64, 270 (1976).

13. J. J. Katz, in *Inorganic Biochemistry*, G. L. Eichhorn, Ed., Vol. 2, p. 1022, American Elsevier, New York (1973).

14. D. W. Krogmann, *The Biochemistry of Green Plants*, Chapters 7 and 8, Prentice-Hall, Englwood Cliffs, NJ (1973).

INDEX

Chemistry of Metalloproteins: Problems and Solutions in Bioinorganic Chemistry, First Edition.
Joseph J. Stephanos and Anthony W. Addison.
© 2014 John Wiley & Sons, Inc. Published 2014 by John Wiley & Sons, Inc.